This item was donated by
The Stark County
Mental Health & Addiction
Recovery Board

www.thesmartstore.org

HANDBOOK OF MILITARY AND VETERAN SUICIDE

HANDBOOK OF MILITARY AND VETERAN SUICIDE

Assessment, Treatment, and Prevention

EDITED BY

BRUCE BONGAR

GLENN SULLIVAN

LARRY JAMES

Oxford University Press is a department of the University of Oxford. It furthers the University's objective of excellence in research, scholarship, and education by publishing worldwide. Oxford is a registered trade mark of Oxford University Press in the UK and certain other countries.

Published in the United States of America by Oxford University Press
198 Madison Avenue, New York, NY 10016, United States of America.

© Oxford University Press 2017

All rights reserved. No part of this publication may be reproduced, stored in a retrieval system, or transmitted, in any form or by any means, without the prior permission in writing of Oxford University Press, or as expressly permitted by law, by license, or under terms agreed with the appropriate reproduction rights organization. Inquiries concerning reproduction outside the scope of the above should be sent to the Rights Department, Oxford University Press, at the address above.

You must not circulate this work in any other form
and you must impose this same condition on any acquirer.

CIP data is on file at the Library of Congress
ISBN 978–0–19–987361–6

1 3 5 7 9 8 6 4 2

Printed by Sheridan Books, Inc., United States of America

Contents

Acknowledgments   vii
About the Editors   ix
Contributors   xi

1. Introduction to Military Suicide   1
   ELVIN SHEYKHANI
   LORI HOLLERAN
   KASIE HUMMEL
   BRUCE BONGAR

2. Why Suicide?   10
   VICTORIA KENDRICK
   LORI HOLLERAN
   DAVID HART
   DANA LOCKWOOD
   TRACY VARGO
   BRUCE BONGAR

3. Suicide and the American Military's Experience in Iraq and Afghanistan   23
   JOSEPH TOMLINS
   WHITNEY BLISS
   LARRY JAMES
   BRUCE BONGAR

4. Suicide in the Army National Guard: Findings, Interpretations, and Implications for Prevention   39
   JAMES GRIFFITH

5. Combat Experience and the Acquired Capability for Suicide   53
   CRAIG J. BRYAN
   TRACY A. CLEMANS
   ANN MARIE HERNANDEZ

6. Combat-Related Killing and the Interpersonal-Psychological Theory of Suicide   64
   LINDSEY L. MONTEITH
   SHIRA MAGUEN

7. Suicide Risk Assessment with Combat Veterans—Part I: Contextual Factors   79
   CHRISTOPHER G. AHNALLEN
   ABBY ADLER
   PHILLIP M. KLEESPIES

8. Suicide Risk Assessment with Combat Veterans—Part II: Assessment and Management   89
   PHILLIP M. KLEESPIES
   ABBY ADLER
   CHRISTOPHER G. AHNALLEN

9. Driving Themselves to Death: Covert and Subintentioned Suicide among Veterans   103
   GLENN SULLIVAN
   PHILLIP C. KROKE
   TIMOTHY B. HOSTLER

10. Identifying MMPI-2 Risk Factors for Suicide   114
    JOHN J. BARRETO
    ROGER L. GREENE

11. Ethical Issues in the Treatment of Suicidal Military Personnel and Veterans   121
    W. BRAD JOHNSON
    GERALD P. KOOCHER

12. Evidence-Based Treatments for PTSD: Clinical Considerations for PTSD and Comorbid Suicidality   131
    AFSOON EFTEKHARI
    SARA J. LANDES
    KATHERINE C. BAILEY
    HANA J. SHIN
    JOSEF I. RUZEK

13. The Collaborative Assessment and Management of Suicidality with Suicidal Service Members   147
    DAVID A. JOBES
    BLAIRE C. SCHEMBARI
    KEITH W. JENNINGS

14. Healing the Hidden Wounds of War: Treating the Combat Veteran with PTSD at Risk for Suicide   166
    HERBERT HENDIN

15. Understanding Traumatic Brain Injury and Suicide Through the Lens of Executive Dysfunction   178
    BEETA Y. HOMAIFAR
    MELODI BILLERA
    SEAN M. BARNES
    NAZANIN BAHRAINI
    LISA A. BRENNER

16. The Problem of Suicide in the United States Special Operations Forces   190
    BRUCE BONGAR
    KATE MASLOWSKI
    CATHERINE HAUSMAN
    DANIELLE SPANGLER
    TRACY VARGO

17. Managing Suicide in the Older Veteran   201
    BAVNA B. VYAS
    LISA M. BROWN
    DAVID DOSA
    DIANE L. ELMORE

18. Person-Centered Suicide Prevention in Primary Care Settings   213
    PAUL R. DUBERSTEIN
    MARSHA WITTINK
    WILFRED R. PIGEON

19. Caring Letters for Military Suicide Prevention   240
    DAVID D. LUXTON

Index   255

Acknowledgments

This book is dedicated to all of our active duty military and veterans. I would also like to acknowledge the contributions of the graduate students in my Clinical Crises and Emergencies Research group - in particular the incredible hard work of our lead graduate students for this book, Danielle Spangler and Catherine Hausman.

Bruce Bongar

In sincere appreciation of my VMI departmental colleagues who proudly wore their country's uniform in time of war: Thomas N. Meriwether, PhD, Colonel, US Army, Vietnam, and Dave I. Cotting, PhD, Captain, US Army, Operation Iraqi Freedom. And to my summer research students, from whom I have learned much: Dave Shaw, Bobby Morris, Hope Hackemeyer, Hannah Granger, Ethan Betts, Phillip Kroke, Nicole Harding, Tim Hostler, and Rachel Kroner.

Glenn Sullivan

I would like to acknowledge and thank all the military personnel and veterans we have lost along the way in service to this great nation.

Larry James

About the Editors

Bruce Bongar, Ph.D., ABPP, FAPM, CPsychol, CSci, is the Calvin Distinguished Professor of Psychology at Palo Alto University, and served as Consulting Professor of Psychiatry and Behavioral Sciences at Stanford University's School of Medicine — as well as Co-Chair and Director of Training for the PGSP-Stanford doctor of psychology program. For over three decades, Professor Bongar's research and published work has focused on the wide-ranging complexities of therapeutic interventions with difficult patients in general, and on suicide and life-threatening behaviors in particular. Dr. Bongar received his Ph.D. from the University of Southern California and served his internship in clinical community psychology with the Los Angeles County Department of Mental Health. Professor Bongar has consulted and published on the topic of suicide risk management and prevention among both active duty military personnel and veteran populations (most recently on the issue of suicide among special operations personnel).

Glenn Sullivan, Ph.D., earned his Ph.D. in clinical psychology at the Pacific Graduate School of Psychology, Palo Alto, California. He completed his clinical internship at the San Francisco Veterans Affairs Medical Center and a postdoctoral residency in postdeployment mental health at the Veterans Affairs Medical Center in Salem, Virginia. Dr. Sullivan is an associate professor of psychology at the Virginia Military Institute, where he received the Thomas Jefferson Teaching Award, which is presented annually to a faculty member "deemed especially talented at inspiring students in the development of their intellect and character." In addition to his numerous publications and presentations, Dr. Sullivan maintains an active private practice in Lexington, Virginia. His clinical specializations include psychological assessment, forensic evaluation, and the treatment of combat veterans.

Larry C. James, Ph.D., ABPP, retired as a colonel from the United States Army, and served as the Chair, Department of Psychology at Walter Reed Army Medical Center, and the Chair, Department of Psychology at Tripler Army Medical Center. Colonel James was awarded the Bronze Star and the Defense Superior Service Medal. He is currently the President & CEO of the Wright Behavioral Health Group, LLC and a professor at Wright State University. Previously he served as the Associate Vice President for Military Affairs at Wright State University in Dayton, Ohio. Prior to that assignment, he served as the Dean, School of Professional Psychology, Wright State University from 2008 to 2013. He received his Ph.D. in Counseling Psychology at the University of Iowa and completed a Post-Doctoral Fellowship in Behavioral Medicine at Tripler Army Medical Center. Dr. James is a recognized expert in psychology, national security, defense issues, clinical psychopharmacology and clinical health psychology. He has lectured internationally on these topics, has published six books (with several others in press), and has over 100 professional papers and conference presentations.

Contributors

Abby Adler, Ph.D.
VA Boston Healthcare System
Boston University School of Medicine
Harvard Medical School

Christopher G. AhnAllen, Ph.D.
VA Boston Healthcare System
Harvard Medical School

Nazanin Bahraini, Ph.D.
Rocky Mountain Mental Illness, Research,
    Education, and Clinical Center (MIRECC)
VA Eastern Colorado Health Care System
VA Salt Lake City Health Care System

Katherine C. Bailey, Ph.D.
National Center for PTSD, Dissemination and
    Training Division
VA Palo Alto Health Care System

Sean M. Barnes, Ph.D.
Rocky Mountain Mental Illness, Research,
    Education, and Clinical Center (MIRECC)
VA Eastern Colorado Health Care System
VA Salt Lake City Health Care System

John J. Barreto, Ph.D.
Palo Alto University

Melodi Billera, LCSW
University of Denver Graduate School of
    Social Work
Rocky Mountain Mental Illness, Research,
    Education, and Clinical Center (MIRECC)
Denver VA Medical Center

Whitney Bliss, M.S.
Palo Alto University

Bruce Bongar, Ph.D., ABPP, FAPM, CPsychol, CSci
Palo Alto University

Lisa A. Brenner, Ph.D.
University of Colorado Denver School of Medicine
Rocky Mountain Mental Illness, Research,
    Education, and Clinical Center (MIRECC)
VA Eastern Colorado Health Care System
VA Salt Lake City Health Care System

Lisa M. Brown, Ph.D., ABPP
Palo Alto University

Craig J. Bryan, Psy.D., ABPP
National Center for Veteran's Studies
University of Utah

Tracy A. Clemans, Psy.D.
National Center for Veteran's Studies
University of Utah

David Dosa, M.D.
Providence VA Medical Center
Brown University

Paul R. Duberstein, Ph.D.
University of Rochester School of Medicine
  and Dentistry
Rochester Health Care Decision Making Group

Afsoon Eftekhari, Ph.D.
National Center for PTSD, Dissemination and
  Training Division
VA Palo Alto Health Care System

Diane L. Elmore, Ph.D., M.P.H.
UCLA–Duke University National Center for Child
  Traumatic Stress

Roger L. Greene, Ph.D.
Palo Alto University

James Griffith, Ph.D.
National Center for Veterans Studies
University of Utah

David Hart, M.S.
Palo Alto University

Catherine Hausman, B.A.
Palo Alto University

Herbert Hendin, M.D.
Suicide Prevention Initiatives

Ann Marie Hernandez, Ph.D.
University of Texas Health Science Center at
  San Antonio

Lori Holleran, M.S.
Palo Alto University

Beeta Y. Homaifar, Ph.D.
Rocky Mountain Mental Illness, Research,
  Education, and Clinical Center (MIRECC)
VA Eastern Colorado Health Care System
VA Salt Lake City Health Care System

Timothy B. Hostler, B.S.
United States Air Force

Kasie Hummel, M.A.
Palo Alto University

Larry James, Ph.D., ABPP
Wright State University

Keith W. Jennings, Ph.D.
The Catholic University of America

David A. Jobes, Ph.D., ABPP
The Catholic University of America

W. Brad Johnson, Ph.D.
United States Naval Academy

Victoria Kendrick, M.S.
Palo Alto University

Phillip M. Kleespies, Ph.D., ABPP
VA Boston Healthcare System
Boston University School of Medicine

Gerald P. Koocher, Ph.D., ABPP
DePaul University

Phillip C. Kroke, B.S.
United States Army

Sara J. Landes, Ph.D.
National Center for PTSD, Dissemination and
  Training Division
VA Palo Alto Health Care System

Dana Lockwood, M.S.
Palo Alto University

David D. Luxton, Ph.D.
University of Washington School of Medicine
National Center for Telehealth & Technology

Shira Maguen, Ph.D.
San Francisco VA Medical Center
University of California, San Francisco

Kate Maslowski, M.A.
Palo Alto University

Lindsey L. Monteith, Ph.D.
Rocky Mountain Mental Illness Research Education and Clinical Center
and
University of Colorado School of Medicine

Wilfred R. Pigeon, Ph.D.
University of Rochester School of Medicine and Dentistry
Canandaiguia VA Medical Center

Josef I. Ruzek, Ph.D.
National Center for PTSD, Dissemination and Training Division
VA Palo Alto Health Care System

Blaire C. Schembari, M.A.
The Catholic University of America

Elvin Sheykhani, M.S.
Palo Alto University

Hana J. Shin, Ph.D.
National Center for PTSD, Dissemination and Training Division
VA Palo Alto Health Care System

Danielle Spangler, M.A., M.S.
Palo Alto University

Glenn Sullivan, Ph.D.
Virginia Military Institute

Joseph Tomlins, Ph.D.
Palo Alto University

Tracy Vargo, M.S.
Palo Alto University

Bavna B. Vyas, M.D.
Bay Pines VA Healthcare System

Marsha Wittink, M.D, M.B.E.
University of Rochester School of Medicine and Dentistry
Rochester Health Care Decision Making Group

HANDBOOK OF MILITARY AND VETERAN SUICIDE

1

Introduction to Military Suicide

Elvin Sheykhani

Lori Holleran

Kasie Hummel

Bruce Bongar

"The solider above all prays for peace, as the soldier must suffer and bear the deepest wounds and scars of war." (MacArthur, 1962, p. 3)

SUICIDE STATISTICS

Each year in the United States more than 30,000 people die by suicide. Over 20% of those decedents are believed to be veterans or current service members (Department of Veterans Affairs, 2012). On average, 22 American veterans die by suicide each day (Department of Veterans Affairs, 2012). In 2008, the suicide rate among active duty personnel exceeded that of the civilian population for the first time in history. The largest increases in completed suicides from 2001 to 2010 within the armed forces were seen in the U.S. Army and Marine Corps (Department of the Army, 2010). Due to the Army and Marine Corps' roles in ground combat, it was postulated that the extended wars in Afghanistan and Iraq may have a disproportionate effect upon suicide rates within these two branches (Hoge, McGurk, Thomas, Cox, Engel, & Castro, 2008). When comparing active duty and reserve components, active duty personnel complete suicide at a disproportionate rate: 57% of all suicide deaths within the U.S. Army involve active duty personnel although they make up less than 49% of the overall force (Greene-Shortridge, Britt, & Castro, 2007; Rusch, Corrigan, Todd, & Bodenhausen, 2010). A recent study conducted jointly by the Department of Defense and the Department of Veteran's Affairs found that 90% of all service members who completed suicide had been diagnosed with a mental health condition. Of those, 50% to 75% are thought to have received inadequate treatment (Department of Defense & Department of Veterans Affairs, 2013).

PREVALENCE OF MENTAL HEALTH CONDITIONS IN OEF/OIF VETERANS

There are currently over 1.6 million veterans of Operation Iraqi Freedom (OIF) and Operation Enduring Freedom (OEF). The majority of these individuals have assimilated back into civilian life with few difficulties, but a sizable minority report adjustment and deployment-related difficulties (Department of Defense Task Force on the Prevention of Suicide by Members of the Armed Forces, 2010; Schell & Marshall, 2008). A recent study conducted of returning OEF and OIF veterans ($n = 1965$) found that 14% screened positive for difficulties associated with posttraumatic stress disorder (PTSD), 14% screened positive for major depressive disorder, and 19% reported probable traumatic brain injury (TBI) related to deployment injuries (Hosek, Kavanagh, & Miller, 2008). It is estimated that over 300,000 OEF/OIF veterans suffer from PTSD, and that 320,000 veterans suffer from TBI. Roughly one-third of all returning veterans report difficulties associated with at least one of these conditions, and 5% report symptoms of both (Hosek et al., 2008; Schell & Marshall, 2008).

Of those reporting probable TBI, 57% report not being evaluated by a physician regarding their symptoms. Of service members experiencing difficulties with major depression disorder or PTSD, 53% report seeking services from a physician or mental health professional (Department of Veterans Affairs, 2012; Hosek et al., 2008; Schell & Marshall, 2008). Of those who seek treatment, it is estimated that only half receive adequate treatment. These deficits in treatment are thought to be due to a multitude of factors known as barriers to care (Schell & Marshall, 2008).

BARRIERS TO CARE

Current research suggests that veterans seek mental health services at a rate similar to that of civilians. Forty percent of veterans report feeling comfortable seeking mental health services versus 41% of the general population (Brown, Creel, Engel, Herrell, & Hoge, 2011). Kim, Thomas, Wilke, Castro, and Hoge (2010) conducted a study in which they examined healthcare behaviors of 15,918 active duty and National Guard soldiers. Of these service members, 10,386 reported being deployed to either Iraq or Afghanistan. The study focused on healthcare utilization and attitudes associated with perceived stigma, access to care, service use, and financial constraints. Kim and colleagues found that 28% of veterans thought that "it would be too embarrassing" to seek mental health services, 40% reported fear that those within their units would have less confidence in their abilities, 45% reported fear of reprisals and consequences from leadership, and 44% feared that they would be seen as weak. Twenty-eight percent of those sampled reported either having difficulty scheduling an appointment or not knowing the appropriate means of contacting a healthcare agency. Furthermore, 20% believed that the financial burden to seek services was too great.

Access to appropriate services remains an obstacle as reported throughout the armed services. Of the 40% who seek services regarding their condition, only half are thought to be provided with adequate and appropriate treatment (Balesco, 2007; Institute of Medicine, 2007). The Institute of Medicine reported that the proportion of veterans and active duty personnel who receive "quality care" is expected to be even smaller than that.

OEF/OIF service members and veterans report that the largest hurdle to receiving behavioral health treatment is often perceived stigma (Hoge et al., 2004; Hoge et al., 2008; Kim et al., 2010). In a study of active duty personnel in the U.S. Army, Warner et al. (2011) found that 51% of the service members believed that seeking mental health treatment would negatively impact their careers. Concerns about appearing weak have historically been a barrier to seeking and receiving care within the military culture (Hoge et al. 2004; Jones, 2002; Jones, 2006; Kim et al., 2010).

HISTORICAL CONTEXT OF MILITARY SUICIDE

War syndromes, which refers to symptoms experienced by servicemen during combat, have been documented for centuries but occurred without being formally acknowledged by the military (Jones 2002; Jones, 2006; Soetekouw et al., 2000). Numerous plausible sources for this lack of military recognition exist, including the broad spectrum of symptoms experienced by servicemen, the subjectivity related to the diagnosis of these symptoms, cultural factors that may impact symptomatic conceptualization, and the ambivalence demonstrated by the military toward psychiatry in general (Jones, 2006; Soetekouw et al., 2000). Here we consider some of the war syndromes experienced during the 1900s and the military's response to these occurrences.

The First World War

While psychological distress associated with war has been documented throughout history, increased recognition of these experiences began during World War I. In the early stages of the fighting, war-related psychological distress, characterized by nervous exhaustion, sleep disturbance, and movement difficulties, was dubbed "shell shock" and was presumed to have an organic etiology (e.g., microscopic brain lesions caused by concussive shock; Jones, 2006; Wessely, 2006).

More psychologically attuned physicians, such as American psychiatrist Thomas Salmon, viewed these same symptoms as responses to combat stress and the result of the mental conflict between self-preservation and the demands of duty (Pols, 2011; Pols & Oak, 2007). Nevertheless, other physicians and military leaders persisted in characterizing these experiences as either malingering or as indicators of weakness (Jones, 2006; Wessely, 2006). Shell shock was often viewed as an attempt to escape military duty (Jones, 2006; Wessely, 2006). At best, it was presumed that only the psychologically weakest and most "unfit" men were disposed to "crack" under the pressure of battle.

The Second World War

The belief that weakness in an individual's morals or constitution contributed to the risk of psychological disability in war carried into World War II (Jones, 2006; Pols, 2011; Wessely, 2006). Roy Halloran, the chief of psychiatry of the U.S. Army Medical Corps. held the belief that battle revealed one's true self, and only those with compromised mental health suffered symptoms during war (Pols, 2011). The United States attempted to avoid sending men into World War II who would not be able to tolerate the intensity of war by implementing psychiatric screening for enlistees (Eagan Chamberlin, 2012). This strategy was supported by the American Psychiatric Association and served as a focus for its annual meetings between 1940 and 1942 (Pols, 2011). Based on this approach, nearly 2.5 million men, or 12% of the men examined, were rejected from enlisting due to emotional or mental defects (Pols, 2011; Pols & Oak, 2007). However, the United States' effort to identify men who were impervious to psychological illness was unsuccessful, with psychological symptoms presenting in more than a third of wounded soldiers in some areas of combat (Pols, 2011).

Soldiers continued to experience psychological and physical symptoms (including fatigue, memory and concentration issues, somatic pains, and sleep disturbances), and these were eventually recognized as the result of "battle exhaustion" or "combat neurosis" (Jones, 2006). During the Second World War, the prevalence rate for combat-related psychological injuries was more than double the rate in World War I (Pols & Oak, 2007). When servicemen experienced severe symptoms, they were regarded as mentally ill and as such were to be repatriated according to policy (Pols, 2011; Pols & Oak, 2007).

In response to the shocking attrition of manpower caused by psychogical injuries, new treatment approaches were implemented. Roy Grinker and John Speigel treated soldiers on the front lines in alignment with ideas first proposed by Salmon during World War I (Pols & Oak, 2007). They injected soldiers with sodium pentothal, inducing a twilight sleep within which the soldiers were encouraged to re-experience and process their trauma in a supportive environment (Pols, 2011; Pols & Oak, 2007). While Grinker and Speigel reported a success rate greater than 70%, military authorities were disappointed that fewer than 2% of the treated soldiers ever returned to combat (Pols, 2011).

Additionally, American psychiatrist Fredrick Hanson delivered a treatment focused on addressing fatigue (Pols, 2011). Soldiers suffering from battle exhaustion would receive "a sedative, warm food, and blankets, and ... be allowed to sleep" (Pols, 2011, p. 317). Psychiatrist Herbert Spiegel considered the impact of group cohesion on soldier morale (Pols & Oak, 2007). This research informed the work of social scientist Samuel Stouffer and his team, which confirmed that soldier morale was significantly related to aspects of the relationships between soldiers, including emotional support, as well as to the relationships between soldiers and their respective commanders (Eagan Chamberlin, 2012; Pols, 2011; Pols & Oak, 2007). Research during World War II contributed to the understanding of psychological responses to war by examining the relationship between physical and psychological symptoms and by demonstrating a positive relationship between these two factors. Additionally, those suffering from psychological difficulties stemming from their service in World War II were asked to consider the source of their symptoms, with 41% indicating their symptoms arose from psychological stress related to their military service (Jones & Wessely, 2005). The research conducted during this period contributed to psychological understanding in two significant ways. First, it shifted "the attention from problems of the abnormal mind in normal times to problems of the normal mind in abnormal times" (Farrell & Appel, 1944, p. 12). Second, the emotional bonds between soldiers and the presence of group cohesion was found to be integral to the overall capabilities of the soldiers to thrive in a wartime setting (Pols, 2011; Pols & Oak, 2007).

While progress occurred in research throughout World War II, popular opinions regarding the symptomatic experience of soldiers remained critical. The prevailing belief of military officials was exemplified during an incident in which Gen. George S. Patton, visiting a field hospital, came upon a soldier with combat fatigue and "slapped him for being a coward" (Pols, 2011, p. 317). While this may seem like an inappropriate treatment approach, it is mild compared to other proposed solutions to widespread combat fatigue. As during World War I, some suggested that these "cowards" and "malingerers" simply be shot (Eagan Chamberlin, 2012) or locked away in mental institutions (Wessely, 2006).

The Vietnam War

Stemming from the knowledge related to the symptomatic experience of psychological illness that had been gained during the previous combat experiences, new regulations were put in place for servicemen during the Vietnam War (Pols & Oak, 2007). Based on suggestions made by psychiatrists examining symptoms and outcomes of World War II soldiers, those serving in Vietnam were limited to one-year nonconsecutive tours of duty, with increased relaxation periods present throughout their tour. Subsequently, it was found that the prevalence of psychological illnesses related to combat were significantly lower than they had been in World War II (Pols & Oak, 2007; Wessely, 2005). While it is difficult to specifically identify how these guidelines influenced outcomes, it is believed that requiring less consecutive time in combat and providing additional respite cycles allowed soldiers to more effectively utilize their existing coping skills and manage combat stressors (Bourne, 1970).

While the symptoms experienced by servicemen in the Vietnam War, mainly somatic in nature, were similar to those seen during prior wars, the Vietnam War represented a shift in how society regarded psychological issues (Eagan Chamberlin, 2012; Jones, 2006; Pols & Oak, 2007). Previously, psychological illness was reflective of more stable, enduring factors related to one's character, typically expected to rectify itself once the individual's setting had changed (Pols & Oak, 2007). However, society became increasingly aware of the potential impact of engaging in combat as servicemen returned from Vietnam with persistent psychological problems, especially violent and suicidal behaviors (Pols & Oak, 2007; Wessely, 2005).

Approximately 15% of American Vietnam veterans experienced chronic psychological issues related to their military service. This high incidence rate slowly moved government and social organizations toward addressing this growing concern (Eagan Chamberlin, 2012; Pols & Oak, 2007). Subsequently, this change in mindset motivated the American Psychological Association to include the diagnosis of PTSD in the third edition of the *Diagnostic and Statistical Manual of Mental Disorders*.

CONTEMPORARY APPROACHES TO RISK MANAGEMENT AND TREATMENT OF SUICIDAL PATIENTS

For the past decade, the rate at which service members have completed suicide has gradually increased, eventually exceeding the rate of the U.S. general public (Department of Defense Task Force on the Prevention of Suicide by Members of the Armed Forces, 2010). This anomaly poses a distinct question to the Department of Defense and the Department of Veterans Affairs—how will mental health practitioners adapt to this apparent increase in risk and treat those who present with suicidal ideation in the hopes of *preventing* suicide? In an effort to address such questions, the Department of Veteran Affairs and the Department of Defense (2013) collaboratively developed guidelines for the evaluation, management, and treatment of suicidal patients.

Initial Risk Management of Suicidal Patients

A mental health practitioner must address three main objectives of managing any suicidal service member or veteran as soon as an individual is deemed at risk of attempting suicide. Regarding patients who are active duty service members, the command element should be included in all aspects of this process (Department of Defense and the Department of Veterans Affairs, 2013).

Safeguarding the patient is the most important factor. A safety plan should be collaboratively developed with the patient. Safety plans are intended to support the patient, manage suicidal behavior, and identify additional resources (Department of Defense and the Department of Veterans Affairs, 2013). The result is a "plan that assists the patient with restricting access to means for completing suicide, problem solving and coping strategies, enhancing social supports and identifying a network of emergency contacts including family members and friends, and ways to enhance motivation" (Department of Defense and the Department of Veterans Affairs, 2013, p. 78).

Second, the patient's access to lethal means should be restricted (Department of Defense and the Department of Veterans Affairs, 2013). The most common lethal means that need to be limited include access to firearms (personal and military issued), medications (prescribed and non-prescribed), alcohol, household poisons, and materials used for hanging. Data indicates that 61% of service members who completed suicide used a firearm (Bush et al., 2013). Moreover, 72% of the firearms used were personal or nonmilitary issue. Other less common methods of suicide that need to be considered are potential drowning, leaping from heights, and vehicular crashes (Department of Defense and the Department of Veterans Affairs, 2013).

Lastly, psychoeducation should be provided to patients, as well as their families, regarding mental health disorders and their association with suicide, potential risk factors and warning signs, protective factors, and treatment options (Department of Defense and the Department of Veterans Affairs, 2013). Patients at increased risk for suicide, should be urged not to use alcohol and nonprescription medications. These patients and their family members should be educated on the possible interactions that may occur between drugs and other substances and how these interactions can increase the likelihood of suicidal behavior. Furthermore, the provider's contact information, information on the resources available through the Veterans and/or Military Crisis Line, and community support resources should also be provided to the patient and their families (Department of Defense and the Department of Veterans Affairs, 2013).

Determining the Appropriate Level of Care

Following the evaluation of the patient and the appraisal of suicidal risk that results, the care setting and level (high, intermediate, or low) should be determined (Department of Defense and the Department of Veterans Affairs, 2013). Care settings can include emergency rooms, inpatient hospital wards, outpatient specialty care clinics, and a variety of options during deployment. If a patient is at a high level of risk, he or she should be evaluated in an emergency room setting, and it should be determined whether hospitalization is appropriate. For intermediate-risk patients, partial hospitalization or intensive outpatient therapy may be most suitable. Low-risk patients have the most flexibility in regard to care settings. In these cases, patient preference, underlying condition, and accessibility should be taken into consideration (Department of Defense and the Department of Veterans Affairs, 2013). The patient's ability to realistically follow a safety plan, his or her level of social support, and the level of reassurance that the patient's access to means can be restricted can permit a shift to a less restrictive level of care.

Treatment of Suicidal Patients

Treatment planning for those who are at risk for suicide must be based on a cost-benefit analysis of each of the potential treatment interventions and on the research that supports each option (Department of Defense and the Department of Veterans Affairs, 2013). However, there appears to be a lack of evidence on such interventions. "The dearth of quality research available on effective suicide prevention practices is mainly due to the difficulty conducting randomized controlled trials (RCTs) in high risk for suicide population and the low base rates of suicide and suicides attempts, even in groups at higher risk for suicide" (Department of Defense and the Department of Veterans Affairs, 2013, p. 88).

Formulating the Treatment Plan

When developing a treatment plan for a veteran or service member who is at risk for suicide, it is important to address the prospective risk factors, mental health and/or medical diagnoses, the available care settings, the nurturing of the therapeutic alliance between patient and provider, and the potential benefits and costs of the various therapeutic interventions (Department of Defense and the Department of Veterans Affairs, 2013). Again, this should be a collaborative process between the patient and provider. If the patient consents, family and unit/command members may also be involved. In general the treatment should address the suicide risk and any mental health diagnoses the patient may be experiencing (Department of Defense and the Department of Veterans Affairs, 2013). Data indicates that of service members who attempted suicide between 2008 and 2010, 25% were diagnosed with an anxiety disorder, 67% with major depression, and 83% with bipolar disorder (Bush et al., 2013). Therefore, it is essential that these symptoms be addressed in conjunction with suicidal thinking and behavior.

Psychotherapy

As stated by the Department of Defense and the Department of Veterans Affairs (2013, p. 90), "Most evidence-based psychotherapy interventions for prevention of suicide can be considered broadly as treatment designed to influence dysfunctional cognitions, emotions, and behaviors through a goal-oriented, systematic procedure." One evidence-based psychotherapy that addresses all of these components and is based on the strong foundations of behavior, learning, cognitive/emotional processing, and interpersonal relationship theories is cognitive-behavioral therapy (CBT). However, according to the Department of Defense and the Department of Veterans Affairs guidelines, CBT can be thought of as an overarching term that includes other therapies that have evolved from the same theoretical foundations. These therapies include dialectical behavior therapy, interpersonal therapy, and psychodynamic therapies, although psychodynamic therapy is not typically used to specifically target suicide. The selection of an efficacious evidence-based psychotherapy option should be based on the care setting, the practitioner's training and confidence in utilizing the method, and the patient's diagnosis (especially with regard to personality and substance use disorders) and preference (Department of Defense and the Department of Veterans Affairs, 2013).

Pharmacotherapy

For many mental health disorders, the use of prescription medications is a common practice in conjunction with psychotherapy. These practices are recognized in the Department of Defense and the Department of Veterans Affairs (2013) guidelines and are recommended for patients who may benefit from this combination. However, the use of pharmacotherapy alone as a means to prevent suicide is contraindicated. According to Bush and colleagues (2013), 43% of service members who attempted suicide between 2008 and 2010 had taken psychotropic medications. Therefore, when a patient is a risk to himself or herself, the medications should be evaluated for potential adverse drug interactions and side effects (i.e., increased suicidal thinking and behavior; Department of Defense and the Department of Veterans Affairs, 2013). Furthermore, when prescribing medications to patients who are at risk for self-harm or suicide, providers need to consider the toxicity of the medications. It may be necessary to limit the quantity dispensed and/or identify a family member or friend who will be responsible for limiting patient access to medications (Department of Defense and the Department of Veterans Affairs, 2013).

Electroconvulsive Therapy

The efficacy of electroconvulsive therapy (ECT) in the prevention and treatment of suicide is mainly attributed to its quick and effective resolution of symptoms of numerous psychological disorders (e.g., major depressive disorder, acute schizophrenia, manic episodes; Department of Defense and the Department of Veterans Affairs, 2013). The decisions to employ ECT should be based on evidence-based recommendations for the patient's specific symptomatology and a cost-benefit analysis. Although ECT is generally safe, side effects can include confusion, memory loss, physical side effects (e.g., jaw pain, nausea, headaches), and medical complications (e.g., heart problems; Mayo Clinic, 2012). There is no research that indicates that this is an effective long-term treatment for reducing suicide risk; therefore, psychotherapy and/or pharmacotherapy is suggested following an acute course of ECT. If ECT is used, experts in care settings that are properly equipped for such services should implement it (Department of Defense and the Department of Veterans Affairs, 2013). Some indications that ECT may be appropriate are if a patient has had success with ECT in the past, has experienced chronic symptoms that do not respond to medications, experiences unbearable side effects with pharmacotherapy, or prefers ECT, or if the risks of other treatments outweigh the risks of ECT.

Following Up

Sixty percent of service members who attempted suicide between 2008 and 2010 had prior outpatient behavioral health care (Bush et al., 2013). Moreover, 9% had received inpatient mental health care within 30 days of their attempt. This data reveals the importance of continued monitoring with the goal of recognizing patients who are at risk and of quickly implementing interventions (Department of Defense and the Department of Veterans Affairs, 2013). Follow-up should begin immediately after the patient is discharged from any acute care setting and can occur in a clinical setting, in a patient's home, or in the community. "The frequency of contact should be determined on an individual basis, and increased when there are increases in risk factors or indicators of suicide risk" (Department of Defense and the Department of Veterans Affairs, 2013, p. 124). However, a patient should be re-evaluated following an inpatient emergency department discharge within seven days. Providers who know the patient and are educated on suicide should establish contact with the patient, either through face-to-face contact, telephone, telemedicine, or other methods. Evaluation of current risk factors and the reinforcement of safety plans should be a focus during such contacts (Department of Defense and the Department of Veterans Affairs, 2013). Last, patients who attempted suicide or are at continued high risk should be monitored for at least one year. Providers should continue to monitor patients at intermediate risk for at least six months. Patients who are low risk may be monitored, but a specific time limit has not been identified (Department of Defense and the Department of Veterans Affairs, 2013).

CONCLUSION

Understanding the historical and cultural context in which guidelines have operated gives some insight into the rates of completed suicides within the U.S. Armed Forces. It has been postulated that the lowering of recruitment standards, the stress of multiple deployments, stigma toward mental health treatment, perceived negative effect on careers, access to care, and service utilization all impact the overall suicide rates (Belasco, 2007; Department of Defense Task Force on Prevention of Suicide, 2013; Green-Shortridge et al., 2007; Hoge et al., 2004; Hoge et al., 2008). Service members operate in an environment of complex stressors that impact functioning as a whole. While suicide continues to be a major problem within the U.S. armed forces, progress has been made in assessment, treatment, and long-term safety planning for individuals. Trends have shown that while overall suicide rates in the U.S. armed forces may have increased since 2001, currently the suicide rates may be beginning to decline (suicide rates among service members in 2013 was 18.7 per 100,000 vs. 22.8 per 100,000 in 2012; Department of the Army 2010; Department of Defense, 2013; Department of Defense and the Department of Veterans Affairs, 2012). An emphasis placed on outreach services in which veterans suffering from mental health conditions and TBI are actively sought out for treatment, education regarding the conditions, and changes in healthcare delivery have had a cumulative impact on overall reducing suicidality (Department of Defense Task Force of Suicide Prevention in the Armed Forces; 2010; Hoge et al., 2004; Hoge et al. 2008). As mental health service allocation has become more commonplace in outpatient primary care settings, individuals suffering from mental disorders are able to be referred to appropriate services. Coupling this advancement with the efforts of the Department of Defense and Department of Veterans Affairs in suicide prevention may have helped curb overall rates. While strides have been made over the past decade to limit suicide within the U.S. military, further research is needed to fully understand suicide dynamics. Additionally, further emphasis may be placed on understanding the course and completion of suicide behaviors. Institutional obstacles still remain in place, as does stigma toward seeking appropriate services. While recommendations have been implemented (i.e., increasing the number of practitioners, offering anonymous counseling, and changing leadership attitudes), suicide continues to be a significant problem within the U.S. armed forces and the nation as a whole.

REFERENCES

Belasco, A. (2007). *The cost of Iraq, Afghanistan, and other global war on terror operations since 9/11*. Washington, DC: Congressional Research Service.

Bourne, P. (1970). Military psychiatry and the Vietnam experience. *American Journal of Psychiatry, 127*, 481–488.

Brown, M. C., Creel, A. H., Engel, C. C., Herrell, R. K., & Hoge, C. W. (2011). Factors associated with interest in receiving help for mental health problems in combat veterans returning from deployment to Iraq. *Journal of Nervous and Mental Disease, 199*(10), 797–801.

Bush, N. E., Reger, M. A., Luxton, D. D., Skopp, N. A., Kinn, J., Smolenski, D., & Gahm, G. A. (2013). Suicides and suicide attempts in the U.S. military, 1998–2010. *Suicide and Life-Threatening Behavior, 43*(3), 262–273. doi:10.1111/sltb.12012

Department of the Army. (2010). *Suicide rates 2001–2010*. Washington, DC: Department of Defense.

Department of Defense. (2013). *Department of Defense suicide events report (DoDSER): Calendar year 2013 annual report*. Washington, DC: Federal Printing Office.

Department of Defense Task Force on the Prevention of Suicide by Members of the Armed Forces. (2010). *The challenge and the promise: Strengthening the force, preventing suicide, and saving lives*. Washington, DC: Department of Defense.

Department of Veterans Affairs. (2012). *Suicide data report 2012*. Washington, DC: Federal Printing Office.

Department of Veterans Affairs, & Department of Defense. (2013). *VA/DoD clinical practice guidelines for assessment and management of patients at risk for suicide*. Washington, DC: Authors. Retrieved from http://www.healthquality.va.gov/guidelines/MH/srb/VADoDCP_SuicideRisk_Full.pdf

Eagan Chamberlin, S. M. (2012). Emasculated by trauma: A social history of post traumatic stress disorder, stigma, and masculinity. *Journal of American Culture, 35*(4), 358–365.

Farrell, M., & Appel, J. (1944). Current trends in military neuropsychiatry. *American Journal of Psychiatry, 101*(1), 12–19.

Greene-Shortridge, T. M., Britt, T. W., & Castro, C. A. (2007). The stigma of mental health problems in the military. *Military Medicine, 172*(2), 157–166.

Hoge, C. W., Castro, C. A., Messer, S. C., McGurk, D., Cotting, D. I.,& Koffman, R. L. (2004). Combat duty in Iraq and Afghanistan, mental health problems and barriers to care. *New England Journal of Medicine, 351*, 13–22.

Hoge, C. W., McGurk, D., Thomas, J. L., Cox, A. L., Engel, C. C. & Castro, C. A. (2008). Mild traumatic brain injury in U.S. soldiers returning from Iraq. *New England Journal of Medicine, 358*(5), 453–463.

Hosek, J., Kavanagh, J., & Miller, L. (2008). *How deployments affect service members*. MG-432-RC 2006. Santa Monica, CA: RAND Corporation. Retrieved from http://www.rand.org/pubs/monographs/MG432

Institute of Medicine, Committee on Treatment of Posttraumatic Stress Disorder, Board on Population Health and Public Health Practice. (2007). *Treatment of posttraumatic stress disorder: An assessment of the evidence*. Washington, DC: National Academies Press.

Jones, E. (2006). Historical approaches to post-combat disorders. *Philosophical Transactions of the Royal Society of London: Series B, Biological Sciences, 361*(1468), 533–542.

Jones, E., & Wessely, S. (2005). War syndromes: The impact of culture on medically unexplained symptoms. *Medical History, 49*(1), 55–78.

Kim, P. Y., Thomas, J. L., Wilke, J. E., Castro, C. A., & Hoge, C. W. (2010). Stigma, barriers to care, and mental services among active duty and National Guard soldiers after combat. *Psychiatric Services, 61*(6), 582–588.

MacAuthur, D. (1962). *Farewell speech at West Point*. Washington, DC: National Archives.

Mayo Clinic. (2012). *Electroconvulsive therapy (ECT) risks*. Retrieved from http://www.mayoclinic.org/testsprocedures/electroconvulsivetherapy/basics/risks/prc-20014161

Pols, H. (2011). The Tunisian campaign, war neuroses, and the reorientation of American psychiatry during world war II. *Harvard Review of Psychiatry, 19*(6), 313–320.

Pols, H., & Oak, S. (2007). War & military mental health: The US psychiatric response in the 20th century. *American Journal of Public Health, 97*(12), 2132–2142.

Rusch, N., Corrigan, P. W., Todd, A. R., & Bodenhausen, G. V. (2010). Implicit self control in people with mental illness. *Journal of Nervous Mental Disorders, 198*(2), 150–163.

Schell, T. L., & Marshall, G. N. (2008). Survey of individuals previously deployed for OEF/OIF. In T. Tamelinian & C. Jaycox (Eds.), *The invisible wounds of war: Psychological and cognitive injuries, their consequences, and services to assist recovery* (pp. 87–115). Santa Monica, CA: RAND Corporation.

Soetekouw, P., de Vries, M., van Bergen, L., Galama, J., Keyser, A., Bleijenberg, G., & van der Meer, J. (2000). Somatic hypotheses of war syndromes.

European Journal of Clinical Investigation, 30(7), 630–641.

Warner, C. H., Appenzeller, G. N., Grieger, T., Belenkiy, S., Breitbach, J., Parker, J., . . . Hoge, C. (2011). Importance of anonymity to encourage honest reporting in mental health screening after combat deployment. *Archives of General Psychiatry*, 68(10), 1065–1071.

Wessely, S. (2005). Risk, psychiatry and the military. *The British Journal of Psychiatry*, 186(6), 459–466.

Wessely, S. (2006). Twentieth-century theories on combat motivation and breakdown. *Journal of Contemporary History*, 41(2), 269–286.

2

Why Suicide?

Victoria Kendrick
Lori Holleran
David Hart
Dana Lockwood
Tracy Vargo
Bruce Bongar

Suicide is a complex phenomenon, which is understood in a number of different ways, involving a multitude of different processes. The complexity of this issue is highlighted when considering suicide on a global level. This multifaceted public health concern can be considerably challenging to address due to the personal nature of the act. Yet understanding suicide has been a significant research endeavor for decades, emphasizing our curiosity regarding aspects of human existence and the meaning of life (Khan & Mian, 2010).

Substantial limitations exist within the field of suicide research, particularly related to the ethical problems that present when examining issues of life and death. However, through the progression of research, significant findings have been discovered regarding suicide. Nevertheless, subtle differences in symptom presentation and reasoning for suicide between cultures, gender, and age create obstacles in reliably predicting with certainty future suicidal behaviors. We begin by considering how this complex behavior has been examined throughout history.

ANTIQUITY

Examining suicides that occurred during antiquity is difficult, as most ancient sources do not detail aspects of psychological or social functioning. In many cases, minimal information is available beyond that strictly pertaining to the individual's motive for killing one's self. The most common documented reason for suicide was shame or dishonor (Lykouras, Poulako-Rebelakao, Tsiamis, & Ploumpidis, 2013). One limitation of the research considering suicide in antiquity is the identification of the individuals committing the acts. There is a dearth of information related to common persons committing suicide, as opposed to famous people and heroes, including characters from ancient texts and regents (Pridmore & McArthur, 2009). However, based on this restricted evidence it is possible to determine probable emotions, such as shame, grief, anger, bereavement, and fear among individuals in literature and history who have committed suicide.

Greeks

Within Greek philosophical schools, the topic of suicide was discussed extensively with strong opinions emerging. Many philosophers, most notably Plato, condemned suicide. Plato believed that to commit suicide was to go against the will of the gods as well as one's obligation to society (Lykouras et al., 2013). In Bury's (1926) translation of Plato's book of laws,

Plato's opinion on suicide was described, stating that individuals who killed themselves should be buried without honor, which included burial in an unmarked grave, and in uninhabited areas far away from their families. Aristotle also condemned suicide, asserting that it was an act of cowardice (Lykouras et al., 2013). Additionally, Aristotle believed that suicide does not only destroy the self but also the community at large (Aristotle, 1853).

Furthermore, Greek physicians examined the dynamics contributing to the experience of depression and its believed final act, suicide. The famous physician Hippocrates described depression as *melancholy* and connected it with a negative outlook toward the outside world in general and life in particular (Lykouras et al., 2013). He also believed that this melancholy would lead to psychological and somatic symptoms, including lack of appetite, insomnia, anxiety, and suicidality.

Romans

According to Roman belief, suicide was regarded as a voluntary death (Minois, 1999). Many in the Roman culture supported suicide, particularly among women who had endured a sexual assault or survived their husband's death. The motives for suicide were separated into six categories, including fury or insanity; advanced age; physical pain; devotion, in cases of spousal loss; shame, particularly after a sexual assault; and sorrow connected to the loss of a close friend of family member (Lykouras et al., 2013). Additionally, the method of suicide was quite important to the Romans, with four prevailing techniques. These methods include death by starvation, symbolizing endurance; death through the use of a weapon, which was considered a gallant and masculine death; death by poison, deemed an easy death; and death by hanging or a fall, which was considered the most cowardly form of suicide (Lykouras et al., 2013).

Group Suicide

While the Romans furthered consideration of contributory factors of suicide, suicidal behavior during antiquity was also seen to expand beyond that of individual actions. During this time, initial occurrences of group suicides were documented. Émile Durkheim (1897/1965) termed this "suicide of the besieged." This group conduct typically occurred in response to cities being captured in war. The largest known group suicide mentioned in historical texts took place in Gamla, where 5,000 residents took their own lives instead of being captured by the Roman army (Lykouras et al., 2013). This group response was widely accepted in Roman culture (Minois, 1999). While not conducted in a unified group, mass suicides were witnessed in Greek history within an elderly population. On the island of Kea, all individuals who reached the age of 70 were legally obligated to drink hemlock, thus ending their lives. This law was eventually abandoned by the late Roman period (Lykouras et al., 2013).

Euthanasia

Much like suicide, active euthanasia was frowned upon because it was viewed as an act against the will of the gods (Papadimitriou et al., 2007). However, passive euthanasia, concerning the ending of life by the deliberate withholding of life-sustaining treatment, was deemed more acceptable. Regarding active euthanasia, most dramatists and philosophers were steadfastly opposed to the act. This is likely because life was viewed as sacred with no one person having the right to violate the sacred trust between humans and the gods (Papadimitriou et al., 2007). Yet there seems to be a tacit acceptance of passive euthanasia throughout antiquity. An illustrative example of this is Hippocrates stating that medicine may not fully assist a severely ill individual, and, as such, passive euthanasia would be acceptable (Papadimitriou et al., 2007).

Although these philosophies represent beliefs held centuries ago, they afford us the opportunity to understand the evolution of how suicide was viewed. Given Pridmore and McArthur's (2009) study findings, many of these themes and motives surrounding suicide remain prevalent today. Among others, factors such as the loss of loved ones, public disgrace, and negative emotions such as shame, guilt, grief, and sorrow continue to influence one's risk for suicide. Although it is true that modern medicine and medical ethics change with time, there appears to exist a stable transmission of attitudes and ideals regarding suicidality and euthanasia (Pridmore & McArthur, 2009).

THE MEDIEVAL PERIOD

Following 1000 AD, documentation of suicidal acts began to increase in frequency and quality (Murray, 2012). Additionally, and perhaps most important, individuals documenting suicides began recognizing and differentiating these acts as responses to situational stressors versus mental health struggles. It is crucial to note that, despite the increased willingness to document acts of self-killing, religion served as an important barrier to public openness (Murray, 2012). Eventually, public strategy transitioned from a focus on punishment to developing preventative measures.

Self-Killing and the Legal System

Termed "self-killing" in the medieval judicial system, suicide was considered a felony punishable by the confiscation of goods and chattels of the deceased from his or her surviving family (Butler, 2006). Using modern language, the estate and assets of an individual who completed suicide would be taken by the court as punishment for the deceased's malice aforethought in killing himself and committing a mortal sin. Essentially, because an individual killed himself with intent and premeditation, his family was punished. Due to the legal proceedings involved in suicide investigations, the best sources of suicide data from the medieval period are legal records (Seabourne & Seabourne, 2001). Unfortunately, the moral stigma and possible punishment associated with self-killing motivated many families to hide a relative's suicide, leading to suspected underreporting (Murray, 2012; Seabourne & Seabourne, 2001). However, based on the information available, researchers have identified a potential base rate of one suicide for every 100,000 people, or .001% (Murray, 2012).

Court records from this time suggest that jurors took into account the existence of a mental defect when returning verdicts in suicide cases (Butler, 2006). Nevertheless, a mental defect was not an affirmative defense to suicide culpability. Instead, juror sympathy appears to have worked along a continuum depending on the age of onset of the claimed mental defect. Jurors were more likely to be sympathetic toward individuals born with mental defects compared to those who developed mental illness later in life. However, the particular level of scrutiny varied by court system, and records indicate immense inconsistencies in how verdicts were rendered when mental health was at issue in a case (Butler, 2006).

Additionally, courts had a difficult time delineating between suicides and self-killings by misadventure (Butler, 2006). Self-killing by misadventure concerns accidental deaths and differs from suicide in the malice aforethought requirement. Grounded in the belief that there was no intent or premeditation involved in the self-killing by misadventure, there was no mortal sin committed and jurors displayed more sympathy in the verdicts they returned (Butler, 2006). It was not uncommon for families to argue self-killing by misadventure in an attempt to protect the assets that were likely to be removed if a guilty verdict was rendered (Butler, 2006; Seabourne & Seabourne).

Gender Differences

The prevalence of female suicide during the medieval period remains a widely contested subject; regardless of these disputes, suicide was the most common felony committed by females throughout this time (Butler, 2006). Historians' beliefs regarding the frequency of suicide behaviors and attempts among females within this period differ greatly. Some historians argue that a female killing herself was unfathomable and that the issue did not arise often, while other scholars argue that the evidence represents a different environment, one in which females undeniably exhibited risk (Butler, 2006). Legal records throughout this period indicate that males were more than twice as likely to complete suicide when compared to females; however, females were twice as likely to attempt suicide when compared to males (Murray, 2012). This disparity is consistent with modern suicide rates. Current risk comparisons between genders indicate that men are four times as likely to commit suicide as compared to women; yet women are three to four times more likely to attempt suicide (Callanan & Davis, 2012; Schrijvers, Bollen, & Sabbe, 2012).

During the medieval period the most common method of self-killing for both males and females was hanging. Additionally, men and women both employed drowning as the second most common method to completing suicide. Overall, men were more likely to complete suicide through the use of weapon, whereas a woman using such a method was extremely rare (Seabourne & Seabourne, 2001). This gender difference has persisted in modern suicide rates. Generally, men are more likely to utilize methods that ensure lethality (firearms, hanging) than women (Callanan & Davis, 2012). Interestingly, suicide through self-poisoning tends to be absent from medieval records (Seabourne & Seabourne, 2001). The scarcity of information is thought to be related to the overwhelming social bias against poisoning as a method of death. Unfortunately, this creates challenges when attempting to examine the utilization of different methods used within the medieval era. In modern times, suicide by poison has become a common method to attempt suicide, specifically among women; fortunately it also provides an increased opportunity for medical intervention and demonstrates diminished lethality (Callanan & Davis, 2012; Schrijvers et al., 2012).

Knight Suicidality

Currently there is a paucity of information regarding suicide among medieval knights; however, historians have recently explored the presence of mental illness among this population. Specifically, it is suggested in ancient writings that some knights suffered from symptoms similar to modern post-traumatic stress disorder (Charny & Kaeuper, 2005; Kaeuper & Kennedy, 1996; Shon, 2011). It was not uncommon for knights to describe instances of fear and hopelessness in their own writings and ultimately experience symptoms related to burnout (Shon, 2011).

While knights and modern military service members both serve in the interest of defense, a distinction exists in how individuals are selected for these roles. Modern service members are recruited regardless of an individual's family legacy in the military. Conversely, knights were born into their nobility, which required rigorous training beginning at a very young age (Kaeuper & Kennedy, 1996; Shon, 2011). Given this formative difference, it is possible that knights had more resilience to the stressors they faced in the line of duty due to their predisposed role in society and understanding of their responsibilities (Shon, 2011). A knight's resilience and familial support presumably served as protective factors in preventing him from acting on the hopelessness he experienced as a result of his duties (Charny, 2005; Shon, 2011).

Medieval Suicide in Historical Context

The medieval period demonstrated a subtle shift in the documentation and public view of self-killing (Murray, 2012). Though still lacking, records of suicidal behavior were more present than in the antiquity era (Murray, 2012; Seabourne & Seabourne, 2001). Additionally, individuals in the court and church began to associate suicide with mental illness and demonstrated more sympathy toward cases involving such issues (Butler, 2006; Murray, 2012). This shift ultimately paved the way for modern society's less punitive view of suicidal acts. Historians believe a large contributing factor to this change was the rising popularity of Christianity (Murray, 2012). Specifically, Christianity bridged the gap between condemning suicide as a mortal sin and practicing its core tenants of kindness and understanding. In fact, some leaders of the Christian church would assist families in concealing suicides to protect them from legal consequences. It was their view that suicide, despite being a mortal sin, should not be an act that imposes legal ramifications on a family that is in mourning (Murray, 2012). Additionally, the medieval period spawned the important discussion of mental illness as it relates to suicidality, leading to reform of legal consequences (Butler, 2006; Murray, 2012). Furthermore, despite the firm legal stance against suicide present in the medieval period, this view eventually transitioned away from punishing the surviving family members of victims toward supporting them (Murray, 2012; Seabourne & Seabourne, 2001). Ultimately, the medieval period made suicide mentionable, thus allowing for manageability in the future.

16TH TO 20TH CENTURY

The limited information about suicides committed during the 16th century comes largely from fragmented, subjective journals with a range of perspectives (Minois, 1999). When describing common motivations for suicide in France during this period, Pierre de L'Estoile wrote that the primary motivations for suicide among members of the general population remained consistent throughout the 16th century. In his writings, he also noted that the primary motivation for suicide among members from a higher social status was despair (Minois, 1999). However, other scholars suggest that the motivation for suicide shifted in accompaniment to the changes in societal values during this period. Specifically, the increased emphasis on individualism and religious anxiety appeared to provide additional motivations for suicide. In addition, literary works endorsed ancient views of heroic suicide as well as suicide as an expression of love and courage, which dispelled some of the shame and fear associated with suicide (Minois, 1999). Other historians suggest that military suicide became more common during the late Renaissance period as a way to avoid dishonor. Honor among military members was established by fulfilling important responsibilities, and capture, or defeat during warfare, was associated with dishonor (Minois, 1999). Thus some preferred death by suicide over living a dishonorable life.

Self-Killing to Suicide

In earlier time periods, the act of ending one's own life was commonly described as "self-killing." However, the word "suicide" was beginning to be used more widely to reference "self-killing" in the 17th and 18th centuries (Shneidman, 1998). With this change in terminology, the perceived motivations for committing suicide also transformed. Some cultures viewed suicide as an honorable action that could serve as repentance for sins. This starkly contrasted with the earlier view of suicide as a shameful act that was condemned and hidden. In fact, in 17th-century Japanese culture, the code of Bushido indicated that an individual could repent for failure or disgrace of one's lord by sacrificing oneself through suicide (Bongar, 2002). Similarly, suicides among English aristocrats throughout the 17th century were labeled "fashionable" suicides regarded as maintaining the honor code (Minois, 1999). Further, suicide was viewed as an honorable response to guilt or love. Meanwhile, others began to suggest severe depression (melancholy) as a potential motivation for suicide. Increased suicide rates during this period were attributed to characteristics of capitalism in Europe, including the movement toward individualism, acceptance of risk, and encouragement of competition. The negative impact of poverty, food shortages, widespread serious medical conditions, and war likely contributed to increased suicide rates. Further, human suffering caused by unrequited love, family conflict, personal losses, shame, and remorse were also associated with suicide (Minois, 1999).

Religion

During the 19th century, theorists began focusing on specific psychological or social reasons for committing suicide rather than the morality of the decision to commit suicide (Durkheim, 1897/1965). Psychological factors included aspects of personality, emotion, and mental illness such as depression, psychosis, and alcohol abuse (Bradatan, 1995/1999). Social features emphasized the nature of group dynamics, interpersonal relationships, and societal values as a whole (Bradatan, 1995/1999). Specifically, Masaryk (1881/1970) asserted that the lack of religiosity in society during this period led to increased unhappiness and social disorganization. He advocated that religion serves as a protective factor against subsequent development of mental illness by structuring and promoting psychological coherence (Masaryk, 1881/1970). Further, because mental illness is associated with suicidal behavior, he argued that strengthening religiosity within society might serve to protect against suicide. This is consistent with more modern perspectives that religion serves as a protective factor against suicide (Dervic et al., 2004; Gearing & Lizardi, 2009). Understanding suicide within this social framework shifted the responsibility from the individual to a broader moral problem of society and further emphasized the individual's suicide as a societal or psychological gesture (Durkheim, 1897/1965).

In the late 19th century, Durkheim (1897/1965) asserted that the structure of social relationships in society or within a subgroup of society influences the motivations for suicide. Specifically he suggested that societies with weaker social relationships offered less protection against impulsive behaviors such as suicide. The level of integration in terms of the quantity and quality of social relationships as well as the level of regulation, or clarity of expectations and sanctions regarding these relationships, were identified as important factors that guided behaviors. Based on these considerations, Durkheim proposed four main types of suicide: egoistic, altruistic, anomic, and fatalistic. Each type was associated with a primary emotional force as well as secondary emotional characteristics (Durkheim, 1897/1965).

According to Durkheim's (1897/1965) theory, egoistic suicide occurs when individuals are not sufficiently integrated into society. This type of suicide is primarily associated with apathy. In addition, melancholy with self-complacence and skeptical disillusionment are associated with egoistic suicide. Altruistic suicide, which is primarily associated with the energy of passion or will, is prevalent in societies characterized by exceptionally high levels of social integration (Durkheim, 1897/1965). The secondary emotional characteristics of calm, sense of duty, mystic enthusiasm, and peaceful courage are also associated with this classification of suicide.

Anomic suicide, which occurs when there is insufficient regulation of social relationships, is primarily associated with irritation and disgust (Durkheim, 1897/1965). The secondary emotional characteristics associated with this type of suicide include retaliations against a particular person (in the case of homicide-suicide) or life in general. Fatalistic suicide occurs when there is excessive regulation of social relationships, meaning that the expectations and sanctions for relationships are too rigid or harsh. Unlike the other types of suicide, this type of suicide was not associated with any specific primary emotional force or secondary emotional characteristics (Durkheim, 1897/1965).

In addition to the four main categories of suicide, three additional combinations including ego-anomic, anomic-altruistic, and ego-altruistic suicide were proposed. Although these types were not associated with any primary emotion, each was described by secondary emotional characteristics. Specifically, ego-anomic suicide was conceptualized as a mixture of agitation and apathy, while anomic-altruistic suicide was described as exasperated effervescence. Ego-altruistic suicide was described as a mixture of melancholy and moral fortitude (Durkheim, 1897/1965).

21ST CENTURY

Although suicide has been documented and studied for decades, those interested in the topic continue to struggle with fully understanding the factors associated with suicide and the motivations for committing the act. The wide body of research examining suicide leading up to the 21st century focused primarily on prediction and prevention (Johnson, Cramer, Conroy, & Gardner, 2014) rather than the rationality of suicide. That is, research focused on how to identify at-risk populations and implement preventative tactics within those populations. The major issues considered in the 21st century, on the other hand, have focused predominately on the rationality of suicide, specifically regarding when suicide is logical, reasonable, and ethical. This research includes physician-assisted suicide as well as the differences found in suicidal ideation and behaviors across varying cultures.

Ethical Perspectives of Suicide

For centuries philosophers have been absorbed in aspects of suicide, presenting innumerable views and hypotheses. Although many of these perspectives originated in ancient times, various interpretations have withstood the test of time and remain present in cultures around the world. These ethical perspectives provide a foundation for opposing understandings of suicide, encompassing various motivations and appropriate responses, and demonstrate how theories have transformed over time.

The moralist perspective of suicide emphasizes the moral obligation to protect life through the prevention of suicide (Cutcliffe & Links, 2011). Therefore, the protection of life constitutes an overriding value, which takes precedence in decision-making (Mishara & Weisstub, 2005). As previously noted, many different philosophers have discussed a moralistic view of suicide including Kant, Plato, and Aristotle, who held beliefs acknowledging that to commit suicide would be to dishonor oneself as well as other people. It was considered a sin and a cowardly choice (Cutcliffe & Links, 2011). Furthermore, Christian philosophy designates suicide as wrong because it goes against the sixth commandment, "Thou shalt not kill" (Mishara & Weisstub, 2005). Although many cultures have adopted a more lenient perspective on self-harming behaviors, cultures that are most predominantly associated with the moralist perspective of suicide include Singapore and Lebanon, both of which have statutes in place recognizing suicide as illegal (Mishara & Weisstub, 2005).

Alternatively, the libertarian perspective considers suicide, if contemplated appropriately, a reasonable response to avoid pain or suffering (Ho, 2014). Libertarians strongly uphold the freedom of choice, and the choice to die by suicide is considered to be within the purview of that freedom (Sartorius, 1983). Philosophers who have written in favor of a libertarian view of suicide include Hume (1789), who stated: "A man who retires from life does no harm to society: He only ceases to do good; which, if it is an injury, is of the lowest kind." Western cultures tend to believe that the individual owns his or her own body and, because of this ownership, people have the right to choose to die by suicide. Therefore the libertarian perspective is most widely held within societies with an emphasis on patient autonomy (Sartorius, 1983). The abolishment of laws forbidding suicide and attempted suicide, such as in Canada in 1972 (Mishara & Weisstub, 2005), indicate the movement away from moralistic views of suicide to a more libertarian view.

Last, in the relativist perspective, suicide is based on a cost-benefit analysis of variables, including situational, cultural, and contemporary factors (Ho, 2014). Within the relativist position exist two subcategories: contextualists and consequentialists. Contextualists maintain that the acceptability of suicide will depend on an analysis considering the needs of the individual, the family, and society. For example, when a person is very old, has a terminal illness, or has a chronic disease; suicide may be morally acceptable, and life-sustaining interventions should not be taken (Mishara & Weisstub, 2010). On the other hand, consequentialists focus on the effects, or outcomes, of the suicide on the individual, their social group, or society (Mishara & Weisstub, 2010).

Cultural Differences

Predictability and prevention are two predominant areas of study within the field of suicide research. Encompassed in the predictability research are studies focused on the reasoning behind an individual deciding to commit suicide. Specific symptomatology, personality factors, and stressors have all been studied to understand how they affect suicidal behaviors in different groups (Colucci & Lester, 2012). Although research mainly focuses on Caucasian samples, it is important to note that suicidal ideation and the symptoms behind the presence of suicide are different for people of varying backgrounds and cultural histories.

The American Psychiatric Association Practice Guidelines recognize 10 categories, comprised of 56 factors, which are linked to an increased risk of suicide (Jacobs et al., 2003). These factors are most predominantly associated with Caucasian Americans. The risk factors include psychological diagnoses such as major depressive disorder and schizophrenia, as well as psychosocial features including recent lack of social support, unemployment, diminishing socioeconomic status, domestic partner violence, and other stressful life events. Additional risk factors include genetic family effects and psychological features such as panic attacks, shame, humiliation, decreased self-esteem, and impulsiveness (Jacobs et al., 2003). It is important to note that although the research focuses on Caucasian Americans, some of these risk factors also pertain to minority groups.

Chu, Chi, Chen, and Leino (2010) classified Asian Americans with suicidal ideation and behaviors into two main subtypes: 48% with a psychiatric suicide construct and 52% with a nonpsychiatric suicide construct. This indicates that relying on psychiatric disorders as a primary indicator of suicidal ideation and behavior may not be sufficient for Asian Americans, as over half have little or no history of mental illness. The nonpsychiatric subtype is consistent with risk factors associated with Asian Americans, including physical and health complications (Chu et al., 2010). Asian Americans rarely distinguish depressive affect from somatization and therefore are more likely to present psychological distress as somatization over depressive factors (Cheng et al., 2010). Additional risk factors include dysfunctional family dynamics, parent–child conflict, lack of family cohesion (Cheng et al., 2010; Chu et al., 2010), high-perceived discrimination, low acculturation, or a combination of these factors (Chu et al., 2010).

Although suicide prevalence within the African American population is lower than all other ethnic minority populations, there has been a definite increase in the suicide rate among African Americans since the 1980s (Lincoln, Taylor, Chatter, & Joe, 2012). Suicide constructs within this population may be characterized by substance abuse and dependence rather than an expressed intent to die (Chu et al., 2010). Findings indicate that emotional support, specifically provided by family members, is associated with lower odds of suicidal ideation and behavior in African Americans (Lincoln et al., 2012). Suicide constructs in Latinos, on the other hand, are associated with acculturation difficulties, alienation, and anger. Further, hopelessness appears to be a more common risk symptom among Latinos than Caucasian or African American populations (Hirsch, Visser, Change, & Jeglic, 2012).

Physician-Assisted Suicide

One of the most controversial aspects of suicide in the 21st century has been the debate surrounding physician-assisted suicide. Western cultures place significant emphasis on patient autonomy, which is inherently contradictory to the society's responsibility to protect suicidal individuals (Mishara & Weisstub, 2013). People are increasingly expressing, due to this autonomy, an interest in controlling the way their lives end, specifically in cases when a terminal illness or chronic condition will provide them with a decreased quality of life (Johnson et al., 2014). In the United States, Oregon first legalized physician-assisted suicide in 1994 with the enactment of the Death with Dignity Act (Mishara & Weisstub, 2005). Since the passing of this act, more than 140 legislative proposals in 27 states have failed. However, at least five states including Oregon, Washington, Montana, Vermont, and Arizona have passed legislation legalizing physician-assisted suicide (Mishara & Weisstub, 2005). The remaining states all have adopted advance directives, which give patients the ability to predetermine the withdrawal of life-sustaining measures if they become incompetent to make medical decisions. Although physician-assisted suicide and advance directives seem relatively similar, they differ in a fundamental quality: physician-assisted suicide hastens death by providing medication to end a life while advance directives withdraw life-sustaining treatment (Menzel & Steinbock, 2013).

The idea of suicide has played a significant role in the debate of physician-assisted suicide, specifically because society has long viewed the desire to die, even a longing for an accelerated death among the terminally ill, as a manifestation of mental illness (Menzel & Steinbock, 2013). Thoughts of death are considered a main criterion for major depressive disorder in the *Diagnostic and Statistical Manual of Mental Disorders* (5th ed.; American Psychiatric Association, 2013) as well as a justification for involuntary commitment (Mayo, 1986). These uses of suicidal ideation substantiate its basis for being an undesirable trait. Therefore, when the Death with Dignity Act was first introduced, there was considerable backlash due to suicidal ideation having a fundamental connection with mental illness and lack of competency (Menzel & Steinbock, 2013). This lack of competency is the principal component of arguments opposing rational suicide, specifically stating that suicidal ideation indicates mental illness and incompetency to make such a life-altering decision. Conversely, proponents of assisted suicide usually associate rational suicide with competence to understand the decision and consequences to commit suicide (Mayo, 1986). This competence, or decisional capacity, is frequently associated with the idea of rationality, which requires logical consistency between one's behaviors and first-order desires or goals, and suicide may arguably be justified to achieve a higher-order goal of reducing suffering (Ho, 2014).

Proponents argue that autonomy gives the individual the right to oversee his or her own goals and destiny, which includes the manner and timing of death (Johnson et al., 2014). Opponents of rational suicide have introduced a number of different options in place of physician-assisted suicide including improving end-of-life care, specifically, broadening doctors' awareness and use of palliative care as well as developing more effective pain management and social support programs for both patient and families (Johnson et al., 2014). The belief that psychology and psychiatry, with the help of hospice and palliative care, can bring adequate care using psychotherapeutic measures is a foundational belief of the opposition's argument (Rich, 2014).

RESEARCH LINEAGE AND ACCOMPLISHMENTS

The evolution of perspective and approach regarding suicide would never have occurred without the dedication and work of psychoanalysts and researchers. From the initial records of suicide during antiquity to more modern beliefs and debates regarding risk, scholars throughout time have helped develop understandings of suicide. Further, they have motivated and mentored suicide research that expands upon their labors. The lineage of suicide research has played an important role in the development of suicide prevention techniques, prevention programs and centers, as well as theories related to suicide.

Freud, Murray, and Shneidman

Sigmund Freud's thoughts about suicide appear as fragments in his theoretical works (Briggs, 2006), but these fragmented ideas helped establish the foundation upon which psychodynamic views of suicide developed (Lee & Stimpson, 2002). One of his early papers titled "Mourning and Melancholia" stated that a suicidal act is a redirection of aggressive and murderous wishes toward another back onto the self (Lee & Stimpson, 2002). His later works confirmed the centrality of the role of aggression in suicide. In "Beyond the Pleasure Principle," Freud introduced the concept of a separate aggressive drive, which he linked with the death instinct (Lee & Stimpson, 2002). This separate drive led to the distinction between life instincts and death instincts. Freud's theory of suicide provided direction for future research and developing ideas in numerous different realms of psychology.

Henry Murray, an American psychologist who pioneered the field of personality theory, was inspired by Freud's work. Although not a suicide researcher, Murray's study of Freud's theories facilitated his development of the "need-press" theory, which later played an important role in the development of Shneidman's (1998) psychache theory. Murray's theory attempts to relate personality and environment to behavioral outcomes. He termed "need" as "a hypothetical process the occurrence of which is imagined in order to account for certain objective and subjective facts" and termed "press" as "a directional tendency in an object or situation" (Murray, 1938, p. 54). Shneidman applied Murray's need-press "personology" to suicidology, which later led to the development of organizations such as the Los Angeles Suicide Prevention Center (LASPC) and the American Association of Suicidology.

Edwin S. Shneidman, known to some as the father of suicidology, became interested in the field of suicide while working at the LA Veteran's Administration in 1949 (Leenaars, 2010). While there Shneidman was asked to write condolence letters to the families of two suicide victims. During his time researching these two cases at the La County Coroner's Office, he stumbled upon hundreds of suicide notes from veterans. This assignment seemed to be a defining moment in Shneidman's career. He ultimately read each of the letters and subsequently committed his career to researching suicide (Leenaars, 2010). Fundamental to Shneidman's work was his belief that suicide is essentially psychological pain, or "psychache." Psychache, as coined by Shneidman refers to "the hurt, anguish, soreness, aching, psychological pain in the psyche, the mind" (Shneidman, 1985, p. 51). Applying Henry A. Murray's need-press theory to suicide and his idea of psychache, Shneidman declared, "suicide is virtually always triggered by the failure to fulfill some need, and the intensity of that need determines the degree of perturbation, which, in turn, leads to lethality" (Sperber, 2011, p. 1).

In 1966, Shneidman took a job at the National Institute of Mental Health in order to draft a proposal for a national program in suicide prevention (Leenaars, 2010). Over the three years that he worked there, the number of suicide prevention centers went from 3 to 200. During that time, he also founded the American Association of Suicidology in Chicago with some of the most influential people in the field of suicide including Erwin Stengel, Karl Menninger, Louis Dublin, Jacques Charon, Paul Friedman, Lawrence Kubie, and Robert Havighurst (Leenaars, 2010).

Mowrer and London

Orval Hobart Mowrer was an influential behaviorist and president of the American Psychological Association in 1954. In 1947, Mowrer proposed a two-factor theory of fear conditioning, which speculated fears are developed through being presented with a neutral stimulus (conditioned stimulus), paired with a fear-eliciting stimulus (unconditioned stimulus; Hofmann, 2009). Furthermore, in a 1939 paper, Mowrer discussed the biological usefulness of conditioned fear and anxiety responses but also alluded to the crisis state that can result when these conditioned responses are no longer adaptive (Mowrer, 1939).

In addition to his work on fear acquisition, Mowrer was also a proponent of integrating insight into treatment (Bixenstine, 2014; Mowrer, 1961). This integration led to Mowrer's recognition by Perry London, a fellow psychologist and researcher, whose areas of study focused on altruism and traumatic stress (Bixenstine, 2014; Lambert, 1992). Despite being essentially shunned from a portion of the psychological community due to his use of morality and religion in sessions, Perry came to Mowrer's defense, interpreting the integration of spirituality in treatment as being one about alleviating suffering, rather than imposing morality on patients (Bixenstine, 2014). At the time, London believed psychotherapy was too entrenched in the medical model, in which morality is distinct from treatment (Bixenstine, 2014; London, 1964). However, in London's 1964 book, he argues psychotherapy has aspects of morality inherent in the areas of insight and action therapies and calls for an integration of the two.

Despite eventual criticisms of Mower's theory, his research on fear acquisition eventually set the stage for the modern behavioral theories of the etiology of crisis states and possible ways to increase coping strategies (Hofmann, 2009). Specifically, the writings of Mowrer and London, though controversial for their time, helped further the field of psychology's understanding of suicidal behavior and protective factors from a behavioral conceptualization, strategies used today in modern suicide prevention and crisis intervention. These crisis interventions and prevention strategies were implemented into prevention centers and intervention programs such as the LASPC and the International Association for Suicide Prevention.

Los Angeles Suicide Prevention Center

Some of the most important and well-known founders of modern thought on suicide are Norman L. Farberow, Robert E. Litman, and Edwin Shneidman. These professionals created the LASPC, the first suicide prevention center in history, in 1958 with help from the National Institute of Mental Health and a grant awarded by the University of Southern California (Leenaars, 2010). Norman Farberow, following his service in the U.S. Air Force and obtaining his PhD, began work with Edwin Shneidman. Their initial research collaboration involving examination of suicide notes would ultimately spawn much of what now defines and shapes suicidology and suicide prevention (Jobes & Nelson, 2006). Robert Litman, who began in the field of psychoanalysis, acted as the executive director of the LASPC. According to Litman, 30 years after the founding of the center (1988), the suicide rate in Los Angeles had been halved (Nelson, 2010). Up until his death in 2010, Litman remained quite active at the LASPC, specifically in training new suicide outreach counselors.

The LASPC embodied a radical idea in mental health in America, specifically because the subject of suicide was often shunned and stigmatized by organizations and broader society. The LASPC helped Americans change their perceptions of suicide by increasing knowledge regarding risk, affording individuals a greater understanding and familiarity with the subject matter. This center was innovative in many ways, including the creation of a 24-hour crisis hotline as well as defining "suicidal crisis," the idea that most suicidal people are suicidal for a relatively short period of time (Litman, Farberow, Shneidman, Heilig, & Kramer, 1965). While utilization of a crisis hotline was novel at the time, the LASPC immediately began to receive calls from those who were suicidal, which provided a new outlet for prevention efforts (Goldney, 2005). The LASPC demonstrated that a community-based agency could provide a useful emergency service for those experiencing risk. Currently, nearly 200 volunteers and staff members field more than 36,000 crisis-related telephone calls per year (Nelson, 2010).

International Association for Suicide Prevention

In 1960, Norman Farberow, a founder of the LASPC, became involved with Austrian psychiatrist and neurologist Erwin Ringel in an international cooperation project to combat suicide on a global scale. This collaboration led to the defining of the term *presuicidal syndrome*, which refers to a syndrome that includes three hallmark symptoms: emotional inhibition, aggression focused on the self, and suicidal fantasies (Ringel, 1976). Despite initial difficulties, Farberow was able to assist in the pioneering of suicide prevention internationally, which later became the International Association for Suicide Prevention (IASP; IASP, 2015).

This organization, which is officially associated with the World Health Organization created a three-fold mission. First, the IASP is dedicated to preventing suicidal behavior worldwide. Second, it has a goal to alleviate the effects of suicide and suicide attempts. Also, the IASP provides a forum for mental health professionals, suicide survivors, crisis workers, and people who are indirectly or directly affected by suicide.

By accessing the organization's website, individuals can access information about suicide and locate resources, such as local crisis centers. Volunteers and professionals in more than 50 countries currently staff the IASP, allowing for global supportive services. Additionally, the IASP sponsors World Suicide Prevention Day annually and hosts an international congress every two years dating back to its inception in 1960 (IASP, 2015). These congresses have historically focused on preventing suicidal behavior internationally through the utilization of new innovations in treatment and intervention (IASP, 2015).

Additionally, the IASP has created four yearly awards, one of which is named in honor of Norman Farberow. This award is given to an individual who has significantly contributed to working with survivors of suicide (IASP, 2015).

CONCLUSION

Curiosity regarding why people commit suicide has afflicted the world for centuries. With the evolution of humanity comes the continued struggle to understand the development of suicide, which is why suicidologists continue to focus their work on understanding the ever-changing perspectives of society. Although there are similar moral and ethical dilemmas throughout history, suicide has also developed to reflect the societal changes in history. For example, with the onset of the technological age came a need to understand suicidal behavior and its relationship with technology and online use. Thus to better understand the "why" behind suicidal behaviors, it is imperative that research continues to evolve with society, in order to competently examine factors associated with suicide and identify people who are at risk for suicidal ideation and behavior.

REFERENCES

American Psychiatric Association. (2013). *Diagnostic and statistical manual of mental disorders* (5th ed.). Washington, DC: Author.

Aristotle. (1853). *The Nicomachean ethics of Aristotle*. London: H.G. Bohn. Retrieved from TheClassics. us

Bixenstine, V. (2014). *O. H. Mowrer's theory of integrity therapy revisited*. New York: Routledge.

Bongar, B. (2002). *The suicidal patient: Clinical and legal standards of care* (2nd ed.). Washington, DC: American Psychological Association.

Bradatan, C. (1995). About some 19th century theories of suicide—interpreting suicide in an east European country. *International Journal of Comparative Sociology, 48*(5), 417–432.

Briggs, S. (2006). "Consenting to its own destruction": A reassessment of Freud's development of a theory of suicide. *Psychoanalytic Review, 93*(4), 541–564.

Butler, S. (2006). Degrees of culpability: Suicide verdicts, mercy, and the jury in medieval England. *Journal of Medieval and Early Modern Studies, 36*(2), 263–290.

Callanan, V., & Davis, M. (2012). Gender differences in suicide methods. *Social Psychiatry and Psychiatric Epidemiology, 47*(6), 857–869.

Charny, G., & Kaeuper, R. (2005). *A knight's own book of chivalry: Geoffroi De Charny*. Philadelphia: University of Pennsylvania Press.

Cheng, J., Fancher, T. L., Ratanasen, M., Conner, K. R., Duberstein, P. R., Sue, S., & Takeuchi, D. (2010). Lifetime suicidal ideation and suicide attempts in Asian Americans. *Asian American Journal of Psychology, 1*(1), 18–30.

Chu, J., Chi, K., Chen, K., & Leino, A. (2010). Ethnic variations in suicidal ideation and behaviors: A prominent subtype marked by nonpsychiatric factors among Asian Americans. *Journal of Clinical Psychology, 70*(12), 1211–1226.

Colucci, E., & Lester, D. (2012). *Suicide and culture: Understanding the context*. Boston: Hogrege.

Cutcliffe, J., & Links, P. (2011). Suicide. In P. Barker (Ed.), *Mental health ethics: The human context* (pp. 260–268). New York: Routledge/Taylor & Francis Group.

Dervic, K., Oquendo, M. A., Grunebaum, M. F., Ellis, S., Burke, A. K., & Mann, J. (2004). Religious affiliation and suicide attempt. *The American Journal of Psychiatry, 161*(12), 2303–2308.

Durkheim, É. (1965). *Suicide: A study in sociology*. Translated by J. A. Spaulding & G. Simpson. New York: The Free Press. (Original work published 1897)

Gearing, R. E., & Lizardi, D. (2009). Religion and suicide. *Journal of Religion and Health, 48*, 332–341.

Goldney, R. D. (2005). The Farberow award: The man. *Crisis, 26*(3) 149–151.

Hirsch, J. K., Visser, P. L., Change, E. C., & Jeglic, E. L. (2012). Race and ethnic differences in hope and hopelessness as moderators of the association between depressive symptoms and suicidal behavior. *Journal of American College Health, 60*(2), 115–125.

Ho, A. O. (2014). Suicide: Rationality and responsibility for life. *Canadian Journal of Psychiatry, 59*(3), 141–147.

Hofmann, S. (2009). Cognitive processes during fear acquisition and extinction in animals and humans: Implications for exposure therapy of anxiety disorders. *Clinical Psychology Review, 28*(2), 199–210. doi:10.1016/j.cpr.2007.04.009

Hume, D. (1789). *An essay on suicide*. Yellow Spring, OH: Kahoe.

International Association for Suicide Prevention. (n.d.). Retrieved from https://www.iasp.info/

Jacobs, D., Baldessarini, R., Conwell, Y., Fawcett, J., Horton, L., Meltzer, H., . . . Simon, R. (2003, November 1). *Practice guidelines for the assessment and treatment of patients with suicidal behaviors.* Retrieved from https://www.ncbi.nlm.nih.gov/pubmed.

Jobes, D. A., & Nelson, K. N. (2006). Shneidman's contributions to the understanding of suicidal thinking. In T. Ellis (Ed.), *Cognition and suicide: Theory, research, and therapy* (pp. 29–49). Washington, DC: American Psychological Association.

Johnson, S. M., Cramer, R. J., Conroy, M. A., & Gardner, B. O. (2014). The role of and challenges for psychologists in physician assisted suicide. *Death Studies, 39*(9), 582–588.

Kaeuper, R., & Kennedy, E. (1996). *The book of chivalry of Geoffroi de Charny: Text, context, and translation.* Philadelphia: University of Pennsylvania Press.

Khan, M. M., & Mian, A. (2010). "The only truly serious philosophical problem:" Ethical aspects of suicide. *International Review of Psychiatry, 22*(3), 288–293.

Lambert, B. (1992, June 22). Perry London, 61, psychologist; Noted for his studies of altruism. Retrieved from www.Nytimes.com.

Lee, J., & Stimpson, Q. (2002). A psychodynamic approach to suicide: A critical and selective review. *British Journal of Guidance and Counselling, 30*(4), 373–382.

Leenaars, A. A. (2010). Edwin S. Shneidman on suicide. *Suicidology Online, 1,* 5–18.

Lincoln, K. D., Taylor, R. J., Chatters, L. M., & Joe, S. (2012). Suicide, negative interaction, and emotional support among Black Americans. *Social Psychiatry and Psychiatric Epidemiology, 47,* 1947–1958.

Litman, R., Farberow, N., Shneidman, E., Heilig, S., & Kramer, J. (1965). Suicide prevention telephone service. *Journal of American Medical Association, 192*(1), 21–25.

Lykouras, L., Poulakou-Rebelakao, E., Tsiamis, C., & Ploumpidis, D. (2013). Suicidal behavior in the ancient Greek and Roman world. *Asian Journal of Psychiatry, 6*(6), 548–551.

London, P. (1964). *The modes and morals of psychotherapy,* 2nd ed. New York: Holt, Rinehart and Winston.

Masaryk, T. G. (1970). *Der selbstmord als sociale massenerscheinung der modernen civilization.* Chicago: University of Chicago Press. (Original work published 1881).

Mayo, D. J. (1986). The concept of rational suicide. *Journal of Medicine and Philosophy, 11*(2), 143–155.

Menzel, P. T., & Steinbock, B. (2013). Advance directives, dementia, and physician-assisted death. *Journal of Law, Medicine, and Ethics, 41*(2), 484–500.

Minois, G. (1999). *History of suicide: Voluntary death in Western culture.* Translated by L. G. Cochrane. Baltimore, MD: Johns Hopkins University Press.

Mishara, B. L., & Weisstub, D. N. (2005). Ethical and legal issues in suicide research. *International Journal of Law and Psychiatry, 28*(1), 23–41.

Mishara, B. L., & Weisstub, D. N. (2010). Resolving ethical dilemmas in suicide prevention: The case of telephone helpline rescue policies. *Suicide and Life Threatening Behavior, 40*(2), 159–169.

Mishara, B. L., & Weisstub, D. N. (2013). Premises and evidence in the rhetoric of assisted suicide and euthanasia. *International Journal of Law and Psychiatry, 36*(5), 427–435.

Mowrer, O. (1939). A stimulus-response analysis of anxiety and its role as a reinforcing agent. *Psychological Review, 46*(6), 553–565.

Mowrer, O. (1961). *The crisis in psychiatry and religion.* Princeton, NJ: Van Nostrand.

Murray, A. (2012). Suicide in the middle ages. *Synergy: Psychiatric Writing Worth Reading, 18*(5), 1–5. Retrieved from http://psychiatry.queensu.ca/assets/Synergy/synergyfall12.pdf

Murray, H. A. (1938). *Explorations in personality.* Oxford: Oxford University Press.

Nelson, V. (2010, February 7). Robert E. Litman dies at 88; co-founder of suicide prevention center. *L.A. Times.* Retrieved from http://articles.latimes.com/2010/mar/07/local/la-me-robert-litman7-2010mar07

Papadimitriou, J. D., Skiadas, P., Mavrantonis, C. S., Polimeropoulos, V. P., Papadimitriou, D. J., & Papacostas, K. J. (2007). Euthanasia and suicide in antiquity: Viewpoint of the dramatists and philosophers. *Journal of the Royal Society of Medicine, 100,* 25–28.

Plato. (1926). *Plato: Laws. Books 1-6.* (Translated by R.G. Bury.). Cambridge, Massachusetts: Harvard University Press.

Pridmore, S., & McArthur, M. (2009). Suicide and Western culture. *Australasian Psychiatry, 17*(1), 42–50.

Rich, B. A. (2014). Pathologizing suffering and the pursuit of a peaceful death. *Cambridge Quarterly Healthcare Ethics, 23*(4), 403–416.

Ringel, E. (1976). The presuicidal syndrome. *Suicide and Life-Threatening Behaviors, 6*(3), 131–149.

Sartorius, R. (1983). Coercive suicide prevention: A libertarian perspective. *Suicide and Life- Threatening Behavior, 13*(4), 293–303.

Schrijvers, D., Bollen, J., & Sabbe, B. (2012). The gender paradox in suicidal behavior and its impact on the suicidal process. *Journal of Affective Disorders*, *138*(1–2), 19–26.

Seabourne, A., & Seabourne, G. (2001). Suicide or accident—self-killing in medieval England: Series of 198 cases from the Eyre records. *The British Journal of Psychiatry*, *178*, 42–47.

Shneidman, E. (1985). *Definition of suicide*. New York: John Wiley.

Shneidman, E. S. (1998). "Suicide" in the encyclopedia Britannica, 1777–1997. *Archives of Suicide Research*, *4*, 189–199.

Shon, E. (2011, December 20). Medieval knights may have had PTSD. Retrieved from http://news.discovery.com/history/archaeology/medieval-knights-ptsd-111220.htm

Sperber, M. (2011). Suicide: Psychache and alienation. *Psychiatric Times*, *28*(11), 10–11.

3

Suicide and the American Military's Experience in Iraq and Afghanistan

Joseph Tomlins
Whitney Bliss
Larry James
Bruce Bongar

OVERVIEW OF KEY OEF/OIF/OND DIFFERENCES

The conflicts in Afghanistan and Iraq have created many novel challenges that have led to significant stress among service members both in the theater of war and at home. Among the many unique factors of Operation Enduring Freedom (OEF) and Operation Iraqi Freedom/Operation New Dawn (OIF/OND) are the presence of two theaters of war, an all-volunteer force, an indefinable battlefield, asymmetrical combat, distinctive wounds, lengthy deployments coupled with short "dwell" periods, and unique demographic characteristics not previously seen in prior conflicts.

Service members in OEF/OIF/OND have been and continue to be confronted with an elusive battlefield. There is often no clear "frontline" or "rear," which, from a military strategy perspective, makes it difficult to target the enemy and creates uncertainty and stress among soldiers (Ressler & Schoomaker, 2014). Tanielian and Jaycox (2008) posit that combat in OEF/OIF is unlike those operational tactics faced in the First Gulf War or trained for in five decades of the Cold War, which has led to new roles for soldiers. For example, service members are often expected to perform myriad duties such as combat, security, humanitarian, and training of local nationals (Hoge, 2010).

Asymmetrical combat compounds the difficulties of an elusive battlefield. In place of distinct battlefields, enemy forces rely on tactics such as improvised explosive devices (IEDs) and other insidious maneuvers that do not directly engage our forces but rather, for example, elicit fear and chaos (e.g., attacking civilians, targeting places of worship). The changed nature of warfare has also had a serious effect on the types of wounds service members face. IEDs, for instance, often cause head injuries such as concussions and traumatic brain injuries (TBIs), which can create and/or exacerbate mental health issues such as depression, loss of identity, and suicidality (Ressler & Schoomaker, 2014).

Tanielian and Jaycox (2008) suggest that one key difference in Iraq and Afghanistan is that they are the United States' first attempt since the end of the Cold War to fight an extended conflict with an all-volunteer force. This is in contrast to prior large engagements (World War I, World War II, Korea, Vietnam), where a cross-sectional draft, despite its inherent problems such as using the judicial system to literally empty out jail cells to put boots on the ground, generally resulted in single tours of duty and at times clearer expectations of what was in store for new soldiers (Ressler & Schoomaker, 2014). Concisely, without a draft, there is no easily accessible pool from which to draw. This, in some ways, forces the United States to rely on the National Guard and Reserve. In fact, of the 2.1 million service members who deployed in support of OEF and OIF before 2011, approximately one-third came from the National Guard and Reserve components (Institute of Medicine, 2013). Some even hypothesize that, by having an all-volunteer force, the nature of the force has changed. For instance, the Department of the Army (2010a) suggests that higher rates of "high-risk" populations have been admitted into the force. Notably, many of the variables defined by the Department of the Army (2010a) as high risk—such as drug and alcohol use, criminal behavior, and mental health concerns (e.g., depression, marital stress)—have all been correlated with suicidality (Bongar & Sullivan, 2013; Fowler, 2013; Nakagawa el al., 2009; Sher, 2009).

Service members often face multiple deployments. For instance, in 2008, there was a mandated benchmark of one year of deployment to a combat theater for every two years outside of combat for the active components of all services and one year of deployment to a theater of war to five years nondeployed for the Reserve components (Tanielian & Jaycox, 2008). Unfortunately, service members often end up with shorter periods back home and longer deployments (Hoge, 2010).

Unique demographic variables among service members also play a role in key OEF/OIF/OND differences. There is a higher prevalence of married military families and female service members than ever before (Hoge, 2010; Urusano et al., 2014). When compounded with lengthy deployment periods and short dwell time, heavy burdens are created such as not being able to spend time with partners or raise children, missing funerals and weddings, and not being able to care for aging parents (Hoge, 2010). Concisely, service members are inherently expected to spend increased, multiple periods of time away from their support system and run the risk of becoming estranged from loved ones.

OEF/OIF/OND Suicide Rates

The 2009 age-adjusted suicide rate for all American males was 19.2[1] for every 100,000 (Centers for Disease Control and Prevention [CDC], 2009). This adjusted rate has been used as a comparison for service member suicide rates in important military suicide documents such as the Department of the Army's Health Promotion, Risk Reduction, and Suicide Prevention Report for 2010 (Department of the Army, 2010a). Using this number as a baseline marker for civilian suicides, the military's active duty suicide rates have historically been lower that for civilians, according to Department of Defense (DOD) service member mortality data (DOD, 2011). From 1980 through the year 2000, the active duty suicide rate fluctuated between 9.6 and 15, with an average rate of 11.2 over that period.

The year 2001 marked the beginning of OEF, and with it came an upward trend in suicide rates across all active duty personnel. From 2001 through 2012, active duty suicide rates increased from 9.9 to 22.7 (DOD, 2011; Luxton et al., 2012; Smolenski et al., 2013). This increase was not only historical because it represented a DOD high but because, for the first time since the military began tallying all suicide events, the active duty suicide rate surpassed the adjusted civilian rate of 19.2. The trend was alarming enough that *Time Magazine* featured the story on its cover with the title "One a Day," signifying that, in the first 155 days of 2012, 154 service members died by suicide (Gibbs & Thompson, 2012). In terms of demographics, most of active duty suicides from 2005 through 2012 were by Caucasian, non-Hispanic, junior enlisted (E1 through E4) males under the age of 30 (Bush et al., 2013; Logan, Skopp, Karch, Reger, & Gahm, 2012; Luxton et al., 2012; Smolenski et al., 2013).

Not only did active duty suicide rates rise, but Reserve and National Guard rates increased precipitously as well. Like with the active duty suicide rates, some of the largest increases occurred between the years 2007 and 2010. Between 2009 and 2010, for example, the number of suicides of National Guard service soldiers doubled from 62 to 112 (Griffith, 2012b). In terms of suicide rates, the combined National Guard and Reserve component rate reached 24.2 in 2012, with the Reserve component at 19.3 and the National Guard at 28.1 (Smolenski et al., 2013). Demographically, the characteristics of those who died by suicide in the Reserve and National Guard components between 2006 and 2012 were similar to those who died on active duty—white, junior enlisted males under 30 (Griffith, 2012a; Luxton et al., 2012; Smolenski et al., 2013).

Although the suicide rate for all users of the Department of Veteran Affairs (VA) medical system did not increase substantially since the beginning of OEF/OIF/OND, an increase in suicide rates was observed within the VA among males under the age of 30 (Kemp, 2014). According to VA suicide surveillance data, the suicide rate for male veterans age 18 to 24 was 79.1 in 2011, compared to 48.3 for males age 25 to 29 and 57.9 for males between 18 to 29 (Kemp, 2014). Ilgen and colleagues (2012) found that the suicide rate for all veterans of OEF/OIF/OND was not significantly higher than veterans from other eras, though they noted that veterans with a psychiatric diagnosis such as depression or posttraumatic stress disorder (PTSD) were at a substantially greater risk for suicide compared to non-OEF/OIF/OND veterans with similar diagnoses (a 4.4-fold increase in risk for OEF/OIF/OND versus a 2.5-fold increase for non-OEF/OIF/OND veterans). Possible explanations for these differences in rates between veteran cohorts are explored throughout this chapter.

Differences Between Services

Significant differences in suicide rates were observed among military services between 2008 and 2012. For instance, in recent years the highest rates were noted in the Army and Marine Corps, while the lowest were found in the Air Force and the Navy (Ramchand, Acosta, Burns, Jaycox, & Pernin, 2011). Historically, however, the Army has had a suicide rate more similar to that of the Navy and Air Force (Ramchand et al., 2011). Data from DOD Suicide Event Reports from 2008 through 2012 show significant differences in suicide rates among the different branches of service, both in terms of the rates themselves as well as the trajectory of those rates. The Army suicide rates for those years went from 18.5 in 2008 to approximately 22 from 2009 to 2011, before rising to a record high 29.7 in 2012. Among all suicide rates recorded during this time period, the Army's rates were consistently higher and also reached the highest rate among all branches. In contrast, the Marine Corps suicide rates started at 19.5 in 2008 and rose to 24.3 in 2012, though the trajectory of the rates varied widely within that timeframe. For example, the 2009 rate rose to 24, then declined in 2010 and 2011 to a low of 14.87 before rapidly rising to 24.3 the following year. The Air Force and Navy suicide rates during this time period also rose from 2008 through 2012, though, like the Marine Corps, both branches showed variable patterns of rate growth. Between 2008 and 2012, the Air Force's suicide rates rose from 12.5 to 15, following a generally consistent growth pattern with a small spike to 15.51 in 2010. The Navy's rates went from 11.6 to 17.8 between 2008 and 2012, with a bumpy trajectory that rose and fell with each subsequent year. (Kinn et al., 2011; Luxton et al., 2012; Luxton et al., 2010; Reger, Luxton, Skopp, Lee, & Gahm, 2009; Smolenski et al., 2013). Considering the relative elevation of rates in the Army and Marine Corps compared to the Air Force and the Navy, it appears then that the service members in the Army and Marine Corps have unique risk factors for suicide that those in the Air Force and Navy do not.

In general, and despite the Marine Corps' drop in 2010/2011, service members in the Army and Marine Corps have had a significantly higher risk for suicide when compared to Air Force, Navy, and civilian populations. One hypothesis for the increase in these two branches is the changed nature of combat in OEF/OIF/OND (e.g., no clear front line, longer and multiple deployments, higher survivability from combat wounds). Service members in the Army may also face unique risk factors due to mental health disorders acquired before deployment. For example, an individual with a substance use disorder that is exposed to combat trauma might be at heightened risk for suicide. This corroborates with LeardMann and colleagues' (2013) prospective longitudinal study (2001–2008) of current and former U.S. military personnel from all service branches, including active and Reserve/National Guard. They found that suicide risk was independently associated with male sex and mental disorders (i.e., depression, bipolar disorder, heavy or binge drinking, and alcohol-related problems) but not with military-specific variables (e.g., combat). Unfortunately, since the data was only collected through 2008, the researchers did not capture suicides in the most recent time period when the rates were at the highest (i.e., 2008–2012). Despite this limitation, the study did include three years of data with statistically significant increases in suicides (i.e., 2006–2008), which highlights the previously mentioned trend in increased suicide rates in the Army and Marine Corps, especially when compared to the Air Force and Navy. Therefore, the increased suicide rate in the Army and Marine Corps may be, in part, related to the prevalence of mental disorders, which were likely exacerbated by military specific stressors both while deployed and when at home over periods of multiple deployments. A more thorough examination of the prevalence of mental disorders in the military is explored later in this chapter.

Summary of Key OEF/OIF/OND Differences

Although the suicide rate in the U.S. Army has traditionally been below demographically matched civilian rates, it has climbed steadily since the beginning of the conflicts in Iraq and Afghanistan. During these tours, suicide rates among service members rose from 9.9 to 22.7 (DOD, 2011; Luxton et al., 2012; Smolenski et al., 2013). In fact, during 2008, and for the first time in U.S. history, service member suicide rates exceeded the rate for demographically matched civilians (Urusano et al., 2014). Suicide rates have increased across all U.S. military service branches—though the Army and Marine Corps have seen heightened numbers when compared to the Air Force and Navy. Trends in the Army's suicide rates generally reflected a steady increase from 2008 to 2012, while the Marine Corps' rates initially climbed with that of the Army's, dropped in 2011, and spiked again in 2012. There are myriad reasons for the increase in suicide rates across all service branches, which are explored more thoroughly later in this chapter; however, there are overarching and deleterious variables that are specific to the conflicts in Iraq and Afghanistan that have likely contributed to heightened stress among service members. For instance, the presence of two theaters of war, an all-volunteer force, an indefinable battlefield, asymmetrical combat, distinctive wounds, lengthy deployments coupled with short dwell periods, and unique demographic characteristics (Hoge, 2010; Ressler, & Schoomaker, 2014; Tanielian & Jaycox, 2008; Urusano et al., 2014) are all challenges specific to OEF/OIF/OND and have likely played a role in increased mental health issues, including suicide.

Relationship Between Deployment and Suicide

Contradictory conclusions. Extant literature on suicide risk among the OEF/OIF/OND cohort often points to the changed nature of deployments that service members must face. OEF/OIF/OND has created a unique context in that service members do not simply deploy for longer periods but also deploy multiple times and have shorter dwell times in between (Gilman et al., 2014). As mentioned earlier, multiple deployments are distinct to OEF/OIF/OND, whereas in other conflicts (e.g., World War II) soldiers were expected to serve one tour (Tanielian & Jaycox, 2008). It seems likely that repeated exposure to high-stress environments would be associated with increased mental health issues, including suicide. In fact, there are a number of studies that have found higher suicide rates among service members who have deployed versus those who have never deployed (Black, Gallaway, Bell, & Ritchie, 2011; Hyman, Ireland, Frost, & Cottrell, 2012; Thomsen, Stander, McWhorter, Rabenhorst, & Milner, 2011). Moreover, multiple deployments (Reger, Gahm, Swanson, & Duma, 2009; Ritchie, 2012; Rona et al., 2007), longer deployments (Rona et al., 2007; Shen, Arkes, & Williams, 2012), and shorter dwell time between deployments (MacGregor, Han, Dougherty, & Galarneau, 2012; Rona et al., 2007) have all been found to increase risk for suicide. Shorter dwell time has also been linked to greater incidence of mental health problems (Department of the Army Office of the Surgeon General, 2009), which, given the link between certain mental illnesses and suicide, suggests a meditational relationship between dwell time and suicide.

Despite research that has connected deployment frequency, dwell time, and deployment duration to suicide, emerging research has begun to question the validity of the relationship of military-specific variables to risk for suicide and posit that deployments do not increase risk for suicide (Gilman et al., 2014; Hyman et al. 2012; LeardMann, 2013; Schoenbaum et al., 2014). For example, Gilman et al. suggest that many of these studies examined deployment history as a whole and thus failed to examine subgroup variations associated with deployment history and suicide. In addition, DOD Suicide Event Reports from 2008 through 2012 indicate that deployment was noted in the history of approximately 50% of service members who died by suicide in those years (Reger et al., 2009; Kinn et al., 2011, Luxton et al., 2012; Luxton et al., 2010, Smolenski, 2013). This suggests that although deployment variables contribute to suicide risk, deployment experience is not a contributing factor in even a majority of suicides.

In order to make sense of the contradictions in findings, it is important to view deployment as an event that is highly variable and that affects several ecological systems in which the service member is encapsulated. For example, Hyman et al. (2012) examined variations associated with deployment history and suicide and found that suicide was associated mental health diagnoses, prescriptions for selective serotonin reuptake inhibitors and sleep medication, a history of deployment, separation or divorce, enlisted rank, and adverse actions leading to reduction in rank. This list alone covers the domains of mental health, interpersonal relationships, pharmaceutical interventions, career status, and disciplinary actions.

In addition to being multifactorial, some of what is known about deployments may seem counterintuitive. For example, Hyman et al. (2012) found that, for Army personnel, multiple deployments actually put service members at reduced risk for suicide, when compared to a single deployment. Hyman et al. suggested that this finding might be due to a "healthy warrior" effect (i.e., service members might be better prepared to face the turmoil's of warfare the second time around). One might argue that the healthy warrior is consistent with historically low suicide rates among service members and may indicate other nonmilitary-specific variables, such as mental health, that have led to the increase in suicide. However, it is still noteworthy that reduced risk for multiple deployments was only observed for the Army and not any other branch of service, lending further to the variability of deployment's effects.

LeardMann et al. (2013) found suicide risk factors among service members are similar to those of civilians. For instance, increased risk was associated with mental disorders (e.g., depression, substance abuse), male gender, and separation from partners. Perhaps the most significant finding was that no military-specific variables such as combat, length of deployment, or number of deployments were associated with increased suicide risk. Schoenbaum et al. (2014) came to a similar conclusion about the contribution of deployment, noting that between 2004 and 2009, suicide rates among service members increased, regardless of whether the individual had deployed or remained at home base—suggesting, again, that military-specific variables such as combat may not be the sole risk factor for suicide.

Interpersonal factors. Military-related stress coupled with a strained support system might also increase risk for suicide. In general, deployment is associated with increased marital strain that occurs before, during, and after deployment (Institute of Medicine, 2013). This is important in the context of suicide, as divorce and separation has been noted as a risk factor for suicide by many researchers (e.g., Hyman et al., 2012; Luxton et al., 2010; Reger et al., 2009). Other studies have found that marriage is linked to low suicide risk among currently deployed service members (Gilman et al., 2014), while others have found that the protective nature of marriage is weaker for service members than it is for the general population (Black et al. 2011; Logan, Skopp, Karch, Reger, & Gahm, 2012). Other important social factors include unit cohesion, which can be described as the ability for an individual to depend on his or her fellow unit members, the degree to which members of the unit cooperate, and whether or not unit members stand up for each other (Mitchell, Gallaway, Millikan, & Bell, 2012). In an investigation of combat exposure, unit cohesion, and suicide risk, Mitchell and colleagues (2012) found that perceived unit cohesion was protective against suicide. In particular, it moderated the relationship between combat exposure suicide risk, where individuals with high combat exposure and high unit cohesion had significantly lower suicide risk than those with high combat exposure and low unit cohesion.

Conditions of the Wars in Afghanistan and Iraq

Improvised explosive devices and TBI. TBI has been referred to as a "signature" wound of the wars in Afghanistan and Iraq (Tanielian & Jaycox, 2008). Indeed, a cursory search of the literature on combat-related TBI yields substantially more results in the context of OEF/OIF/OND service members and veterans compared to Vietnam or other-era veterans. Compared to Vietnam, overall survival rates for wounded OEF/OIF service members were greater, including the rates for the most severely wounded (Goldberg, 2007). Credit for this increased survival is partially due to advancements in body armor, combat trauma medicine, and medevac technologies (Goldberg, 2007). If TBI is the signature wound of OEF/OIF/OND, it may be fair to say that the IED is the signature enemy weapon of these conflicts. In one early exploration of battlefield injuries from 2001 to 2005, a substantial 78% of wounds were due to explosions, the highest percentage recorded in any conflict to date (Owens et al., 2008). Additionally, fewer thoracic injuries were noted compared to other conflicts, attesting to the protective qualities of body armor (Owens et al., 2008). Another examination of explosion injuries sustained during OIF from 2004 to 2007 found that nearly 80% of these injuries were attributable to

IEDs (Eskridge et al., 2012), further suggesting the prevalence of this weapon's use.

Perhaps nowhere has the impact of IEDs been most deeply felt than in the realm of TBI. Once again, improvements in body armor such as Kevlar helmets have shaped the nature of TBI in combat, leading to fewer penetrating brain injuries and increases in nonpenetrating brain injuries (Ling, Bandak, Armonda, Grant, & Ecklund, 2009). There are several classifications of blast injuries that can contribute to TBI. The first is a primary injury caused by a wave of pressure emanating from the explosive device (CDC, 2013). This blast wave can cause injury to several bodily systems, and it is thought that this wave of pressure may be partially responsible for the clinical and neurobiological presentation of TBI, though the precise mechanisms are not yet fully understood (Ling et al., 2009). Injuries classified as secondary involve damage from fragments or debris that become airborne during the explosion, which can result in what is known as penetrating TBI (CDC, 2013; Ling et al., 2009). Damage can also be caused by the force of the body being thrown by the explosive blast, resulting in what is known as a tertiary injury (CDC, 2013). The force of being thrown into structures, vehicles, other people, or the ground can result in the brain slamming against the inside of the skull, which can also result in tissue damage (Magnuson, Leonessa, & Ling, 2012).

Between the years 2000 and the first quarter of 2014, the DOD recorded more than 300,000 TBIs in service members, most from the wars in Iraq and Afghanistan (Defense and Veterans Brain Injury Center [DVBIC], 2014). Among these injuries classified by DVBIC, 82.4% were considered mild TBI (mTBI), which refers to an injury to the head that results in less than 24 hours of disorientation and/or memory loss and less than 30 minutes of unconsciousness. It is worth noting that each case recorded represents a distinct person and is based on the first injury he or she reported. Because of this, these data cannot account for individuals who may have sustained another, more severe TBI after their initial mTBI was recorded. Therefore, the actual number of TBIs sustained in this 14-year period is difficult to ascertain from these data. It also important to note that these numbers constitute only reported TBIs and do not account for the many mild brain insults that were not reported by service members for reasons such as not wanting to be taken off duty or not seeing the need to report due to the mildness of symptoms.

The links between TBI and suicide, particularly in military and veteran populations, are not entirely clear.[2] Studies of civilians have noted individuals with TBI are at an increased risk for suicidal ideation and attempts (Simpson & Tate, 2002), as well as increased risk for completed suicide (Teasdale & Engberg, 2001). Simpson and Tate suggested that this increased risk might be due to the psychiatric disturbances that happen in the wake of such an injury, such as depression. A review of studies of TBI in service members and veterans has yielded some support for the role of mental illness in TBI and suicide. For example, in one study of soldiers in Iraq referred for assessment and treatment related to mTBI, suicidal ideation was associated with comorbid depression or depression and PTSD combined (Bryan, Clemans, Hernandez, & Rudd, 2013). In another sample of OEF/OIF/OND veterans referred for outpatient VA PTSD treatment, researchers did not find a statistically significant difference in suicidality between those who had comorbid mTBI and those who did not (Barnes, Walter, & Chard, 2012). However, they did find that mTBI was associated with more severe PTSD symptoms on the Clinical Administered PTSD Screener, suggesting that mTBI may be related to increased PTSD symptom severity. This is important because greater severity of PTSD symptoms is related to an increase in suicide risk (Freeman, Roca, & Moore, 2000). Interestingly, there is some evidence for a dose-response effect related to TBI, symptoms of depression and PTSD, and suicidal ideation and behavior. One study conducted by Bryan and Clemans (2013) on a sample of soldiers in an outpatient TBI clinic indicated that the greater the number of TBIs sustained, the greater the symptoms of PTSD and depression, as well as the risk for past-year suicidal ideation. The effect on suicide risk was observed even after depression and PTSD symptom severity was controlled.

The relationship between TBI and suicide has also been shown to be unmediated by psychiatric comorbidities. In one study of veterans who died by suicide and who utilized VA medical services between 2001 and 2006, individuals with TBI were 1.5 times more likely to die by suicide than veterans without TBI (Brenner, Ignacio, & Blow, 2011). These effects were not accounted for by demographic factors or co-occurring disorders that may impact suicidality. In contrast, another investigation of airmen who died by suicide between 2001 and 2009 did not find any association between the diagnosis of a TBI and death by suicide (Skopp, Trofimovich, Grimes, Oetjen-Gerdes, & Gahm, 2012). Given the correlational nature of studies investigating the relationship among TBI, suicide, and possible mediating psychiatric variables, as well as the conflicting results obtained by other studies finding unmediated or absent ties between TBI and suicide, it is clear that more research is needed to understand whether or not TBI contributes suicidality and, if so, how and to what degree.

PTSD. Greater survival in the wake of traumatic, life-threatening events may also contribute to another common "invisible wound" of war: PTSD. Other chapters in this volume focus on the relationships among combat exposure, trauma, and suicide. However, in the context of the wars in Iraq and Afghanistan, it is important to note that, among veterans in the VA medical system, depression and PTSD are associated with increased risk for suicide (Bossarte et al., 2012; Jakupcak et al., 2009). In a substantial sample of soldiers and Marines who deployed to Iraq and Afghanistan, Hoge and colleagues (2004) found that war theater deployments were related to increases in both depression and PTSD symptoms from pre- to postdeployment, indicating that deployment experiences—especially combat violence—contributed to these increases in reporting.

There have also been observable differences in postdeployment mental health outcomes when comparing active duty to National Guard service members. In one study Milliken, Auchterlonie, and Hoge (2007) gave both active duty and Reserve/Guard component soldiers who had just returned from deployment a postdeployment health assessment and a reassessment three to six months later. When comparing the mental health of National Guard service members to active duty service members on the follow-up assessment, the researchers found that soldiers in the National Guard reported more mental and physical health complaints than their active duty counterparts. Specifically, they reported higher levels of PTSD, depression, and overall mental health risk. One proposed reason for this increased risk was lack of unit support postdeployment, the result of returning to a one weekend a month, two weeks a year training schedule (Milliken et al., 2007).

In general, OEF/OIF/OND veterans seeking services in the VA have a higher prevalence of psychiatric diagnoses compared to non-OEF/OIF/OND veterans (Ilgen et al., 2012). Specifically, in terms of PTSD in OEF/OIF/OND veterans, the diagnosis has been related to increased suicidal ideation (Guerra & Calhoun, 2011). Another study of OEF/OIF/OND veterans found that those with a PTSD diagnosis are four times more likely to report suicidal ideation compared to those without the diagnosis, and veterans with two or more psychiatric comorbidities were 5.7 times more likely to have suicidal ideation. These studies confirm a meta-analysis of 16 studies on PTSD and suicide in veterans from multiple cohorts, which notes that although the association between PTSD and suicide is complex and not entirely clear, PTSD is associated with greater suicidal ideation and behavior in veterans (Pompili et al., 2013).

Sleep. An additional contributor to suicide in OEF/OIF/OND service members and veterans is poor sleep. One study of the relationship between sleep and mental health outcomes by Luxton et al. (2011) found that less than six hours of sleep in theater and feelings of tiredness that substantially interferes with work activities were both associated with TBI, more severe PTSD symptoms, suicidal ideation, and suicide attempts. Depressive symptoms were also significantly related to short sleep duration. These findings are important because they tie an OEF/OIF/OND deployment factor (i.e., disturbed sleep) to suicide and mental health conditions associated with suicide. They are also important because disturbance in sleep may persist even after returning from deployment (Luxton et al., 2011), which implies that these sleep difficulties may disturb these service members as they transition to civilian life and veteran status.

Access to care. One factor that may explain greater observed mental health symptoms—including suicidality—in OEF/OIF/OND service members is access to care. In their comparison study of active duty versus National Guard postdeployment mental health, Milliken et al. (2007) proposed that a push to access time-limited medical benefits may have accounted for greater reporting of mental health problems in National Guard soldiers. At the time the study was conducted, Reserve and National Guard service members had only six months to access TRICARE and 24 months to access VA benefits after transitioning back to civilian status (Milliken et al., 2007). It was not until the National Defense Authorization Act of 2008 that the VA changed its policies to allow five years of free care for veterans with combat theater–related mental and physical health problems, provided that those veterans served during and after the Persian Gulf War (Department of Veterans Affairs, 2009). It is possible that relatively elevated rates of mental health diagnoses and reported suicidal ideation and behavior in OEF/OIF/OND veterans are related to better access to care rather than a greater incidence of these factors compared to veterans from other eras. This may mean that more veterans are seeking care, including younger, more at-risk veterans. Another hypothesis explaining the increase in suicides in OEF/OIF/OND veterans is the VA's enhanced focus on suicide event tracking and reporting, which began in earnest in 2007 (Kemp, 2014).

Although an increase in the length of time afforded to attain medical benefits may partially explain an increase in mental health diagnoses, it is important to note that the VA has also struggled to provide timely medical and mental health services to veterans. Such difficulties have included failure to fill clinical mental health job positions in psychiatry, psychology, social work, and nursing (*Honoring the Commitment*, 2013), difficulties scheduling timely medical appointments (Department of Veterans Affairs Office of the Inspector General, 2013), and intentional alteration of electronic wait lists to obfuscate unacceptably long wait times (Department of Veterans Affairs Office of the Inspector General, 2014). Although not definitively substantiated, is possible that the compromised quality of care observed at certain VA medical centers may have contributed to the exacerbation of mental health conditions in veterans seeking care, including suicide ideations, attempts, and completions.

Other Risk Factors

In addition to the stressors endured throughout the deployment cycle, research has suggested other OEF/OIF/OND era factors that may have contributed to the rise in suicides. The Army Study to Assess Risk and Resilience in Service members (Army STARRS) is a multicomponent epidemiological study designed to generate evidence-based recommendations to reduce Army suicides and increase knowledge about risk and resilience factors for suicidality and its psychopathological correlates. Using data from the Army STARRS study, Nock et al. (2014) studied suicidal behavior among service members before and after serving in the Army and found that about 13.9% of soldiers had considered suicide at some point in their lifetime, 5.3% had made a suicide plan, and 2.4% had attempted suicide. Importantly, 47% to 60% of these outcomes first occurred prior to enlistment. This study was consistent with Schoenbaum et al. (2014) and LeardMann et al. (2013) in that service member suicidal risk factors mirrored that of the civilian population (i.e., suicidality was not independently related to military specific variables). Moreover, Ursano et al. (2014) noted that about 25% of nondeployed U.S. Army personnel met criteria for a substance use, depression, anxiety, or disruptive disorder. It is important to note that three-fourths of these disorders had pre-enlistment onset, which highlights the importance of early identification. Notably, suicidal behavior among the service members with mental disorders matched those of civilian populations with similar disorders and did not exceed those of the civilian population until they had entered the Army (Ursano et al., 2014). This suggests that perhaps mental disorders compounded by military-specific stress may be a leading contributor to suicidal behavior among OEF/OIF/OND service members and veterans.

In the 2010 Department of the Army report on health promotion, risk reduction, and suicide prevention in the force, leaders expressed concern about the incidence of predeployment, pre-enlistment risk factors that may contribute to suicidal ideation, behavior, and completion. This report indicated that between 2004 and 2007, the number of waivers the Army granted for drug and conduct offenses increased significantly. A waiver essentially grants a recruit entry into the military despite having a criminal or drug use background that would have otherwise disqualified them from service. Simultaneously, the report indicated that the number of Chapter 11 discharges decreased, meaning that the Army increasingly retained recruits who demonstrated a lack of capability, effort, and/or ability to adapt to military life. In addition to not discharging unfit recruits, the Army also allowed a substantial number of soldiers with two or more felony charges to remain in service. This was possible because, at the time of the report, there was no regulation requiring that non-drug-related felonies result in discharge. For drug use and possession felonies, which constituted the overwhelming majority of felonies recorded by the Army from 2001 to 2009, the report indicated that as many as 40% of drug-positive urinalyses were never referred for felony investigation. The report also found that misdemeanor crimes rose significantly between 2004 and 2009.

In terms of suicide risk, the recruitment and retention of high-risk soldiers—particularly those with criminal histories—is important. In the same 2010 report that outlined the decline in so-called "good discipline" of the force, the Army also noted that for individuals who died by suicide between 2005 and 2009, approximately 25% had a history of investigation for a founded misdemeanor or felony. In the context of OEF and OIF, the report directly cited increased operations tempo as a reason substandard recruits were not discharged in a timely fashion. Although it is unclear if recruitment waivers and retention of felony-level offenders was also due to increased deployments in support of OEF/OIF, it is not unreasonable to assume that the Army's extensive involvement in these wars precipitated at least some of these practices.

As researchers attempt to elucidate the cause of the increased suicide rates in OEF/OIF/OND, more are led to tease apart the extent to which suicide risk is due to pre-existing psychopathology such as substance abuse. For instance, the Department of the Army (2010) found that in 2009 there were 146 deaths in the Army due to high-risk behavior such as drug overdose. Furthermore, approximately 29% of those who died by suicide had drugs and/or alcohol in their systems at the time of death (Department of the Army, 2010). This is consistent with LeardMann et al.'s (2013) finding that mental health problems, including alcohol-related problems, were significantly associated with an increase in the risk of suicide among service members. As we know with the civilian population, substance use and life stressors (particularly related to family and work) are a deadly cocktail for suicide risk (Bush et al., 2013; Dobscha et al., 2014; Kochanski-Ruscio, Carreno-Ponce, DeYoung, Grammer, & Ghahramanlou-Holloway, 2014). In sum, the increase of suicide rates may be related to prior mental health diagnoses—primarily those associated with risky behaviors such as substance and alcohol abuse.

Summary of Reasons for Increased Suicide in OEF/OIF/OND Service Members

It appears that suicide risk for service members is independently associated with variables that are commonly associated with suicide—such as depression, alcohol/substance use, male gender, loss of face, and marital strife—and not as strongly associated with military-specific variables as previously thought (e.g., deployments). However, despite the fact that the contribution of deployment variables to suicide is not as prominent or simple as once believed, it is still important to consider these variables as possible risk factors. Veterans and military personnel who served in the conflicts in Iraq and Afghanistan survived battlefield injuries at rates unprecedented in recent military history. In addition, service members in these conflicts faced frequent deployments, low dwell time, long duty hours, and frequent combat engagements that often consisted of IED attacks of various forms. These wartime circumstances led to an increase in afflictions such as TBI, PTSD, and depression, all of which have a relationship to suicide. However, analysis of the many studies attempting to investigate these relationships yields unclear relationships among these and other variables, such as premilitary psychiatric morbidity, criminal background, unit cohesion, recruitment and retention practices, and social support. There appears in many cases to be a diathesis-stress effect, where previously existing mental health disorders and psychosocial stressors in service members are compounded by military-specific stressors within both the deployed (e.g., combat) and garrison environments. In order to better understand how these factors interact and contribute to suicidality, more research exploring mediating and moderating relationships will be especially important.

SUICIDE PREVENTION IN THE DEPARTMENT OF DEFENSE

In the midst of the rise in military suicides in the mid- to late 2000s, the DOD and its branches expanded pre-existing suicide prevention programs in an effort to stave off the rise in deaths (Ramchand, et al., 2011). One of the DOD's most substantial decisions was to streamline suicide event reporting from all branches of service into a central reporting system, a change from previous practices of keeping such reporting compartmentalized by branch (Ramchand et al., 2011). Through the Defense Centers of Excellence for Psychological Health and Traumatic Brain Injury (2013), the DOD also created the Real Warriors initiative, a multimedia campaign featuring stories of service members and veterans who experienced psychological health challenges. One aim of this campaign was to normalize these challenges, as well as help-seeking behaviors, and demonstrate to service members that help-seeking does not inherently result in career consequences such as loss of promotions or security clearances. In 2009, the DOD also expanded its annual suicide prevention conferences to include the VA—an effort to bridge the gap between these separate but entwined organizations (Ramchand et al., 2011).

All branches of service made swift and significant suicide prevention policy revisions to address the rise in suicide rates. Although each branch has separate and varied suicide prevention tactics and strategies, there are similarities that extend across all programs. One of the most common is the emphasis on peer intervention. All branches of service have adopted peer intervention models such as Ask, Care, Escort (Department of the Army, 2010b), Ask, Care, Treat (Department of the Navy, 2011), and React, Ask, Care, Escort (Werbel, 2010). The Army and the Marine Corps in particular have emphasized peer intervention, with the Marine Corps suicide prevention policy indicating that a failure to help a suicidal fellow Marine is in direct conflict with the Marine Corps ethos (Department of the Navy, 2012; Ramchand et al., 2011).

All branches of service have also emphasized the importance of leadership, which includes making suicide prevention a priority, setting a tone that encourages help-seeking, and demonstrating skill and willingness to intervene with suicidal subordinates and colleagues (Department of the Air Force, 2012; Department of the Army, 2010b; Department of the Navy, 2009, 2012). In addition to ensuring that peers and leaders are trained to successfully intervene with suicidal service members, the Army also advanced its gatekeeper training program, which was designed to provide more advanced skills training to select individuals. By regulation, gatekeepers include first-line military and civilian supervisors, chaplains, medical professionals, attorneys, and other individuals with a high likelihood of encountering service members in medically and psychosocially vulnerable situations (Department of the Army, 2010b).

The branches of service also turned attention toward expanding programs to engender resilience in the ranks. Navy and Marine Corps Combat and Operational Stress Control, for example, organizes levels of stress into a four-colored classification system (green, yellow, orange, red). This coding system is used to assess the status of self and others and can serve as a guide for various degrees of intervention, depending on the severity of the stress state (Department of the Navy, 2010). This functions as a resiliency-developing model, because it encourages checking in with oneself, identifying current states of stress, and acting to reduce stress using the classification system to inform next steps. In a similar effort to prevent suicides by increasing overall well-being, the Army developed resilience programs such as Comprehensive Soldier and Family Fitness (CSF2), which taps into physical, emotional, social, spiritual, and family aspects of a soldier's life (Department of the Army, 2013). CSF2 is intended to increase baseline psychological health and flexibility, with the ultimate goal of decreasing mental health symptoms and diagnoses (Cornum, Matthews, & Seligman, 2011). In an evaluation of the program, CSF2 was shown to indirectly lead to reductions in mental health diagnoses through improvements in optimism and adaptability (Harms, Herian, Krasikova, Vanhove, & Lester, 2013).

It is important to note that many of these advances were developed or expanded in rapid response to the rise in suicide rates among service members serving in the OEF/OIF/OND era. Despite the aggressive bolstering of these programs from 2008 and beyond, the raw number of suicides in the four branches of service from 2010 to 2012 decreased only for the Air Force (59 in 2010 to 50 in 2011 and 2013). In the other branches of service, the suicide numbers actually increased in that time period (Smolenski et al., 2013). Simultaneous to these suicide prevention program modifications, the DOD also made a substantial commitment to preserving and increasing the daily quality of life for service members and families, especially in the context of the protracted combat engagements in Iraq and Afghanistan. These efforts included improvements in educational benefits for service members and families, recreation, financial assistance, housing, and timely access to mental health care (DOD, 2009). Although an empirically valid evaluation of the efficacy of these latest efforts has not yet emerged in the scientific literature, it seems from the increases in both raw suicide numbers and rates of suicide that these programs have not been as effective as was hoped and expected. This supposition points to the complexity of suicide and its prevention, where even the most heartfelt and stringent program improvements may not yield desired reductions in suicides.

CONCLUSION

Sustained military engagement in Afghanistan and Iraq brought over 2 million service members into unpredictable combat environments fraught with such dangers as IEDs and enemy combatants who blend in seamlessly with the civilian population. For many service members, deployments to these theaters of war were long, frequent, and coupled with insufficient rest time in-between. At the same time, suicide rates in all branches of service began to climb to record levels, with the highest rates seen in the branches with the greatest percentage of forces devoted to combat operations: the Army and the Marine Corps. This trend was also noted by the VA, which recorded a similar uptick in suicides for veterans of the OEF/OIF/OND era relative to veterans of other cohorts. Given the operational stressors inherent in these conflicts, it was natural for laypeople, the media, and researchers to look to deployment and combat stressors as primary reasons for these increases in suicide rates.

However, the research on the role of deployment and wartime mental health sequelae such as PTSD, TBI, and other mental health outcomes suggests that the connections between these variables and suicide is, in many cases, more complicated than originally expected. Studies exploring these relationships were sometimes contradictory, such as those investigating the effects of deployment on suicide rates or those looking at the role of TBI in suicidal ideation and behavior. Often these studies yielded even more questions about mediating and moderating factors that explain the relationships between these variables and suicide. Some relationships were clearer, such as that between PTSD and suicidality in veterans, though even these relationships also appear to be affected by factors such as psychiatric comorbidity, symptom severity, social support, unit cohesion, sleep quality, and resiliency.

Attempts to uncover the contributors to the rise in suicide rates also revealed information suggesting that suicide contributors for some service members, such as mental health problems, may have predeployment or even pre-enlistment origins. Additionally, the Army reported that for several years in the mid-2000s, recruitment standards were lowered and unfit soldiers were retained, many of whom had histories of criminal and substance abuse behaviors that placed them at increased risk for suicide. Research in this vein suggests that many service members may enter the military with predisposing suicide risk factors or may have been allowed to stay in service when, in fact, discharge may have been more appropriate. It is possible that military stressors such as war-zone deployment and combat may have compounded these nonmilitary-related risk factors, placing already vulnerable individuals at an even greater risk for suicide. However, it is important to note that psychological autopsies of suicide-deceased service members indicate that approximately 50% of service members who died by suicide from 2008 to 2012 had no history of deployment. This indicates that many suicidal service members in the OEF/OIF/OND era may have been affected by the same problems with living that are faced by civilians.

Despite the complications inherent in mapping out contributors to military suicide, researchers have uncovered information that could be invaluable for helping veterans of these conflicts as they serve out their enlistments and move once again into the civilian sector. However, the value of this research will depend on how well it is translated into practice within the DOD, the VA, and other medical and mental health care systems that serve veterans and military personnel. Time will tell if these current research findings and the questions they evoke will prove beneficial in assisting these veterans, as well as other veterans from future conflicts who will undoubtedly face similar complex psychosocial challenges both on and off the battlefield.

FUTURE DIRECTIONS

1. There remains a need to better understand how previously existing mental health disorders and psychosocial stressors in service members are compounded by military-specific stressors within both the deployed (e.g., combat) and garrison environments and how these factors interact and contribute to suicidality. More research exploring mediating and moderating relationships will be especially important.

2. Psychological autopsies of suicide-deceased service members indicate that approximately 50% of service members who died by suicide from 2008 to 2012 had no history of deployment. This indicates that many suicidal service members in the OEF/OIF/OND era may have been affected by the same problems with living that are faced by civilians. Further research will need to be done to disentangle this finding.
3. There remains a need to investigate the effects of deployment on suicide rates or those looking at the role of TBI in suicidal ideation and behavior. Often these studies yield even more questions about mediating and moderating factors that explain the relationships between these variables and suicide.

NOTES

1. For the duration of this chapter, suicide rates are presented using the rate alone (e.g., 19.2), under the assumption that this number is out of 100,000 individuals.

2. An extensive overview of TBI and suicide can be accessed in chapter 16 of this volume. However, because the current chapter focuses on service member experiences in OEF/OIF, and because TBI is such a hallmark of these conflicts, a focused summary of the topic is included here.

REFERENCES

Barnes, S. M., Walter, K. H., & Chard, K. M. (2012). Does a history of mild traumatic brain injury increase suicide risk in veterans with PTSD?. *Rehabilitation Psychology*, 57(1), 18-26. doi:10.1037/a0027007

Black, S. A., Gallaway, M., Bell, M. R., & Ritchie, E. C. (2011). Prevalence and risk factors associated with suicides of Army soldiers 2001–2009. *Military Psychology*, 23(4), 433–451. doi:10.1037/h0094766

Bongar, B., & Sullivan, G. R. (2013). Treating and managing care of the suicidal patient. In G. P. Koocher, J. C. Norcross, & B. A. Greene (Eds.), *Psychologists' desk reference* (3rd ed., pp. 185–190). New York: Oxford University Press.

Bossarte, R. M., Knox, K. L., Piegari, R., Altieri, J., Kemp, J., & Katz, I. R. (2012). Prevalence and characteristics of suicide ideation and attempts among active military and veteran participants in a national health survey. *American Journal of Public Health*, 102(Suppl. 1), S38–S40. doi:10.2105/AJPH.2011.300487

Brenner, L. A., Ignacio, R. V., & Blow, F. C. (2011). Suicide and traumatic brain injury among individuals seeking Veterans Health Administration services. *The Journal of Head Trauma Rehabilitation*, 26(4), 257–264. doi:10.1097/HTR.0b013e31821fdb6e

Bryan, C. J., & Clemans, T. A. (2013). Repetitive traumatic brain injury, psychological symptoms, and suicide risk in a clinical sample of deployed military personnel. *JAMA Psychiatry*, 70(7), 686–691. doi:10.1001/jamapsychiatry.2013.1093

Bryan, C. J., Clemans, T. A., Hernandez, A. M., & Rudd, M. D. (2013). Loss of consciousness, depression, posttraumatic stress disorder, and suicide risk among deployed military personnel with mild traumatic brain injury. *The Journal of Head Trauma Rehabilitation*, 28(1), 13–20. doi:10.1097/HTR.0b013e31826c73cc

Bush, N. E., Reger, M. A., Luxton, D. D., Skopp, N. A., Kinn, J., Smolenski, D., & Gahm, G. A. (2013). Suicides and suicide attempts in the U.S. military, 2008–2010. *Suicide and Life-Threatening Behavior*, 43(3), 262–273. doi:10.1111/sltb.12012

Centers for Disease Control and Prevention. (2009). *United States suicide injury deaths and rates per 100,000*. Retrieved from http://webappa.cdc.gov/sasweb/ncipc/mortrate10_us.html

Centers for Disease Control and Prevention. (2013). *Explosions and blast injuries: A primer for clinicians*. Retrieved from http://emergency.cdc.gov/masscasualties/explosions.asp

Cornum, R., Matthews, M. D., & Seligman, M. E. P. (2011). Comprehensive soldier fitness: Building resilience in a challenging institutional context. *American Psychologist*, 66(1), 4–9. doi:10.1037/a0021420

Defense and Veterans Brain Injury Center. (2014). *DoD numbers for traumatic brain injury worldwide—totals*. Retrieved from http://dvbic.dcoe.mil/sites/default/files/uploads/Worldwide%20Totals%202000-2014Q1.pdf

Defense Centers of Excellence for Psychological Health and Traumatic Brain Injury. (2013). *About us*. Retrieved from http://www.realwarriors.net/aboutus

Department of the Air Force. (2012). *Suicide prevention program* (Air Force instruction 90-505). Retrieved from http://static.e-publishing.af.mil/production/1/af_sg/publication/afi90-505/afi90-505.pdf

Department of the Army. (2010a). *Army health promotion, risk reduction, suicide prevention report 2010*. Retrieved from http://csf2.army.mil/downloads/HP-RR-SPReport2010.pdf

Department of the Army. (2010b). *Health promotion, risk reduction, and suicide prevention* (Department of the Army pamphlet 600-24). Retrieved from http://www.apd.army.mil/pdffiles/p600_24.pdf

Department of the Army. (2013). *Comprehensive soldier and family fitness program* (Army directive 2013-07). Retrieved from http://armypubs.army.mil/epubs/pdf/ad2013_07.pdf

Department of the Army, Office of the Surgeon, Multinational Force–Iraq and Office of the Surgeon General, US Army Medical Command, Mental Health Advisory Team. (2009). *Operation Iraqi Freedom 07-09.* Retrieved from http://armymedicine.mil/Documents/MHAT-VI-OIF-Redacted.pdf

Department of Defense. (2009). Report of the 2nd quadrennial quality of life review. Retrieved from http://www.militaryonesource.mil/12038/MOS/Reports/Quadrennial%20Quality%20of%20Life%20Review%202009.pdf

Department of Defense. (2011). *Active duty military death rates per 100,000 serving.* Retrieved from https://www.dmdc.osd.mil/dcas/pages/report_number_serve.xhtml

Department of the Navy. (2009). *Suicide prevention program* (OPNAV instruction 1720.4A). Retrieved from http://www.med.navy.mil/sites/nmcphc/Documents/health-promotion-wellness/psychological-emotional-wellbeing/opnav-inst-1720-4a-navy-suicide-prevention-policy.pdf

Department of the Navy. (2010). *Combat and operational stress control.* Retrieved from http://www.med.navy.mil/sites/nmcsd/nccosc/coscConference/Documents/COSC%20MRCP%20NTTP%20Doctrine.pdf

Department of the Navy. (2011). *Commanding officer's suicide prevention and response toolbox.* Retrieved from http://www.public.navy.mil/bupers-npc/support/21st_Century_Sailor/suicide_prevention/Documents/Suicide%20Prevention%20Commander%20Toolbox%2015%20Nov.pdf

Department of the Navy. (2012). *Marine Corps suicide prevention program* (Marine Corps order 1720.2). Retrieved from http://www.marines.mil/Portals/59/Publications/MCO%201720.2.pdf

Department of Veterans Affairs. (2009). *Combat veteran health care benefits and co-pay exemptions post-discharge from military service* (VHA directive 2008-054). Retrieved from http://www.va.gov/vhapublications/ViewPublication.asp?pub_ID=1758

Department of Veterans Affairs Office of Inspector General. (2013). *Healthcare inspection: Appointment scheduling and access patient call center VA San Diego Healthcare System San Diego, California* (Report No. 12-04108-96). Retrieved from http://www.va.gov/oig/pubs/VAOIG-12-04108-96.pdf

Department of Veterans Affairs Office of Inspector General. (2014). *Veterans Health Administration interim report: Review of patient wait times, scheduling practices, and alleged patient deaths at the Phoenix Health Care System* (Report No. 14-02603-178). Retrieved from http://www.va.gov/oig/pubs/vaoig-14-02603-178.pdf

Dobscha, S., Denneson, L., Kovas, A., Corson, K., Helmer, D., & Bair, M. (2014). Primary care clinician responses to positive suicidal ideation risk assessments in veterans of Iraq and Afghanistan. *General Hospital Psychiatry, 36*(3), 310–317. doi:10.1016/j.genhosppsych.2013.11.007

Eskridge, S. L., Macera, C. A., Galarneau, M. R., Holbrook, T. L., Woodruff, S. I., MacGregor, A. J., . . . Shaffer, R. A. (2012). Injuries from combat explosions in Iraq: Injury type, location, and severity. *Injury, 43*(10), 1678–1682. doi:doi:10.1016/j.injury.2012.05.027

Fowler, J. (2013). Core principles in treating suicidal patients. *Psychotherapy, 50*(3), 268–272. doi:10.1037/a0032030

Freeman, T. W., Roca, V., & Moore, W. M. (2000). A comparison of chronic combat-related posttraumatic stress disorder (PTSD) patients with and without a history of suicide attempt. *Journal of Nervous and Mental Disease, 188*(7), 460–463. doi:10.1097/00005053-200007000-00011

Gibbs, N., & Thompson, M. (2012, July 23). The war on suicide? *Time Magazine, 180*(4). Retrieved from http://content.time.com/time/magazine/article/0,9171,2119337,00.html

Gilman, S. E., Bromet, E. J., Cox, K. L., Colpe, L. J., Fullerton, C. S., Gruber, M. J., . . . Kessler, R. C. (2014). Sociodemographic and career history predictors of suicide mortality in the United States Army 2004–2009. *Psychological Medicine, 44*(12), 2579–2592. doi:10.1017/S003329171400018X.

Goldberg, M. S. (2007). *Projecting the costs to care for veterans of U.S. military operations in Iraq and Afghanistan: Statement before the Committee on Veterans' Affairs.* Retrieved from http://www.cbo.gov/sites/default/files/cbofiles/ftpdocs/87xx/doc8710/10-17-va-admin_testimony.pdf

Griffith, J. (2012a). Correlates of suicide among Army National Guard soldiers. *Military Psychology, 24*(6), 568–591. doi:10.1080/08995605.2012.736324

Griffith, J. (2012b). Suicide in the Army National Guard: An empirical inquiry. *Suicide and Life-Threatening Behavior, 42*(1), 104–119. doi:10.1111/j.1943-278X.2011.00075.x

Guerra, V., & Calhoun, P. S. (2011). Examining the relation between posttraumatic stress disorder and suicidal ideation in an OEF/OIF veteran sample. *Journal of Anxiety Disorders, 25*(1), 12–18. doi:10.1016/j.janxdis.2010.06.025

Harms, P. D., Herian, M. N., Krasikova, D. B., Vanhove, A., & Lester, P. B. (2013). *The Comprehensive Soldier and Family Fitness Program evaluation report #4: Evaluation of resilience training and mental and behavioral health outcomes.* Retrieved from http://www.ppc.sas.upenn.edu/csftechreport-4mrt.pdf

Hoge, C. W. (2010). *Once a warrior—always a warrior: Navigating the transition from combat to home.* Guilford, CT: GPP Life.

Hoge, C. W., Castro, C. A., Messer, S. C., McGurk, D., Cotting, D. I., & Koffman, R. L. (2004). Combat duty in Iraq and Afghanistan, mental health problems, and barriers to care. *The New England Journal of Medicine 351*(1), 13–22. doi:doi:10.1056/NEJMoa040603

Honoring the commitment: Overcoming barriers to quality mental health care for veterans. (2013, February 13). Statement of Office of Inspector General, Department of Veterans Affairs. Retrieved from http://www.va.gov/oig/pubs/statements/VAOIG-statement-20130213.pdf

Hyman, J., Ireland, R., Frost, L., & Cottrell, L. (2012). Suicide incidence and risk factors in an active duty US military population. *American Journal of Public Health, 102*(Suppl. 1), S138–S146. doi:10.2105/AJPH.2011.300484

Ilgen, M. A., McCarthy, J. F., Ignacio, R. V., Bohnert, A. S. B.,; Valenstein, M., Blow, F. C., & Katz, I. R. (2012). Psychopathology, Iraq and Afghanistan service, and suicide among Veterans Health Administration patients. *Journal of Consulting and Clinical Psychology, 80*(3), 323–330. doi:10.1037/a0028266

Institute of Medicine. (2013). *Returning home from Iraq and Afghanistan: Assessment of readjustment needs of veterans, service members, and their families.* Washington, DC: National Academies Press.

Jakupcak, M., Cook, J., Imel, Z., Fontana, A., Rosenheck, R., & McFall, M. (2009). Posttraumatic stress disorder as a risk factor for suicidal ideation in Iraq and Afghanistan war veterans. *Journal of Traumatic Stress, 22*(4), 303–306. doi:10.1002/jts.20423

Kemp, J. E. (2014). *Suicide rates in VHA patients through 2011 with comparisons with other Americans and other veterans through 2010.* Retrieved from http://www.mentalhealth.va.gov/docs/Suicide_Data_Report_Update_January_2014.pdf

Kinn, J. T., Luxton, D. D., Reger, M. A., Gahm, G. A., Skopp, N. A., & Bush, N. E. (2011). *Department of Defense suicide event report (DoDSER) calendar year 2010 annual report.* Retrieved from http://www.t2.health.mil/sites/default/files/dodser/DoDSER_2010_Annual_Report.pdf

Kochanski-Ruscio, K., Carreno-Ponce, J., DeYoung, K., Grammer, G., & Ghahramanlou-Holloway, M. (2014). Diagnostic and psychosocial differences in psychiatrically hospitalized military service members with single versus multiple suicide attempts. *Comprehensive Psychiatry, 55*(3), 450–456. doi:10.1016/j.comppsych.2013.10.012

LeardMann, C. A., Powell, T. M., Smith, T. C., Bell, M. R., Smith, B., Boyko, E. J., . . . Hoge, C. W. (2013). Risk factors associated with suicide in current and former US military personnel. *JAMA: Journal of the American Medical Association, 310*(5), 496–506. doi:10.1001/jama.2013.65164

Ling, G., Bandak, F., Armonda, R., Grant, G., & Ecklund, J. (2009). Explosive blast neurotrauma. *Journal of Neurotrauma, 26*(6), 815–825. doi:10.1089/neu.2007.0484

Logan, J., Skopp, N. A., Karch, D., Reger, M. A., & Gahm, G. A. (2012). Characteristics of suicides among US Army active duty personnel in 17 US states from 2005 to 2007. *American Journal of Public Health, 102*(Suppl.), S40–S44. doi:10.2105/AJPH.2011.300481

Luxton, D. D., Greenburg, D., Ryan, J., Niven, A., Wheeler, G., & Mysliwiec, V. (2011). Prevalence and impact of short sleep duration in redeployed OIF soldiers. *Sleep: Journal of Sleep and Sleep Disorders Research, 34*(9), 1189–1195. doi:10.5665/sleep.1236

Luxton, D. D., Osenbach, J. E., Reger, M. A., Smolenski, D. J., Skopp, N. A., Bush, N. E., & Gahm, G. A. (2012). *Department of Defense suicide event report (DoDSER) calendar year 2011 annual report.* Retrieved from http://www.t2.health.mil/sites/default/files/dodser/DoDSER_2011_Annual_Report.pdf

Luxton, D. D., Skopp. N. A., Kinn, J. T., Bush, N. E., Reger, M. A., & Gahm, G. A. (2010). *Department of Defense suicide event report (DoDSER) calendar year 2009 annual report.* Retrieved from http://www.t2.health.mil/sites/default/files/dodser/DoDSER_2009_Annual_Report.pdf

MacGregor, A. J., Han, P. P., Dougherty, A. L., & Galarneau, M. R. (2012). Effect of dwell time on the mental health of US military personnel with multiple combat tours. *American Journal of Public Health, 102*(Suppl. 1), A55–A59. doi:10.2105/AJPH.2011.300341

Magnuson, J., Leonessa, F., & Ling, G. S. F. (2012). Neuropathology of explosive blast traumatic brain injury. *Current Neurology and Neuroscience Reports, 12*(5), 570–579. doi:10.1007/s11910-012-0303-6

Milliken, C. S., Auchterlonie, J. L., & Hoge, C. W. (2007). Longitudinal assessment of mental health problems among active and reserve component soldiers returning from the Iraq War. *JAMA: Journal of the American Medical Association*, 298(18), 2141–2148. doi:10.1001/jama.298.18.2141

Mitchell, M. M., Gallaway, M. S., Millikan, A. M., & Bell, M. (2012). Interaction of combat exposure and unit cohesion in predicting suicide-related ideation among post-deployment soldiers. *Suicide and Life-Threatening Behavior*, 42(5), 486–494. doi:10.1111/j.1943-278X.2012.00106.x

Nakagawa, A., Grunebaum, M. F., Oquendo, M. A., Burke, A. K., Kashima, H., & Mann, J. (2009). Clinical correlates of planned, more lethal suicide attempts in major depressive disorder. *Journal of Affective Disorders*, 112(1–3), 237–242. doi:10.1016/j.jad.2008.03.021

Nock, M., Stein, M., Heeringa, S., Ursano, R., Colpe, L., Fullerton, C., . . . Kessler, R. (2014). Prevalence and correlates of suicidal behavior among soldiers: Results from the Army Study to Assess Risk and Resilience in Servicemembers (Army STARRS). *JAMA Psychiatry*, 71(5), 514–522. doi:10.1001/jamapsychiatry.2014.30

Owens, B. D., Kragh, J. F. Jr., Wenke, J. C., Macaitis, J., Wade, C. E., & Holcomb, J. B. (2008). Combat wounds in Operation Iraqi Freedom and Operation Enduring Freedom. *Journal of Trauma: Injury, Infection & Critical Care*, 64(2), 295–299. doi:10.1097/TA.0b013e318163b875

Pompili, M., Sher, L., Serafini, G., Forte, A., Innamorati, M., Dominici, G., . . . Girardi, P. (2013). Posttraumatic stress disorder and suicide risk among veterans: A literature review. *Journal of Nervous and Mental Disease*, 201(9), 802–812. doi:10.1097/NMD.0b013e3182a21458

Ramchand, R., Acosta, J., Burns, R. M., Jaycox, L. H., & Pernin, C. G. (Eds.) (2011). *The war within: Preventing suicide in the U.S. military.* Santa Monica, CA: RAND Corporation.

Reger, M. A., Gahm, G. A., Swanson, R. D., & Duma, S. J. (2009). Association between number of deployments to Iraq and mental health screening outcomes in US Army soldiers. *Journal of Clinical Psychiatry*, 70(9), 1266–1272. doi:10.4088/JCP.08m04361

Reger, M. A., Luxton, D. D., Skopp, N. A., Lee, J. A., & Gahm, G. A. (2009). *Department of Defense suicide event report (DoDSER) calendar year 2008 annual report.* Retrieved from http://www.t2.health.mil/sites/default/files/dodser/DoDSER_2008_Annual_Report.pdf

Ressler, K., & Schoomaker, E. (2014). Commentary on "The Army Study to Assess Risk and Resilience in Servicemembers (Army STARRS)": Army STARRS: A Framingham-like study of psychological health risk factors in soldiers. *Psychiatry*, 77(2), 120–129. doi:10.1521/psyc.2014.77.2.120

Ritchie, E. (2012). *Suicide and the United States army: Perspectives from the former psychiatry consultant to the army surgeon general.* Retrieved from http://dana.org/Cerebrum/Default.aspx?id=39471

Rona, R., Fear, N., Hull, L., Greenberg, N., Earnshaw, M., Hotopf, M., & Wessely, S. (2007). Mental health consequences of overstretch in the UK armed forces: First phase of a cohort study. *BMJ*, 335(7620), 603. doi:10.1136/bmj.39274.585752.BE

Schoenbaum, M., Kessler, R., Gilman, S., Colpe, L., Heeringa, S., Stein, M., . . . Cox, K. (2014). Predictors of suicide and accident death in the Army Study to Assess Risk and Resilience in Servicemembers (Army STARRS): Results from the Army Study to Assess Risk and Resilience in Servicemembers (Army STARRS). *JAMA Psychiatry*, 71(5), 493–503. doi:10.1001/jamapsychiatry.2013.4417

Shen, Y., Arkes, J., & Williams, T. V. (2012). Effects of Iraq/Afghanistan deployments on major depression and substance use disorder: Analysis of active duty personnel in the US military. *American Journal of Public Health*, 102(Suppl. 1), S80–S87. doi:10.2105/AJPH.2011.300425

Sher, L. (2009). A model of suicidal behavior in war veterans with posttraumatic mood disorder. *Medical Hypotheses*, 73(2), 215–219. doi:10.1016/j.mehy.2008.12.052

Simpson, G., & Tate, R. (2002). Suicidality after traumatic brain injury: Demographic, injury and clinical correlates. *Psychological Medicine: A Journal of Research in Psychiatry and the Allied Sciences*, 32(4), 687–697. doi:10.1017/S0033291702005561

Skopp, N. A., Trofimovich, L., Grimes, J., Oetjen-Gerdes, L., & Gahm, G. A. (2012). Relations between suicide and traumatic brain injury, psychiatric diagnoses, and relationship problems, active component, U.S. Armed Forces, 2001–2009. *Medical Surveillance Monthly Report*, 19(2), 7–11. Retrieved from http://www.afhsc.mil/viewMSMR?file=2012/v19_n02.pdf

Smolenski. D. J., Reger, M. A., Alexander, C. L., Skopp, N. A., Bush, N. E., Luxton, D. D., & Gahm, G. A. (2013). *Department of Defense suicide event report (DoDSER) calendar year 2012 annual report.*

Retrieved from http://www.t2.health.mil/sites/default/files/dodser_ar2012_20140306-2.pdf

Tanielian, T., & Jaycox, L. H. (Eds.). (2008). *Invisible wounds of war: Psychological and cognitive injuries, their consequences, and services to assist recovery.* Santa Monica, CA: RAND Corporation.

Teasdale, T. W., & Engberg, A. W. (2001). Suicide after traumatic brain injury: A population study. *Journal of Neurology, Neurosurgery, and Psychiatry, 71*(4), 436–440. doi:10.1136/jnnp.71.4.436

Thomsen, C. J., Stander, V. A., McWhorter, S. K., Rabenhorst, M. M., & Milner, J. S. (2011). Effects of combat deployment on risky and self-destructive behavior among active duty military personnel. *Journal of Psychiatric Research, 45*(10), 1321–1331. doi:10.1016/j.jpsychires.2011.04.003

Ursano, R., Colpe, L., Heeringa, S., Kessler, R., Schoenbaum, M., & Stein, M. (2014). The Army Study to Assess Risk and Resilience in Servicemembers (Army STARRS). *Psychiatry, 77*(2), 107–119. doi:10.1521/psyc.2014.77.2.107

Werbel, A. D. (2010). *Marine Corps suicide prevention program (MCSPP) update for the annual military suicide prevention conference.* Retrieved from http://www.dcoe.mil/content/navigation/documents/SPC2010/0830%20-%201030/Werbel%20-%20Services%20Update.pdf

4

Suicide in the Army National Guard

Findings, Interpretations, and Implications for Prevention

James Griffith

THE EMERGENT PROBLEM

One of the unique aspects of Operation Enduring Freedom (OEF) and Operation Iraqi Freedom (OIF) has been the deployment of large numbers of U.S. military reserve personnel. As of 2011, about 1.7 million U.S. Army military personnel had been deployed to Iraq or Afghanistan, with about one-third from the Army National Guard (ARNG) and the U.S. Army Reserve (USAR; Baiocchi, 2013). Subsequently, the reserves have been a focus of study. Recent studies suggest that reservists, compared to active component personnel, are at greater risk for developing posttraumatic stress disorder (PTSD) and related behavioral health problems (Griffith, 2010; Thomas et al., 2010). During postdeployment, reservists have reported higher rates of PTSD and related symptoms relative to deployed active component personnel (Hoge, Auchterlonie, & Milliken, 2006; Hourani et al., 2007; Milliken, Auchterlonie, & Hoge, 2007; Schell & Marshall, 2008). This is particularly true for reservists who also experienced stressful life events after deployment, such loss of significant relationships, financial troubles, civilian employment difficulties, and so on (Jacobson et al., 2008). Furthermore, longitudinal studies have shown that PTSD was more prevalent among reserve component than among active component members three to six months after having returned from deployments (Milliken et al., 2007) and sometimes even up to four years later (Jacobson et al., 2008; Seal, Bertenthal, Miner, Sen, & Marmar, 2007).

Suicide rates have also been observed to be higher among reservists, especially among ARNG soldiers (Griffith, 2012d).

Military suicide rates were first observed to have increased in 2004, when the United States was engaged in the Iraq and Afghanistan wars. In 2008, the military suicide rate surpassed the civilian age-adjusted rate (20 versus 19 suicides per 100,000 personnel). Within the military services, rates for the Marine Corps and the Army first showed increased rates in 2002, with the ARNG having the highest rate in 2010 (31 suicides per 100,000 personnel; U.S. Army Office of the Chief of Public Affairs, 2010). The Navy and Air Force rates similarly increased during this time frame, although the increase had been much less pronounced. The Marine Corps and Army bore the burden of major ground operations, leading to the preliminary assessment that participation in combat likely caused greater personal distress and thus placed soldiers at greater risk for suicide (U.S. Army of the Office of Chief of Public Affairs, 2010, p. 16). Other hypotheses have been offered as to why more ARNG soldiers than others have symptoms of PTSD, display suicidal behaviors, and report other stress-related outcomes, including difficulties returning to and retaining civilian employment, adjustment problems from having left civilian life (both family and civilian employment) and having to return a year or more later, and fewer supports and resources afforded to reserve military personnel when no longer recalled to active duty (Griffith, 2015a).

PURPOSE

This chapter summarizes the findings of various studies that investigated suicide in the ARNG population. (For a summary of suicide findings for both the active component Army and the ARNG, see Griffith, 2012a; Griffith, 2012b). Findings reported here were taken from a series of studies conducted when the author had been called to active duty during 2010–2012 to conduct analyses that responded to questions of the director of the National Guard Bureau and the Vice Chief Staff of the Army concerning suicide in the Army. Analyses relied primarily on ARNG soldiers who had committed suicide from calendar years (CYs) 2007 through 2012 combined with annual random samples of nonsuicides (N = 6,511, of which 513 were suicides). Additionally, findings from analyses of two archival data sets are summarized here: (a) ARNG soldiers who had recently returned from deployment and who had responded to questions about combat experiences, postdeployment stressors, and suicidal behaviors (N = 4,567) and (b) ARNG soldiers who had not been called to active duty and mainly served one weekend a month and had answered questions about current stressors, suicidal behaviors, and earlier lifetime experiences (N = 15,597). Findings provide direction for policy change to address suicide risk in the ARNG by addressing the following questions:

- What is the suicide risk among ARNG soldiers? Has the risk changed over time?
- What factors place soldiers at risk for suicide?
- Do suicide completers represent a homogenous group, or are two or more subgroups identifiable?
- What are the differences in suicide rates among the ARNG, USAR, and active component Army? How are noted differences explained?

SUMMARY OF FINDINGS

What Is the Suicide Risk Among ARNG Soldiers?

Suicide risk in populations is traditionally assessed as rates. Rates are derived by the number of suicides in a given population or subpopulation divided the total number of individuals in that given population or subpopulation. Due to the small size of such quotients, they are multiplied by 100,000 to represent rates per 100,000. Such rates are referred to as crude rates, as they do not take into account other factors, such as gender and age adjustments.

Historically, the Army (about 12 suicides per 100,000 personnel) has had lower suicide rates than comparable age-adjusted civilian rates (about 20 per 100,000) (Ramchand, Acosta, Burns, Jaycox, &

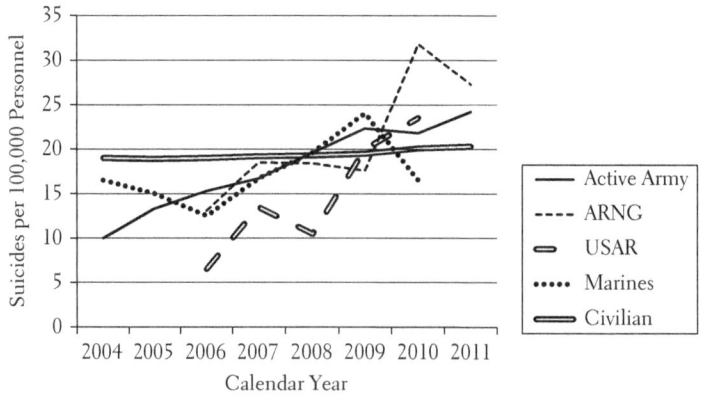

FIGURE 4.1 Suicide Crude Rates per 100,000 for Various Military Services, 2004–2011.
Sources of rates: Active Army ARNG, USAR, and civilian rates are taken from U.S. Army Public Health Command (USAPHC) (May, 2015). Surveillance of suicidal behavior: Summary of rates. Aberdeen, MD: Behavioral and Social Health Outcome Program (BSHOP), Epidemiology and Disease Surveillance (EDS), Army Institute of Public Health (AIPH), United States Army Public Health Command (USAPHC).. Marine rates are taken from Military personnel and veterans suicide prevention, American Foundation for the Prevention of Suicide. Retrieved July 29, 2015 at https://www.afsp.org. Civilian rates are age-adjusted. Rates for USAR and Marines for 2011 were not available.

Pernin, 2011). Since 2004, ARNG suicide rates have fluctuated above and below the active component Army and USAR suicide rates (see Figure 4.1).

Rates for the ARNG increased starting in 2006 and in 2010 exceeded the rate of the active component Army (32 per 100,000 compared to 24 per 100,000) and the USAR (23 per 100,000), as well as the most available civilian age-adjusted rate (20 per 100,000 in 2007).[1] The ARNG suicide rate was first reported as having increased in 2010, at about 32 suicides per 100,000 personnel. Since that time, the rate has remained from 27 to 32 per 100,000 (see Table 4.1). These rates were higher than the most available suicide estimate from the Centers for Disease Control and Prevention (CDC). The 2010 national suicide rate for 20- to 49-year-olds (all races and both sexes) was 15.5 suicides per 100,000 people, much lower than that observed for the ARNG in 2010, at 32 suicides per 100,000 personnel. (The civilian rate was derived from the web application WISQARS [2011].)

Since 2010, the overall ARNG suicide rate has remained high, compared to the other services, and stable. Subpopulations have had higher or lower suicide rates than the ARNG overall, suggesting, respectively, more or less risk for suicide among specific groups of soldiers. Subpopulations having increased rates higher than the ARNG overall rate included those of younger ages, males, whites, singles, those with alternate high school degrees, and those serving part-time in the ARNG (M-Day status). Soldiers who had received enlistment waivers showed noticeably increased rates for 2011 and 2012. Waiver counts among suicides and the ARNG

TABLE 4.1 Suicide Crude Rates (per 100,000), Overall and by Demographic Groups, for Calendar Years 2007–2012

Row Demographic Group	2007	2008	2009	2010	2011	2012	2012 Standard Error (SE)
Total	18.52	17.26	17.36	31.50	27.28	30.75	2.93
Age 17–24	20.10	20.13	18.53	42.18	29.47	32.01	5.00
Age 25–29	11.23	11.82	21.04	40.22	39.87	26.48	6.07
Male	20.48	19.77	18.93	34.13	31.24	34.15	3.35
White	22.91	21.69	22.17	35.61	30.07	37.26	3.76
Single	20.04	16.94	19.69	40.83	31.94	31.74	4.20
Married	18.33	16.40	13.64	18.56	22.44	28.43	4.29
Alt HS	5.70	31.86	33.14	54.05	42.76	44.65	11.53
AFQT <= 50	16.79	13.40	11.24	44.54	28.25	19.54	5.05
Privates (E1–E3)	26.45	19.04	20.71	48.55	22.56	24.85	5.70
Nonprior service	18.29	16.83	17.75	37.08	26.34	31.15	3.81
M-Day	21.61	20.42	21.04	36.32	31.22	29.04	2.98
Combat MOS	17.04	22.26	19.53	34.22	27.66	25.38	3.63
Ever Deploy	15.00	17.95	19.89	26.00	28.20	32.67	4.33
Never Deploy	21.04	16.74	15.11	36.99	26.37	28.95	3.98
Not in training	19.50	17.12	18.93	32.77	28.12	33.23	3.21
Enlist waiver	15.77	20.38	13.06	21.39	36.19	54.84	12.58
Reside in west U.S.	29.44	24.43	23.98	51.94	30.59	38.26	7.36

Notes: 2012 SE = estimated standard error for calendar year 2012 rates (standard error is the quotient—rate/square root [suicide count]). Alt HS = alternative high school (including categories of other nontraditional high school credential, other than Student Recruit program, high school degree in 365 days, home study diploma, high school certificate of attendance, test-based equivalency diploma, overseas GED, occupational program certificate, and correspondence school diploma). AFQT = Armed Forces Qualification Test. AFQT <= 50 = soldiers with mental category IIIB and below. Nonprior service = soldiers who had no previous military service before entering the Army National Guard. M-day = mobilization day asset, a reservist who serves only one weekend a month and 15 days annual training. Combat MOS = combat arms military occupational speciality; for males, duty MOSs of 11, 13, 19, 21, 25, 31, 68, 79, 88, 89, and 91, and for females, duty MOSs of 15, 21, 25, 31, 68, and 92. These typically are MOSs involving combat knowledge and skill. Ever deployed = the soldier has been mobilized and served full-time in military service. Never deployed = the soldier has not been mobilized and served full-time in military service. Not in training = training categories of in or awaiting warrant officer basic course, individual active duty training status, awaiting or in basic officer leader course, and in split phase training. Enlist waiver = soldiers who received a waiver to enter military service (usually a medical or moral waiver). Due to the small counts, caution should be used in interpreting these rates. Reside in west U.S. = soldier's residency was west of the Mississippi River.

overall are low, contributing to large error variance, so some caution should be exercised in their interpretation. Waivers mostly involve medical or moral conduct (e.g., pre-enlistment criminal convictions). Other subpopulations showed increased rates but were lower than the ARNG overall rate. These included soldiers who were married and soldiers in training status. Civilian studies have shown marriage as a protective factor against suicide and more so for men than for women (Kraut & Walld, 2003). Lower suicide risk for those in training likely relates to selection criteria for entry into training and/or being full-time in demanding and accountable environments. Finally, of particular note is that soldiers having been deployed or not and having combat arms military occupational specialities (MOSs) showed comparable risk to all ARNG soldiers. These two factors, then, seemingly had not conferred additional suicide risk.

What Factors Place Soldiers at Risk for Suicide?

The previous discussion draws on rates depicted in Table 4.1. Rates apply to the ARNG population overall and to specific sociodemographic subpopulations. While informative, such rates do not consider risk factors all at once or, in other words, determine the risk factor of one demographic characteristic relative to another. To examine the simultaneous associations of sociodemographic characteristics with suicide, logistic regression analyses were routinely performed. Each year, suicide cases were combined with a random sample for 1,000 nonsuicide cases. Personnel information on suicides and nonsuicides was then used to predict variables most associated with having committed suicide. Table 4.2 displays the logistic regression of several demographic characteristics to predict whether someone had committed suicide or not.

TABLE 4.2 Prediction of 2007–2012 Army National Guard Suicides by Soldier Characteristics (Logistic Regression)

Predictor Variable	r with Suicide	Regress Coefficient	Odds Ratio	R^2 Added	% of Total R^2
Male	.04*	1.08**	2.93		
White	.08*	1.00**	2.71		
17–24 years	.02	1.84**	6.29		
25–29 years	−.00	.17+	1.18	.062	60.2
Single	.18*	.91**	2.49	.009	8.7
In training	−.08*	−2.52**	0.08	.	
Previously deployed	−.12*	.00	0.51		
Combat MOS	.05	.10	1.05	.017	16.5
Reside in west U.S.	.17*	.54**	2.85	.010	9.7
Year 2007		−.69**	0.50		
Year 2008		−.58*	0.56		
Year 2009		−.49*	0.61		
Year 2010		−.12	0.88		
Year 2011		.18	1.20	.005	4.9
Constant		−4.53	0.01		
$X^2 (5,14) =$	705.82				
Total R^2				.103	100.0

Notes: N = 6,511, 2007–2012 suicides plus random sample of soldiers each calendar year for comparison. Variables that were highly correlated with predictor variables were not considered to avoid multicollinearity problems.

Table 4.1 provides the nomenclature for the predictor variables. Each predictor variable here was coded as 1s and all other categories of the variables as zeros (i.e., dummy coded).

Calendar year 2011 served as the reference category.

Odds ratio = amount of times the variable value (e.g., male) is more likely to commit suicide than the reference group (e.g., female); for example, males are 2.93 more likely than females to commit suicide. MOS = military occupational specialty. Reside in west U.S. = soldier's residency was west of the Mississippi River.

* $p < .01$; ** $p < .001$.

Since 2010, primary risk factors associated with suicide have remained the same. Using the CYs 2007 to 2012 suicide–nonsuicide data, variables most associated with suicide were being young (17 to 29 years of age), white, and male. About 10% variance of having committed suicide can be explained by variables considered in the analyses.[2] The low amount of explained variance is not uncommon in epidemiological studies of physical and mental disorders (Warheit, Holzer, & Schwab, 1973). Many variables likely contribute to suicide risk, including previous life experiences and genetic makeup, yet were lacking in the analyses.

Of the 10% variance explained, two-thirds (60.2% plus 8.7%) was accounted for by soldier demographic characteristics (i.e., age, gender, and race), about 5% by year-to-year fluctuations in suicide, and the remaining 17% by military-related variables, such as having been deployed, having combat arms MOS, having prior military service, being in some form of training, and so on. Additionally, residing in the western United States conveyed suicide risk, similar to civilian suicides. (This risk is probably related to residential instability patterns that diminish social integration; see Barkan, Rocque, & Houle, 2013.) In the present study, residing in the western United States imparted a 2.85-fold increased risk for suicide (R^2 added = 9.7%). When analytic variables were treated together, soldiers 17 to 29 years of age were 6.29 times more likely than other ages to commit suicide, white soldiers were 2.71 times more likely to commit suicide than other races, and males were 2.93 times more likely to commit suicide than females. Single soldiers were 2.49 times more likely to have committed suicide compared to married ones. When treated together (three-way interaction), soldiers who were 17 to 24 years old, male, and white had 10 times the risk for suicide (odds ratio = 10.05, reported in Griffith, 2015b). Having been previously deployed and having combat arms MOS were not associated with suicide risk, consistent with other military studies (LeardMann et al., 2013; Millikan et al., 2011).

Primary risk factors identified here have been identified by both active component Army (Millikan, Spiess, Mitchell, Watts, & Porter, 2011; U.S. Army Office of the Chief of Public Affairs, 2010) and civilian research studies (Pagliaro, 1995). In addition, analyses of Army suicides have continued to report deployment-related experiences to be weakly or unrelated to suicide risk (Griffith, 2015b; Leardmann et al., 2013; cf. Schoenbaum et al., 2014). Most recently, Bryan et al. (2015) performed a meta-analysis of 22 published studies to examine the relationship of deployment and combat experiences with suicidal behaviors. Combined effects were small and positive, though with much heterogeneity, conveying greater risk to those deployed. Having been exposed to killing or atrocity showed the largest combined effect, with much less heterogeneity.

Do Suicide Completers Represent a Homogenous Group, or Are Two or More Subgroups Identifiable?

One aspect of the study of suicides in the ARNG was to examine whether suicides represent one group having similar demographic backgrounds and events surrounding suicide as compared to members of other groups. To accomplish this, cluster analyses were performed on soldiers' demographic background—for 2007–2010 ARNG suicides (summarized in Griffith, 2012d) and then again for 2007–2012 ARNG suicides (summarized later). First, an agglomerative clustering method was used to ascertain how many groupings best characterized the suicides. Second, a k-means clustering method was used, specifying the number of groupings and comparing the groupings on the variables used for clustering.

Suicides from 2007 through 2010 were best described as "first-termers" and "careerists." Members of the first-termer group were defined as younger in age, male, white, single, privates first class and specialists, with nonprior military service, fewer years of service, mobilization day (M-day; part-time military status), and to be in training and never before deployed. This group accounted for about two-thirds of the 2007–2010 suicides. Members of the careerist group were defined as older in age, male, white, married, staff sergeants and sergeants first class, with prior military service, more years of service, M-day status, and to have already been deployed. This group accounted for about one-third of the 2007–2010 suicides. There were important differences reflected between the two groups when considering events surrounding the suicides, though not always meeting traditional levels of statistical significance (reported in Griffith, 2012d). Army Regulation 15-6 (sometimes called "37-liner" information) required gathering information surrounding suicides after their occurrence and was first reliably captured in 2009.

First-termers reported more suicidal thoughts, isolation, mood, and anxiety problems, and thus they were thought to be less recognizable as being at risk for suicide. Thus careful screening of such soldiers was recommended (discussed later in this chapter). Careerists, on the other hand, reported problems that could be more readily observed (e.g., prior suicide attempts, interpersonal problems, antisocial behavior, PTSD, physical illnesses, and legal issues such as DUI/DWI). Many of these problems were readily observable, thus allowing for recognizing these soldiers as probable suicide risks. Members of both groups displayed past behavioral health problems, alcohol/substance abuse, and loss of significant relationships.

When suicides were available for 2007–2012, cluster analysis was performed again. Three groups rather than two were identified. The first group was called "first-termers." As before, this group comprised nearly two-thirds of the 2007–2012 ARNG suicides. These suicides were generally 17 to 24 years old, mostly single, male, white, privates and corporals; were less likely to have been deployed; and had M-day status and on average four years of military service. The second group was called "re-up soldiers," comprising about a quarter of the 200–2012 ARNG suicides. (The term *re-up* was used because these soldiers were higher in rank and had more years of service, indicating they had reenlisted and/or extended their service.) These suicides were slightly older (more likely 24 to 29 years old or 30 years old), generally married, mostly male, white, sergeants, and less likely to have been deployed or have M-day status; on average they had 12 years of military service. The third group was called "careerists," comprising about one-tenth of the 2007–2012 ARNG suicides. These suicides were the most married group, mostly male, white, 30 years of age and older; were more likely to have been deployed and less likely to have M-day status; and on average had 25 years of military service.

The significance of the three groups was more than simply differences in soldier background characteristics. First, the three groups showed trends in the percentage of all ARNG suicides over time (CYs 2007–2011):

- First-termers increased from 2008 to 2010–2011 and then leveled off. This trend likely explains the "jump" in suicide counts in 2010.
- Re-up soldiers increased from 2007 to 2009 and then leveled off.
- Careerists decreased from 2008 to 2011 and then steadily rose. This trend reflects the slight shift in suicides to the older ages during this time period. This increase may relate to the depressed economy at the time. A recent CDC-sponsored analysis of suicide rates and business cycles in the United States from 1928 to 2007 (Luo, Florence, Quispe-Agnoli, Ouyang, & Crosby, 2011) reported suicide rates rose during recessions and fell during expansions, with effects being greatest among age groups in prime working years (e.g., 30 to 45 years of age).

Second, the three groups systematically differed in events surrounding the suicide. First-termers, and to some extent re-up soldiers, were characterized by family, school, income, performance, and behavioral health problems and substance abuse. Careerists were characterized by relationship problems, such as divorced or separated, and loss of significant others, in addition to job dissatisfaction. These differences suggest possible ways to more readily identify soldiers at-risk, that is, look for the persistent appearance of these circumstances among specific soldier groups. Cluster analyses conducted on ARNG suicides (CYs 2007–2014) have shown very similar findings, with more recent results showing, once again, two groups (first-termers and careerists), rather than three groups of suicides (Griffith, 2015b).

How Are Differences in Suicide Rates Among the Army Components Explained?

Differences in suicide rates and trends have been observed among the various Army components—that is, the active component Army and the reserve component Army. The reserve component consists of the USAR and the ARNG. ARNG suicide rates have fluctuated above and below the active Army and USAR suicide rates and, before 2007, were below the civilian age-adjusted suicide rate. In 2006, the ARNG suicide rate surpassed the USAR and became more consistent with the active component Army suicide rate and increased in 2010.

Differences between the ARNG and USAR suicide rates were likely *not* due to differences in the way soldiers experience military service. In the recent Department of Defense Status of Forces surveys (Williams, 2013), junior-ranking enlisted soldiers (representing most suicides) in the ARNG and in the USAR did not differ in responses about military experiences (e.g., cohesion, leadership, training, and unit administration). In fact, on some survey items, ARNG soldiers were more satisfied with military service than USAR soldiers. Higher suicide rates for ARNG compared to the USAR likely have to do with the force structure of the ARNG—which involves more combat arms MOSs and results in proportionally more young males in the ARNG than in the USAR. Two primary risk factors for suicide in studies of civilian and military suicides are being young in age and male (Kessler, Berglund, Borges, Nock, & Wang, 2005; Millikan et al., 2011). Differences between the ARNG and active component Army suicide rates likely relate to the nature of reserve military service compared to active duty military service. Traditional ARNG soldiers are predominantly part-time soldiers, serving one weekend a month and 15 days annually. Contrasted to active component Army soldiers, ARNG soldiers, when not on active duty, receive diminished surveillance of behaviors and well-being, in addition to the complexities involved with offering health care to ARNG soldiers. Over three-quarters of all ARNG suicides have occurred in non-drill, "civilian" status (Griffith, 2012d, 2015b).

INTERPRETATION OF FINDINGS

Findings described here suggest that suicides in the ARNG have more to do with *who the person is* (e.g., age, gender, and race) and how this intervenes in the relationship of stressful events and subsequent coping, *rather than direct effects of what he or she experiences*. Studies of members of the active component Army, Department of Defense, and civilian populations (Kessler et al., 2005) have shown three primary factors associated with being at risk for suicide: age (17–24 years and 25–29 years), race (white), and gender (male). By comparison, other variables, such as having a combat arms MOS, having been deployed, having prior service, being in-training, and so on, generally have shown little or no association with suicide (LeardMann et al., 2013; Millikan et al., 2011). Military experience variables—in particular, having been deployed, number and length of deployments, and so on—have shown little to no association with suicide, and the military-related experience variables all together accounted for little explained variance in suicide. Findings are consistent with those reported by the U.S. Army Office of the Chief of Public Affairs (2010, 2012) and a recent Department of Defense large-scale study of service member suicides (LeardMann et al., 2013).

Why Are Age, Gender, and Race Suicide Risk Factors?

The extensive sociological literature on suicide suggests at-risk factors relate to age-specific tasks of identity and relationship development, *contextualized by gender and race*, which likely define the type and amount of support and coping resources. To elaborate, *age* necessitates age-specific tasks concerning development of self-identity and the quality of interpersonal relations. Army suicides occur mostly from 17 years of age through the mid-20s. This age span prescribes specific tasks for the individual, which defines who the individual is and how his or her identity relates to others. Erikson (1968) called these, respectively, identity versus role confusion and intimacy versus isolation. Such themes are evident in adolescent and teen suicide studies (Conner & Goldston, 2007; Portes, Sandhu, & Longwell-Grice, 2002). Major suicide theorists, such as Joiner (Joiner, VanOrden, Witte, & Rudd, 2009) and Durkheim (1897/1951), have noted the importance of self-identity (or lack thereof) in the context of others. Self-identity provides the individual with a sense of worth and meaning—characteristics often absent in suicide cases. Often, life events, such as personal relationship problems, behavioral health conditions, alcohol and substance abuse, job problems, and feelings of loneliness, make these tasks difficult, if not impossible, to accomplish effectively. Many of these problems are evident among Army suicides (Millikan et al., 2011).

Race and *gender* define the context of coping. The extended support network afforded to African Americans is documented in the general literature (Early, 1992; Gibbs, 1997; Kubrin & Wadsworth, 2009; Taylor, Chatters, Tucker, & Lewis, 1990). Whites would be expected to have more negative consequences related to stressful circumstances than would African Americans, due to the benefits of indigenous social supports (Cohen & Wills, 1985). Some have also described African Americans' higher level of participation in religion as an additional inhibition against self-harm (Early, 1992; Kubrin & Wadsworth, 2009). Still others have described the greater resiliency among African Americans in adapting to adverse life circumstances, such as discrimination, unemployment, poverty, urban living, and so on (McIntosh & Santos, 1981; Seiden, 1981).

Male gender is associated with differences in the benefit of support, in addition to being socialized to be competitive and aggressive, and having familiarity and comfort with weapons of violence. Males' aggressiveness and competiveness, along with greater exposure, familiarity, and comfort with weapons, often lead to greater availability and reduced inhibition to use them for self-harm (Kubrin & Wadsworth, 2009). These behaviors have particular significance because about 70% of the ARNG suicides involved firearms (Griffith, 2015b). As described by Maris and colleagues (2000), males are more likely to engage in risky behaviors such as excessive alcohol consumption, unsafe use of firearms, and so on, including suicidal behaviors. Males are also less likely to engage in protective behaviors, such as seeking help for problems, being aware of signs of personal distress, having flexible coping skills, and maintaining social supports. Many of these associations result from different socialization patterns between males and females, in particular, interpersonal behaviors (Wilson, 1987). Evidence suggests that men benefit more than women from social integration and are more vulnerable to the negative effects of stressful circumstances in its absence (Cohen, 2004; House, Landis, & Umberson, 1988).

Evidence for Dispositional Risk

Along with these primary demographic risk factors there is accumulating evidence that individuals have inherent *dispositions* to suicide risk. This risk likely interacts with troubled personal circumstances, resulting in elevated overall suicide risk. Several findings from analyses of ARNG soldiers' survey responses support this notion. Survey responses were obtained from the 2010 Reintegration Unit Risk Inventory, given to about 5,000 soldiers who had recently returned from deployment (Griffith, 2012c, 2012d), and the 2010 Unit Risk Inventory, given to about 12,500 of soldiers who were mainly serving in part-time military status (Griffith, 2014).

- *Persistent suicide risk.* Suicide risk was observed for the same soldiers during and after deployment. Soldiers reporting suicide risk behaviors during deployment represented about two-thirds of those with suicide risk behaviors after deployment (Griffith, 2012c).
- *Underlying traits.* Negative affectivity appears to play a role in suicide risk among deployed, combat-exposed ARNG soldiers. Negative affectivity is associated with habitually experiencing negative cognitions and emotions about oneself and surrounding life events, often accompanied by persistent depressive symptoms and conditions, imparting a long-term risk for suicide (Connor & Ilgen, 2011). Among a large sample of deployed, combat-exposed ARNG soldiers, Griffith (2012c) observed that combat exposure had no direct relationships to postdeployment suicidal behaviors. However, combat exposure was indirectly related to suicidal behaviors through PTSD symptoms and negative mood.
- *Negative childhood events.* Childhood abuse may impart harmful understandings of oneself, events, and others, along with ineffective coping (Pompili et al., 2011). ARNG soldiers who reported childhood abuse were three to eight times more likely to report suicidal behaviors (i.e., thought about suicide, made plans, or had attempted), with the highest likelihood of such behaviors for self-reported physical abuse (Griffith, 2014). Similarly, suicide attempts among Canadian military personnel were more strongly associated with past sexual and interpersonal trauma (e.g., sexual assault, spousal abuse, child abuse) than with more recent deployment and combat experiences (Belik, Stein, Asmundson, & Sareen, 2009).

In summary, suicide risk is likely defined by age-specific tasks, which are more difficult for some soldiers to successfully accomplish than others. Some researchers have described difficulties in terms of earlier adverse life experiences, which defined particular paths or trajectories of unsuccessful coping (Pompili et al., 2011; Seguin, Renaud, Lesage, Roberts, & Turecki, 2011). Race and gender influence the amount of available social support and integration, both of which are important to consider when coping with stressful life events. Indeed, several theories of suicide have emphasized the importance of social context in suicide (Durkheim, 1897/1951; Joiner et al., 2009). Greater sensitivity should be shown toward how the individual negotiates age-specific tasks, in particular, identity formation and the development of intimate relationships. Life events associated with personal distress should be mitigated to the extent possible, allowing positive self-identity and development of interpersonal relationships. Informal supports should be available and offered to the individual. Unit climate and leadership likely play a role in this process. Positive effects of leadership on soldier well-being, particularly reducing the negative effects of stressors on individual strain, is evident in the military and health literature (Griffith & West, 2010; Mental Health Advisory Team, 2009; Solomon, Mukilincer, & Hobfoll, 1986).

IMPLICATIONS FOR SUICIDE PREVENTION

The findings described here imply several directions for strategies of suicide prevention. These include the following.

Screen Soldiers at Risk

Groups of soldiers having specific characteristics were identified as being at greater risk for suicide, namely those young in age, male, and white, with a history of behavioral and substance abuse problems (Millikan et al., 2011), and, among the suicides, first-termers versus careerists. Certain circumstances, such as relationship problems, loss of significant others, financial troubles, legal problems, and job dissatisfaction, were associated with increased suicide risk. Many of these were found more among careerists than first-termers. Making gatekeepers, such as family and friends, first-line supervisors, civilian coworkers, clergy, and physicians, aware of the risks associated with these circumstances could help identify at-risk soldiers (Mann et al., 2005).

There is a need to screen prospective recruits as well. Nock et al. (2014) examined variables associated with suicidal behaviors among a representative cross-sectional sample of nondeployed active duty Army soldiers ($N = 5,428$). Approximately one-third of postenlistment suicide attempts were associated with pre-enlistment mental conditions, which led to the recommendation to screen for pre-enlistment mental conditions. At present, screening methods to assess psychological adjustment do not lend themselves to consistent administration, scoring, and follow-up. The initial induction interview, conducted by medical personnel at the Military Entry Processing Stations, consists of a few open-ended questions, with no standard scoring or criteria for referral. Similarly, soldiers respond to questions about stress levels, suicidal thoughts, and substance abuse during the Post Deployment Health Assessment and Post Deployment Health Reassessment (which is usually conducted annually), but again, scoring and criteria for referral are not standardized. There is no systematic screening in between these assessments.

Improvements might be made in current screening assessments, in particular, content, standardization, and frequency of administration. Current assessments for suicide risk ask soldiers broad questions regarding suicidal thoughts, plans, and attempts. Positive responses to these questions are generally considered by suicide experts to be among the most important "warning signs" or short-term indicators of imminent risk for suicide. Research findings suggest that more than one-half of individuals who have died by suicide actually had denied suicidal ideation and/or intent during their most recent screening (Busch, Fawcett, & Jacobs, 2003; Coombs et al., 1992; Hall, Platt, & Hall 1999; Kovacs, Beck, & Weissman, 1976). The majority of individuals who subsequently attempt suicide or die by suicide were therefore "missed" by current screening methods focused on self-reporting of suicidal thoughts and behaviors.

To tap underlying risk for suicide, assessments might include content of negative cognitions, akin to negative affectivity. A possible starting point might be Rudd's (2006) fluid vulnerability theory. The theory proposes that the underlying cognitions of suicidal individuals are characterized by hopelessness, perceived burdensomeness, self-hatred, distress, and intolerance. These cognitions, which are not explored by existing screeners, make individuals chronically vulnerable to suicide and become more evident for specific life events, increasing suicide risk. Assessments of underlying cognitions might better identify soldiers who are at risk but who deny symptoms (e.g., suicidal thoughts or plans) on general suicide screenings. In this regard, the Social Cognitions Scale has shown its usefulness in screening for suicide risk (Bryan et al., 2014).

Develop Protocol for Determined Follow-Up

After screening and identification of those at risk, a formal protocol should be followed for referral and follow-up. After some initial indication of suicide risk, many soldiers subsequently are not referred for further assessment. Among the 2006 to 2009 suicides, 52% had reported two depressive items and 44% reported one PTSD symptom, yet very few (6%) were referred and followed up on (U.S. Army Public Health Command, 2010). Among 2005–2009 suicides, about one-half (48%) had received outpatient care for behavioral health disorders. Additionally, a majority (82%) of the suicides had some evidence of personal distress; the most common sources of distress were relationship problems (56%) and military or work-related stressors (50%). There is a clear need for more determined referral and follow-up of soldiers who are screened as being at risk for suicide. This includes a designated person responsible for referral and follow-up and a mechanism whereby the at-risk individual can receive medical treatment. For ARNG military personnel, this protocol should include mechanisms for personnel to receive readily available mental health treatment in the military health care system regardless of their military status. At present, ARNG soldiers not on active duty orders must rely on their private health care insurance. Most young soldiers lack such insurance, or, if they have insurance, it is fairly limited in mental health treatment coverage.

Train in the Handling of Firearms

Nearly three-quarters of the ARNG suicides have died by personal firearms (Griffith, 2015b). The research literature suggests that restricting suicide means, in particular, those with lethal consequences such as firearms, might reduce suicides in the military (Kubrin & Wadsworth, 2009; Mann et al., 2002; Ramchand et al., 2011). More deliberation should be given to developing education and training for soldiers (and their family members) on the handling and storage of personally owned firearms and to recognizing when reduced access to firearms is needed. Such weapons safety training could reduce suicide in the U.S. military. Similar safety programs concerning drinking and driving have been instituted in the U.S. military with reasonable success (e.g., has reduced automobile accidents and fatalities associated with DWI/DUI).

Train Those At Risk to Recognize Symptoms

Training should be developed explicitly for soldiers who are at risk for suicide. The current Army suicide prevention programs (ACE education program) is directed primarily at those who might recognize suicide intentions in others and *not in themselves*. Little education and training have been directed to the individual at risk. Such training should better inform soldiers of (a) events that may cause personal distress (e.g., "Dear John" letters); (b) possible psychological and behavioral reactions to stressful events, in particular, adverse reactions (e.g., sense of hopelessness); and (c) effective coping responses.

Refocus Policy on Empirically Supported Risk Factors for Suicide

The military and public continue to see war and its consequences as an explanation for the current rise in suicide among U.S. military personnel. Yet the results summarized here showed deployment and combat exposure have little to no association with suicide risk (LeardMann et al., 2013; Millikan et al., 2011; U.S. Army Public Health Command, 2010). Findings suggest that suicide risk has more to do with who one is rather than what one experiences. Variables most associated with suicide risk are basic demographic characteristics—namely, age, gender, and race, in addition to having a past or current behavioral health condition and substance abuse. Findings also suggest that suicide risk is persistent. When examining survey data of deployed ARNG soldiers, suicidal behaviors were observed largely among the same soldiers during and after deployment. After having returned from deployment, combat exposure, in combination with postdeployment stressors, was associated with increased suicide risk. Thus, for some soldiers, deployment and combat experiences likely intervene in the relationship of negative effects of stressors and suicide risk.

AUTHOR'S NOTE

The findings and views presented here are solely those of the author and do not reflect the position of any entity, public or private. This chapter was originally drafted in spring 2012 and edited again in summer 2015.

NOTES

1. Trends depicted in Figure 4.1 have continued into 2015, with some fluctuation for each military service (personal communication, Defense Suicide Prevention Office staff, 2015). For example, in 2013, the ARNG rate was 34 suicides per 100,000 personnel and then the rate dropped to 22 suicides per 100,000 personnel in 2014, only to increase by some 20% or more in 2015, likely to be similar to the rates previous to 2014 (Griffith, 2015b; Klimas, 2015).

2. It should be noted that logistic regression (used here) does not have a direct equivalent to the total R^2 or added R^2 that is found in ordinary least squares (OLS) regression, which represents the proportion of variance explained in the criterion variable by the predictor variables. There are, however, for logistic regression models several analogues, pseudo R^2, which mimic the OLS R^2 in evaluating the goodness-of-fit of the model. Of these analogues, the Nagelkerke's pseudo R^2 is preferred and can be used to determine how well the model improves (accuracy of prediction) over the null model. Some have used Nagelkerke's pseudo R^2 similar to the R^2 in OLS regression to determine goodness-of-fit. So, even though predictor variables are significant for a given model, the model may have little practical predictive value (i.e., low values of pseudo R^2 values; see Meyers, Gamst, & Guarino, 2006; see also University of Strathclyde, 2015).

REFERENCES

Baiocchi, D. (2013). *Measuring Army deployments to Iraq and Afghanistan*. Washington, DC: RAND Corporation.

Barkan, S., Rocque, M., & Houle, J.(2013). State and regional suicide rates: A new look at an old puzzle. *Sociological Perspectives, 56*, 287–297.

Belik, S., Stein, M. B., Asmundson, G. J. G., & Sareen, J. (2009). Relation between traumatic events and suicide attempts in Canadian military personnel. *Canadian Journal of Psychiatry, 54*(2), 93–104.

Bryan, C., Griffith, J., Pace, B. T., Hinkson, K., Bryan, A. O., Clemens, T. A., & Imel, Z. E. (2015). Combat exposure and risk for suicidal thoughts and behaviors among military personnel and veterans: A systematic review and meta-analysis. *Suicide and Life-Threatening Behavior Journal, 45*(5), 633–649.

Bryan, C. J., Rudd, M. D., Wertenberger, E., Etienne, N., Ray-Sannerud, B. N., Morrow, & Peterson, A. L. (2014). Improving the detection and prediction of suicidal behavior among military personnel by measuring suicidal beliefs: An evaluation of the Suicide Cognitions Scale. *Journal of Affective Disorders, 159*, 15–22.

Busch, K. A., Fawcett, J., & Jacobs, D. G. (2003). Clinical correlates of inpatient suicide. *Journal of Clinical Psychiatry, 64*, 4–19.

Cohen, S. (2004). Social relationships and health. *American Psychologist, 59*(8), 676–684.

Cohen, S., & Wills, T. A. (1985). Stress, social support, and the buffering hypothesis. *Psychological Bulletin, 98*, 310–357.

Conner, K. R., & Goldston, D. B. (2007). Rates of suicide among males increase steadily from age 11 to 21: Developmental framework and outline for prevention. *Aggression and Violent Behavior, 12*(2), 193–207.

Connor, K. R., & Ilgen, M. A. (2011). Substance use disorders and suicidal behavior. In R. C. O'Connor, S. Platt, & J. Gordon, J. (Eds.), *International handbook of suicide prevention: Research, policy and practice* (pp. 93–108). West Sussex, UK: Wiley.

Coombs, D. W., Miller, H. L., Alarcon, R., Herlihy, C., Lee, J. M., & Morrison, D. P. (1992). Presuicide attempt communications between parasuicides and consulted caregivers. *Suicide and Life-Threatening Behavior, 22*, 289–302.

Durkheim, É. (1951). *Suicide: A study in sociology*. New York: Free Press. (Original work published 1897)

Early, K. (1992). *Religion and suicide in the African American community*. Westport, CT: Greenwood Press.

Erikson, E. H. (1968). *Identity, youth and crisis*. New York: Norton, 1968.

Gibbs, J. T. (1997). African-American suicide: A cultural paradox. *Suicide and Life-Threatening Behavior, 27*, 68–79.

Griffith, J. (2010). Citizens coping as soldiers: A review of postdeployment stress symptoms among deployed reservists. *Military Psychology*, 22, 176–206.

Griffith, J., & West, C. (2010). The Army National Guard in OIF/OEF: Relationships among combat exposure, postdeployment stressors, social support, and risk behaviors. *Applied and Preventive Psychology*, 14(1–4), 86–94.

Griffith, J. (2012a). Army suicides: "Knowns" and an interpretative framework for future directions. *Military Psychology*, 24(5), 488–512.

Griffith, J. (2012b). Correlates of suicide in the Army National Guard. *Military Psychology*, 24(6), 568–591.

Griffith, J. (2012c). Suicide and war: The mediating effects of negative mood, posttraumatic stress disorder symptoms and social support among Army National Guard soldiers. *Suicide and Life-Threatening Behavior*, 42(4), 453–469.

Griffith, J. (2012d). Suicide in the Army National Guard: An empirical inquiry. *Suicide and Life-Threatening Behavior*, 42(1), 104–119.

Griffith, J. (2014). Prevalence of childhood abuse among Army National Guard soldiers and its relationship to suicidal behavior. *Military Behavioral Health Journal*, 2(2), 114–122.

Griffith, J. (2015a). Homecoming of soldiers who are citizens: Re-employment and financial status of returning OIF/OEF Army National Guard Soldiers. *Work: A Journal of Prevention, Assessment & Rehabilitation*, 50(1), 85–96.

Griffith, J. (2015b, January). *Synopsis: Analysis of the CY2007-2014 ARNG suicides*. Paper presented at the 2015 VA/DoD Suicide Prevention Conference, Dallas, TX.

Hall, R. C., Platt, D. E., & Hall, R. C. (1999). Suicide risk assessment: A review of risk factors for suicide in 100 patients who made severe suicide attempts. Evaluation of suicide risk in a time of managed care. *Psychosomatics*, 40, 18–27.

Hoge, C. W., Auchterlonie, J. L., & Milliken, C. S. (2006). Mental health problems, use of mental health services, and attrition from military service after returning from deployment to Iraq or Afghanistan. *JAMA: Journal of the American Medical Association*, 295(9), 1023–1032.

Hourani, L. L., Bray, R. M., Marsden, M. E., Witt, M., Vandermaas-Peeler, R., Scheffler, S., . . . Strange, L. (2007). *2006 Department of Defense survey of health related behaviors among the Guard and Reserve force*. Report for the Assistant Secretary of Defense, Health Affairs. Retrieved from http://www.dodwws. rti.org/guardres/2006RCHlthBevSurvey18sept.ppt

House, J. S., Landis, K. R., & Umberson, D. (1988). Social relationships and health. *Science*, 241, 540–545.

Jacobson, I. G., Ryan, M. A. K., Hooper, T. I., Smith, T. C., Amoroso, P. J., Boyko, E. J., . . . Bell, N.S. (2008). Alcohol use and alcohol-related problems before and after military combat deployment. *JAMA: Journal of the American Medical Association*, 300(6), 663–675.

Joiner, T. E., VanOrden, K. A., Witte, T. K., & Rudd, M. D. (2009). *The interpersonal theory of suicide: Guidance for working with suicidal clients*. Washington, DC: American Psychological Association.

Kessler, R. C., Berglund, P. A., Borges, G., Nock, M., & Wang, P. S. (2005). Trends in suicide ideation, plans, gestures, and attempts in the United States 1990–92 to 2001–03. *JAMA: Journal of the American Medical Association*, 293(2), 2487–2495.

Klimas, J. (2015, July 10). Suicides up among Army, Army National Guard. *The Washington Times*.

Kovacs, M., Beck, A. T., & Weissman, A. (1976). The communication of suicidal intent. *Archives of General Psychiatry*, 33, 199–201.

Kraut, A., & Walld, R. (2003). Influence of lack of full-time employment on attempted suicide in Manitoba, Canada. *Scandinavian Journal of Work, Environment & Health*, 29(1), 15–21.

Kubrin, C. E., & Wadsworth, T. (2009). Explaining suicide among Blacks and Whites: How socioeconomic factors and gun availability affect race-specific suicide rates. *Social Science Quarterly*, 90(5), 1203–1227.

LeardMann, C. A., Powell, T. M., Smith, T. C., Bell, M. R., Smith, B., Boyko, E. J., . . . Hoge, C. W. (2013). Risk factors associated with suicide in current and former US military personnel. *JAMA: Journal of the American Medical Association*, 310(5), 496–506.

Luo, F., Florence, C. S., Quispe-Agnoli, M., Ouyang, L., & Crosby, A. E. (2011). Impact of business cycles on U.S. suicide rates, 1928–2007. *American Journal of Public Health*, 101(6), 1139–1146.

Mann, J. J., Apter, A., Bertolote, J., Beautrais, A., Currier, D., Haas, A, . . . Hendin, H. (2005). Suicide prevention strategies: A systematic review. *JAMA: Journal of the American Medical Association*, 294(16), 2064–2074.

Maris, R. W., Berman, A. L., & Silverman, M. M. (2000). *Comprehensive textbook of suicidality*. New York: Guilford Press.

McIntosh, J. L., & Santos, J. L. (1981). Suicide among minority elderly: A preliminary investigation.

Suicide and Life-Threatening Behavior, 11(3), 151–166.

Mental Health Advisory Team. (2009, May 8). *Operation Iraqi Freedom 07-09 report.* Washington, DC: Office of the Army Surgeon General, US Army Medical Command.

Meyers, L. S., Gamst, G., & Guarino, A. J. (2006). *Applied multivariate research: Design and interpretation.* Thousand Oaks, CA: SAGE.

Millikan, A., Spiess, A., Mitchell, M., Watts, C., & Porter, P. (2011). *Analyses of Army suicides, 2005–2010* (Epidemiological Report No. 14-HK-0DS8-10p). Aberdeen Proving Ground, MD: US Army Public Health Command.

Milliken, C. S., Auchterlonie, J. L., & Hoge, C. W. (2007). Longitudinal assessment of mental health problems among active and reserve component soldiers returning from the Iraq War. *JAMA: Journal of the American Medical Association, 298*(18), 2141–2148.

Nock, M. K., Stein, M. B., Heeringa, S. G., Ursano, R. J., Colpe, L. J., Fullerton, C. S., . . . Kessler, R. C. (2014). Prevalence and correlates of suicidal behavior among soldiers: Results from the Army Study to Assess Risk and Resilience in Servicemembers (Army STARRS). *JAMA Psychiatry, 71*(5), 514–522.

Pagliaro, L. A. (1995). Adolescent depression and suicide: A review and analysis of the current literature. *Canadian Journal of School Psychology, 11*(2), 191–201.

Pompili, M., Innamorati, M., Szanto, K., DiVittorio, C., Conwell, Y., Lester, D., . . . Amore, M. (2011). Life events as precipitants of suicide attempters among first-term suicide attempters, repeaters, and non-attempters. *Psychiatry Research, 186*, 300–305.

Portes, P. R., Sandhu, D. S., & Longwell-Grice, R. (2002). Understanding adolescent suicide: A psychosocial interpretation of developmental and contextual factors. *Adolescence, 37*(148), 805–814.

Ramchand, R., Acosta, J., Burns, R. M., Jayjox, L. H., & Pernin, C. G. (2011). *The war within: Preventing suicide in the U.S. military.* Santa Monica, CA: RAND Corporation.

Rudd, M. D. (2006). Fluid vulnerability theory: A cognitive approach to understanding the process of acute and chronic suicide risk. In T. E. Ellis (Ed.), *Cognition and suicide* (pp. 355–368). Washington, DC: American Psychological Association.

Schell, T. L., & Marshall, G. N. (2008). Survey of individuals previously deployed to OEF/OIF. In T. Tanielian, & L. H. Jaycox (Eds.), *Invisible wounds of war: Psychological and cognitive injuries, their consequences, and services to assist recovery* (pp. 87–161). Santa Monica, CA: RAND Corporation.

Schoenbaum, M., Kessler, R. C., Gilman, S. E., Colpe, L. J., Heeringa, S. G., Stein, M. B., . . . Army STARRS Collaborators. (2014). Predictors of suicide and accident death in the Army Study to Assess Risk and Resilience in Servicemembers (Army STARRS): Results from the Army Study to Assess Risk and Resilience in Servicemembers (Army STARRS). *JAMA Psychiatry, 71*(5), 493–503.

Seal, K. H., Bertenthal, D., Miner, C. R., Sen, S., & Marmar, C. (2007). Bringing the war back home. *Archives of Internal Medicine, 167*, 476–482.

Seguin, M., Renaud, J., Lesage, A., Roberts, M., & Turecki, G. (2011). Youth and young adult suicide: A study of life trajectory. *Journal of Psychiatric Research, 45*, 863–870.

Seiden, R. H. (1981). Mellowing with age: Factors influencing nonwhite suicide rate. *The International Journal of Aging and Human Development, 13*(4), 265–284.

Solomon, Z., Mukilincer, M., & Hobfoll, S. E. (1986). Effects of social and battle intensity on loneliness and breakdown during combat. *Journal of Personality and Social Psychology, 51*(6), 1269–1276.

Taylor, R. J., Chatters, L. M., Tucker, M. B., & Lewis, E. (1990). Developments in research on Black families: A decade in review. *Journal of Marriage and the Family, 52*, 993–1014.

Thomas, J. L., Wilk, J. E., Riviere, L. A., McGurk, D., Castro, C. A., & Hoge, C. A. (2010). Prevalence of mental health problems and functional impairment among active component and National Guard soldiers 3 and 12 months following combat in Iraq. *Archives of General Psychiatry, 67*(6), 614–623.

University of Strathclyde. (2015, July). *Goodness of fit measures.* Retrieved from https://www.strath.ac.uk/aer/materials/5furtherquantitativeresearchdesignandanalysis/unit6/goodnessoffitmeasures/

US Army Office of the Chief of Public Affairs. (2010). *Army health promotion, risk reduction, and suicide prevention report.* Report prepared for the US Army Vice Chief of Staff. Washington, DC: The Pentagon.

US Army Office of the Chief of Public Affairs. (2012). *Army 2020: Generating health and discipline in the force: Ahead of the strategic reset.* Report 2012. Report prepared for the US Army Vice Chief of Staff. Washington, DC: Department of the Army.

US Army Public Health Command. (2010, April). *Analyses of Army suicides 2003–2009* (Epidemiological Report no. 14-HK-0BW9-10c). Aberdeen, MD: Author.

Warheit, G. J., Holzer, C. E., & Schwab, J., J. (1973). An analysis of social class and racial differences in depressive symptomatology: A community study. *Journal of Health and Social Behavior*, 14(4), 291–299.

Williams, K. (2013). *The 2012 Status of Forces Survey of Reserve Component Members*. Rosslyn, VA: Defense Manpower Data Center.

Wilson, W. J. (1987). *The truly disadvantaged: The inner city, the underclass, and public policy*. Chicago: University of Chicago Press.

WISQARS (2011). *WISQARS Fatal Injury Reports, National and Regional, 1999 –2009*. Retrieved from http://webappa.cdc.gov/sasweb/ncipc/mortrate10_us.html

5

Combat Experience and the Acquired Capability for Suicide

Craig J. Bryan
Tracy A. Clemans
Ann Marie Hernandez

INTRODUCTION

Suicide among members of the U.S. Armed Forces has increased dramatically during the past 10 years and is now the second most common cause of death among active duty military personnel. Suicide rates have risen across all branches of service since 2004, and in 2008 the active duty military suicide rate surpassed the age- and gender-adjusted national average (Kang & Bullman, 2008) despite historical trends for decreased suicide risk among military personnel. Furthermore, suicide attempts (defined as self-inflicted, potentially injurious behaviors with nonfatal outcomes for which there is evidence, whether explicit or implicit, of intent to die; Silverman, Berman, Sanddal, O'Carroll, & Joiner, 2007) also appear to be increasing in frequency, although estimates of suicide attempts are believed to be much less reliable than deaths by suicide (Ramchand, Acosta, Burns, Jaycox, & Pernin, 2011) because many suicide attempts go unreported.

In addition to increased suicide rates, many military personnel have experienced increased exposure to various forms of stress due to ongoing military operations in Iraq and Afghanistan. These stressors include exposure to traumatic events associated with combat operations, such as death, killing, injury, and life-threatening situations, but also include chronic and persistent nontraumatic stressors such as increased work demands and responsibilities, extended separations from family, and frequent moves. To date, research on military suicide has predominantly focused on exposure to traumas and major life stressors, especially combat exposure, due in large part to the temporal relationships between the rise in military suicides following the initiation of combat operations in Iraq and Afghanistan (Bryan et al., 2015). Much less attention has been given to the role of persistent, relatively "benign" stressors, however, which might be just as important for understanding military suicide (Heron, Bryan, Dougherty, & Chapman, 2013). When considering the impact of combat exposure on suicide risk among military personnel, a well-known and frequently cited model that proposes to explain this association is the interpersonal-psychological theory of suicide (IPTS; Joiner, 2005; Van Orden et al., 2010). In this chapter, we dismantle and expand on one particular dimension of the IPTS, the acquired capability for suicide, which has specifically been proposed as a mechanism for understanding the combat–suicide association. The acquired capability for suicide is examined within the context of military training and combat exposure, with an eye toward integrating recent research with theoretical propositions.

THE INTERPERSONAL-PSYCHOLOGICAL THEORY OF SUICIDE

The IPTS (Joiner, 2005) is a relatively newer theoretical model of suicidal behaviors that has garnered a growing amount of empirical support. The model proposes that three factors must be present simultaneously for an individual to engage in lethal suicidal behavior: perceived burdensomeness, thwarted belongingness, and acquired capability (see Figure 5.1).

Perceived burdensomeness is the sense that one does not contribute to, but rather detracts from, the well-being and/or security of those around him or her and society more generally. It is important to note that this sense of burden persists even in the face of disconfirming evidence. As a result, the individual may perceive his or her death as necessary to decrease this burden on others, thereby ensuring the ongoing well-being and safety of the group. Supporting this assertion are a number of studies showing that higher levels of perceived burdensomeness are associated with more severe suicidal ideation and suicide attempts in both civilian (Joiner et al., 2009; Van Orden, Witte, Gordon, Bender, & Joiner, 2008) and military samples (Anestis, Bryan, Cornette, & Joiner, 2009; Brenner et al., 2008; Bryan, 2011; Bryan, Clemans, & Hernandez, 2012; Bryan, Morrow, Anestis, & Joiner, 2009). Furthermore, nearly half of active duty soldiers interviewed after making a suicide attempt reported feeling "like a burden to others" on the day they made their suicide attempt (Bryan & Rudd, 2012), and 16.9% specifically reported their suicide attempt was motivated in part "to make others better off" (Bryan, Rudd, & Wertenberger, 2013). Indeed, perceived burdensomeness has become one of the most robust and reliable predictors of suicidal ideation and suicide attempts, regularly outperforming other common risk factors such as depression and hopelessness.

Thwarted belongingness is conceptualized as either the actual or the perceived lack of meaningful social connections despite efforts to develop and deepen relationships with others. More specifically, individuals may genuinely lack a social support system or feel disconnected from their existing social support network. The need for interpersonal attachment is believed to be a fundamental and necessary condition for human emotional and physical adaptation and well-being (Baumeister & Budnick, 1995). Likewise, the inability

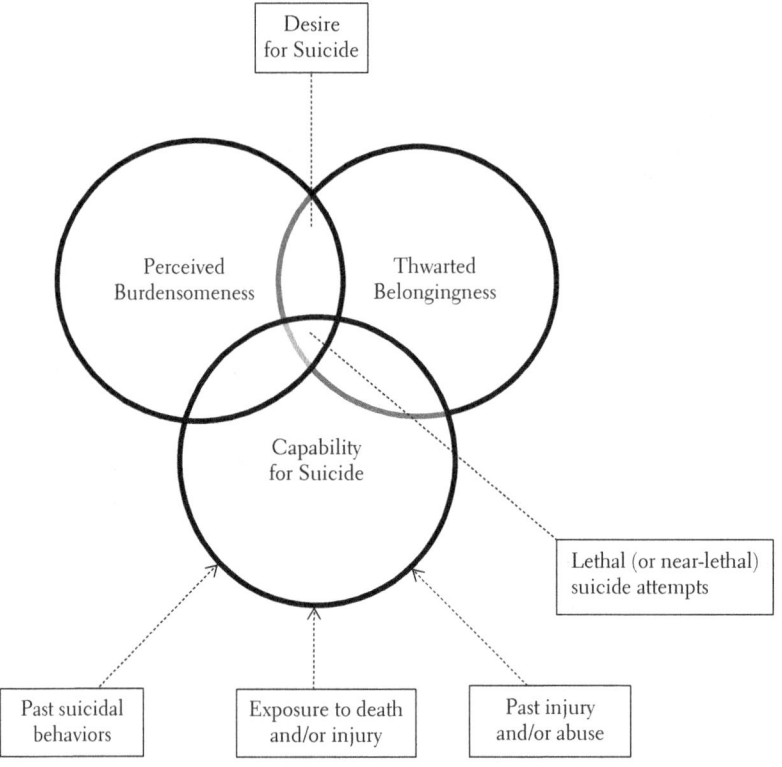

FIGURE 5.1 Assumptions of the interpersonal-psychological theory of suicide.
Sources: Bryan & Cukrowicz, 2011; Joiner, 2005.

or lack of opportunity to relate or belong to a social network negatively impacts physical and emotional health. In fact, research suggests that thwarted belongingness and social isolation is associated with suicidal ideation, attempts, and death by suicide (Conner, Britton, Sworts, & Joiner, 2007; Joiner, Hollar, & Orden, 2006; Van Orden et al., 2008). Among military personnel, thwarted belongingness is correlated with suicidal ideation while deployed (Bryan, 2011), and feeling lonely or isolated is reported by almost two-thirds of soldiers on the day of their suicide attempts (Bryan & Rudd, 2012). Furthermore, the desire to relieve feelings of aloneness, emptiness, or isolation is reported as a motivation for nearly half of soldiers' suicide attempts (Bryan, Rudd, et al., 2013).

According to the IPTS, the combined presence of perceived burdensomeness and thwarted belongingness underlie the desire to die by suicide, which is theoretically distinct from the ability to engage in suicidal behaviors. In other words, the combined presence of perceived burdensomeness with thwarted belongingness explains *why* military personnel would want to die by suicide. This assertion is supported by studies confirming that a significant proportion of the general population (15%) expresses a desire to engage in suicidal behaviors at some point in their lives (i.e., suicidal ideation; Nock et al., 2008), a considerably smaller proportion (2.7%) of the general population actually make suicide attempts (Nock et al., 2008), and even fewer (.01%) die by suicide (Centers for Disease Control and Prevention, 2012). The IPTS therefore seeks to explain *who*, out of the larger number of individuals who desire suicide, will actually die by suicide.

The *acquired capability for suicide* is the degree to which an individual can endure the fear and/or pain associated with a lethal suicide attempt. From the perspective of the IPTS, suicidal behavior is intrinsically fear-inducing and painful and therefore requires decreased fear of death and elevated pain tolerance. The IPTS postulates that individuals acquire this fearlessness and pain tolerance (i.e., the capability for suicide) through repeated exposure to painful and fearful situations such as nonsuicidal self-injury, previous suicide attempts, and violence (Van Orden et al., 2008; Van Orden et al., 2010). This process is particularly relevant in the military, which routinely and repeatedly exposes service members to fear- and pain-inducing situations, some of which can involve extreme violence and aggression, as part of combat training exercises. The repeated exposure to pain, adversity, and aggression is believed to explain why active duty military personnel, even those who have just completed basic training but have not yet deployed, score higher than civilians on measures of acquired capability for suicide (Bryan, Cukrowicz, West, & Morrow, 2010; Bryan, Morrow, Anestis, & Joiner, 2009). Because higher levels of fearlessness and pain tolerance are associated with more frequent suicide attempts (Joiner et al., 2009; Van Orden et al., 2008) and death by suicide (Holm-Denoma et al., 2008), these findings that military personnel may be more "capable" for engaging in suicidal behaviors are of particular concern and highlight one aspect of military service that might serve as a unique risk factor for suicide. The acquired capability for suicide is therefore discussed in greater detail.

Military Service and the Acquired Capability for Suicide

As noted, the IPTS differentiates the desire for suicide from the capability to engage in suicidal behaviors (Joiner, 2005; Van Orden et al., 2010). Simply desiring suicide (i.e., experience thwarted belongingness and perceived burdensomeness) in and of itself is not sufficient to make a lethal or near-lethal suicide attempt; one must also be able to do it. Likewise, having the capability to engage in suicidal behaviors (i.e., low fear of death and high pain tolerance) by itself is insufficient for suicidal behavior, since an individual must also want to die by suicide in order to put this capability into practice. The IPTS therefore provides an elegant model for explaining why some military personnel think about suicide but do not make suicide attempts, whereas others who are capable of lethal self-inflicted injury do not do so.

The two main components underlying acquired capability—fearlessness about death and pain insensitivity—are thought to accumulate as a result of prolonged, repeated exposure to painful and/or fear-inducing stimuli. As described in the learning and motivation literature, the initial reaction an individual has to an aversive or provocative stimuli (i.e., fear) lessens over time with repeated, prolonged exposure to the same fear-inducing stimulus (Charney, 2004). This decrease in fear after repeated exposure to the fear-inducing stimulus is known as habituation. A simple, nonclinical example of habituation is the service member who, during his or her first parachuting training exercise, experiences a high level of fear and discomfort. Upon landing safely on the ground, however, the service member typically experiences an opposing, *positive* emotion (e.g., exhilaration). With each subsequent jump, the fear and discomfort declines as a result of habituation, and the exhilaration becomes magnified through the opponent process, until eventually parachuting is primarily experienced as a positive event. Research has demonstrated that prolonged exposure to emotional stressors or fear-inducing stimuli often leads to behavioral adaptation and habituation and has identified structural, neurochemical, and functional alterations in the emotional processing regions of the brain as a result of prolonged, repeated exposure to aversive stimuli (Charney, 2004; Shin et al., 2006; Tischler et al., 2006). In this same way, individuals develop the capability to overcome the fear of death and acquire greater capacity to tolerate pain. Among military personnel, fearlessness of death and high pain tolerance paradoxically serve as both a strength and valued trait of an effective warrior *and* as a risk factor for suicide.

Previous suicide attempts and other forms of self-injurious behaviors are considered one of the most direct methods for acquiring the capability for suicide. Consistent with this hypothesis, one of the most robust predictors of future suicidal behavior is past suicide attempts, which predict intensity of subsequent suicidal ideation (Joiner, 2005), suicide attempts (Maser et al., 2002), and death by suicide (Brown et al., 2000; Maser et al., 2002). These findings extend to military personnel as well (Rudd, Joiner, & Rajab, 1996). In many ways, suicide attempts function as "practice runs" that reduce the individual's fear of death and aversion to self-inflicted injury by providing a desired and intrinsically reinforcing emotional response. Supporting this perspective are research findings demonstrating that emotional relief is the most frequently reported outcome of self-inflicted injury (Brown, Comtois, & Linehan, 2002; Bryan, Rudd, et al., 2013; Nock & Prinstein, 2004). Among active duty soldiers who had attempted suicide, emotional relief was the most commonly reported motivation for making suicide attempts, with 100% of soldiers reporting they attempted suicide "to stop bad feelings" (Bryan, Rudd, et al., 2013).

Less direct pathways to acquired capability occur through the experience of other painful, albeit nonsuicidal, stimuli such as drug use, aggression, and physical or sexual abuse (Darke & Ross, 2002; Whitlock & Broadhurst, 1969). Individuals who have attempted suicide tend to report greater exposure to violence and aggression during their lives as compared to those who have never attempted suicide (Van Orden et al., 2008). Along these same lines, military personnel who have been the victims of childhood physical or sexual abuse, or physical abuse or battering as an adult, are significantly more likely to experience suicidal ideation than military personnel who have not experienced these assaults (Bryan, McNaughton-Cassill, Osman, & Hernandez, 2013). Suicide attempts are more frequent among military personnel who reported unwanted sexual experiences or violent assault (including rape) as an adult (Bryan, McNaughton, et al., 2013), suggesting that victimization can increase suicide risk.

Military service can also indirectly impact the acquired capability for suicide in several ways, such as routine military training and combat exposure (Selby et al., 2010). For instance, basic military training (a.k.a. boot camp, basic training) across the branches is typically experienced as highly aversive and stressful, and physical injury is common. Within the Army, for instance, between 40% and 60% of female recruits and 15% to 35% of male recruits sustain an injury of some form (Knapik, Cuthie, Canham, Hewitson, & Laurin, 1997). Recruits are also introduced to weapons and trained to use them effectively, practice aggressive and violent actions such as hand-to-hand combat, and participate in simulated combat environments. Indeed, the primary goal of basic military training is to intentionally induce stress so that service members learn to tolerate this psychological state and respond calmly to fearful situations. Supporting the possibility of enhanced fearlessness and pain tolerance following basic training is research showing that U.S. Air Force personnel who had recently graduated from basic military training scored higher on measures of acquired capability compared to a civilian clinical sample that contained multiple suicide attempters, a group believed to have especially high levels of acquired capability (Bryan et al., 2010).

Selby and colleagues (2011) have argued that differences in training and operational demands between branches of service might also contribute differentially to the acquired capability for suicide, which in turn may account for the observed differences in suicide rates between branches. For instance, relative to the Air Force and Navy, the Army and Marine Corps have experienced the most dramatic rises in suicides over the past decade and have also experienced the highest suicide rates during basic military training (Scoville, Gardner, Magill, Potter, & Kark, 2004). Because of the more explicit emphasis on ground combat operations in the Army and Marine Corps, it is possible that soldiers and Marines have acquired a greater capability for suicide relative to airmen and sailors. It is noteworthy, however, that self-selection bias could confound this issue. Specifically, individuals with higher levels of acquired capability to begin with may be more inclined to join the Army or Marine Corps. Unfortunately, no studies to date have explicitly considered between-branch differences.

The Role of Combat

Of the many proposed reasons for increased suicide rates among military personnel, the issue that has arguably received the greatest amount of attention and discussion is the potential role of combat exposure. With over 10 years of sustained military operations in Iraq and Afghanistan, military personnel are now routinely exposed to combat-related traumas, with many military personnel deploying multiple times. Indirect evidence of the proposed link between combat exposure and suicide risk arise from studies investigating the link among posttraumatic stress disorder (PTSD) and suicide risk. For instance, when compared to combat veterans without PTSD, those with PTSD have significantly higher rates of death by suicide (Boscarino, 2006; Drescher, Rosen, Burling, & Foy, 2003; Farberow, Kang, & Bullman, 1990), suicide attempts (Freeman, Roca, & Moore, 2000; Krammer, Lindy, Green, Grace, & Leonard, 1994; Nad, Marcinko, Vuksan-Eusa, Jakovljeviç, & Jakovljevic, 2008), and suicidal ideation (Butterfield et al., 2005). Comorbid psychological problems, especially depression, appear to augment risk even further (Bryan, Clemans, et al., 2012; Lehmann, McCormick, & McCracken, 1995; Rudd, Goulding, & Bryan, 2011; Waller, Lyons, & Costantini-Ferrando, 1999). Although these data suggest an indirect link between combat and suicide risk, none of these studies directly measured combat exposure, thereby limiting the conclusions that can be made.

Based on these findings, some have argued that combat exposure increases suicide risk through repeated exposure to traumatic events, which facilitates the acquired capability for suicide (e.g., Anestis et al., 2009; Selby et al., 2010). While deployed to combat zones, service members may engage in violent actions (e.g., firefights) or witness the consequences of these actions (e.g., wounds, dead bodies), which could habituate service members to death and increase their capacity to tolerate pain and suffering. Several studies have investigated this proposed link and have found that, consistent with the tenets of IPTS, combat exposure is associated with significantly increased levels of fearlessness and pain tolerance but not perceived burdensomeness or thwarted belongingness (Bryan, Cukrowicz, et al., 2010). This suggests that combat exposure contributes more to the capability for suicide than to the desire for suicide. Furthermore, although all combat-related experiences seem to contribute to fearlessness and pain tolerance, combat experiences marked by violence and aggression (e.g., firing at the enemy, being shot at, being attacked) show the strongest relationship (Bryan & Cukrowicz, 2011).

A number of studies have also explored the link between combat exposure and suicide risk among military personnel and veterans. Among active military personnel, results are mixed, with studies alternately finding that combat exposure is weakly associated with suicidal ideation (Griffith, 2012) or is *not* related to suicidal ideation (Bryan, Hernandez, Allison, & Clemans, 2013). A study of Canadian military personnel found that exposure to atrocities such as mass killings is associated with suicidal ideation but general combat action is not (Sareen et al., 2007), further contributing to the inconsistent findings. Among military veterans, however, findings are much more consistent, with several studies demonstrating significant indirect associations of combat exposure, and killing in particular, with increased suicidal ideation and intent (Fontana et al., 1992; Maguen et al., 2012; Sareen et al., 2007; Thoresen & Mehlum, 2008).

Despite the fact that research has confirmed a relationship between combat exposure with higher levels of fearlessness and pain tolerance, and higher levels of fearlessness and pain tolerance is, in turn, related to increased suicidal ideation, these separate segments of the chain have not yet been linked into one, continuous chain. In other words, no studies have demonstrated that the association of combat exposure with increased suicide risk occurs through fearlessness and pain tolerance. In fact, the only study to date that has explicitly sought to test this pathway failed to support the combat–fearlessness–suicide risk chain (Bryan, Hernandez, et al., 2013), indicating that combat exposure was neither directly nor indirectly associated with suicide risk. Along these same lines, Department of Defense statistics indicate that less than half of military personnel who die by suicide have ever been deployed (although considerable variability exists between branches), and deployment history is not overrepresented among military suicides (Department of the Army, 2012; Department of Defense, 2010), which has further contributed to the general confusion about any possible link between combat and suicide.

This seeming disparity in results can be understood when one considers the role of PTSD and temporal factors. PTSD clearly plays a central role in suicide risk among many military personnel, especially when it is combined with depression. Reexperiencing symptoms, in particular, are most predictive of suicidal ideation (Nye & Bell, 2007), which aligns with other research findings that violent daydreaming (Selby, Anestis, & Joiner, 2007) and nightmares (Agargun, Boysan, & Hanoglu, 2004) predict suicidal ideation and behaviors. Interestingly, reexperiencing symptoms of posttraumatic stress have been found to contribute to higher levels of fearlessness and pain tolerance (Bryan & Anestis, 2011), suggesting that military personnel who repeatedly re-live or re-experience a violent or traumatic event are, in essence, acquiring a greater capability for suicide. As time passes and the service member continues to re-experience these violent or traumatic events, they habituate to the fear of death and subsequently their suicide risk increases, which is consistent with the IPTS. Indeed, close examination of the studies summarized here suggest that time does, indeed, play an important role when considering the combat–suicide link. Specifically, while service members are deployed, combat appears to be unrelated to suicide risk (Bryan, Hernandez, et al., 2013), but within the first six months after their return from deployment, an indirect, though very small, effect emerges (Griffith, 2012). As time continues to pass, certain types of combat experiences emerge as more significant factors for understanding suicidal ideation. Specifically, witnessing death or killing is a pernicious experience (Sareen et al., 2007) that persists as the service member transitions from active military duty to veteran status (Fontana et al., 1992; Maguen et al., 2012; Rudd, 2015; Thoresen & Mehlum, 2008). Likewise, the association of combat experiences with suicide risk strengthens among military personnel and veterans with PTSD, especially those individuals who repeatedly re-live violent or aggressive combat experiences.

Research seems to suggest that not all combat experiences are equal (Bryan et al., 2015), and different dimensions of combat can affect service members in very different ways based on such variables as proximity to violence (e.g., hand-to-hand combat vs. artillery fire), severity of violence (e.g., nonhostile patrols vs. firefights), and personal responsibility (e.g., killing enemy vs. witnessing others being killed). It is also important not to overlook the impact of combat exposure on personnel with noncombat roles, such as medical professionals, chaplains, and mortuary affairs personnel, who may be exposed to the *consequences* of violent combat actions (e.g., dead bodies, severe injury), even though they are not the direct agents of this violence.

Implications for Prevention and Clinical Intervention

From a clinical perspective, the acquired capability for suicide provides a general framework that can help to guide and structure clinical work with suicidal military personnel. When conducting risk assessments and determining level of risk, clinicians should first and foremost always assess a service member's suicide attempt history, since previous suicide attempts are the most reliable and robust predictor of current and future suicidal behaviors (Bryan & Rudd, 2006). Identifying patterns of suicide-related behaviors and trends in risk over time is one of the simplest and most straightforward methods for improving the accuracy of one's clinical decision-making. This is especially true and important for service members with a history of multiple suicide attempts.

Given that military service intrinsically requires repeated exposure to painful and provocative experiences, a fact that is unlikely to change at any point in the foreseeable future, prevention efforts and clinical interventions must adapt to this reality. Nonetheless, early intervention with empirically supported strategies for conditions associated with the acquired capability for suicide, such as prolonged exposure or cognitive processing therapy for PTSD, could prevent the eventual onset or worsening of suicidal ideation. Supporting this possibility are recent data from clinical trials conducted with active duty military personnel, which indicate that both prolonged exposure and cognitive processing therapy are associated with reductions in suicidal ideation (Clemans et al., 2010; McLean et al., 2012), which could be due to reductions in re-experiencing symptoms. Along these same lines, it is important to keep in mind that although the very nature of military service appears to facilitate an acquired capability for suicide, the IPTS posits that service members will die by suicide only when they possess both the capability *and* the desire for suicide. Treatments such as brief cognitive behavioral therapy to prevent suicide attempts (Rudd et al., 2015) directly reduce service members' desire for suicide by targeting their maladaptive belief systems, including perceived burdensomeness and thwarted belongingness.

Restriction of access to lethal means for suicide (e.g., firearms, medications) is another strategy designed to directly target the acquired capability for suicide. Service members' familiarity and training with firearms is believed to be one reason a greater proportion of military suicides are by self-inflicted gunshot wounds relative to the general population (Department of Defense, 2010). Consistent with the IPTS, this familiarity habituates service members to the fear of firearms, and repeated use of weapons could contribute to a sense of exhilaration when handling and/or using them. Engaging service members' friends or family in assisting with the restriction of lethal means, such as firearms, can significantly reduce the likelihood of an adverse outcome during a suicidal crisis. As described by Bryan, Stone, and Rudd (2011), means restriction is one of the only effective methods for preventing death by suicide and can be successfully accomplished with service members. Military clinicians have the additional benefit of being able to work with a suicidal service member's chain of command to facilitate restriction of means, although it is recommended that this be done collaboratively with the patient's involvement and buy-in. Additional information about firearms safety and means restriction counseling can be found in Bryan, Stone, et al. (2012) and Britton, Bryan, and Valenstein (2016).

CONCLUSION

Military personnel are explicitly trained to overcome their fear of injury and death, typically through repeated exposure to scenarios and environments that increasingly mimic actual combat situations, which habituates them to fear and eventually replaces this fear with exhilaration and/or other positive emotions (i.e., the opponent process). Fearlessness is an essential quality of a service member; retreating from danger and life-threatening situations are generally not conducive to an effective fighting force. Yet at the same time, fear of death is a well-known protective factor for suicide, given that individuals who are afraid to die are unlikely to attempt suicide. Consequently, fearlessness about death paradoxically serves both as a necessary strength and asset for military personnel as well as a risk factor for suicide. Given that military personnel seem to have a greater capability for suicide relative to other groups, it is critical that prevention efforts and clinical interventions work to reduce service members' desire for suicide using early and empirically supported strategies.

FUTURE DIRECTIONS

Although our understanding of suicide among military personnel is advancing rapidly and continues to expand, several important questions remained unanswered:

1. Are the current methods we have for measuring fearlessness about death and pain tolerance reliable and valid?
2. Do some types of traumas and adverse life events (e.g., child abuse, sexual assault, combat) contribute more to the acquired capability for suicide and subsequent suicidal behaviors in comparison to others?
3. Why does combat exposure explain only a very small amount of suicidal ideation among service members?

4. Why does combat exposure seem to contribute to suicidal ideation but not suicide attempts or death by suicide among service members?
5. Can fearlessness about death or pain tolerance be modified through clinical interventions?
6. How does fearlessness about death and pain tolerance change over the course of military training and combat deployments?

AUTHORS' NOTE

The views expressed in this chapter are those of the authors and do not reflect the official position or policy of the Department of Defense, the Department of Veterans Affairs, or the U.S. government.

REFERENCES

Agargun, M. Y., Boysan, M., & Hanoglu, L. (2004). Sleeping position, dream emotions, and subjective sleep quality. *Sleep Hypnosis*, 6(1), 8–13.

Anestis, M. D., Bryan, C. J., Cornette, M. M., & Joiner, T. E. (2009). Understanding suicidal behavior in the military: An evaluation of Joiner's interpersonal psychological theory of suicidal behavior in two case studies of deployed veterans. *Journal of Mental Health Counseling*, 31, 60–75.

Baumeister, R. F., & Budnick, C. (1995). The need to belong: Desire for interpersonal attachments as a fundamental human motivation. *Psychological Bulletin*, 117(3), 497–529.

Boscarino, J. A. (2006). External-cause mortality after psychologic trauma: The effects of stress exposure and predisposition. *Comprehensive Psychiatry*, 47, 503–514.

Brenner, L. A., Gutierrez, P. M., Cornette, M. M., Betthauser, L. M., Bahraini, N., & Staves, P. J. (2008). A qualitative study of potential suicide risk factors in returning combat veterans. *Journal of Mental Health Counseling*, 30, 211–225.

Britton, P. C., Bryan, C. J., & Valenstein, M. (2016). Motivational interviewing for means restriction counseling with patients at risk for suicide. *Cognitive and Behavioral Practice*, 23(1), 51–61.

Brown, G. K., Beck, A. T., Steer, R. A., & Grisham, J. R. (2000). Risk factors for suicide in psychiatric outpatients: A 20-year prospective study. *Journal of Consulting and Clinical Psychology*, 68, 371–377.

Brown, M. Z., Comtois, K. A., & Linehan, M. (2002). Reasons for suicide attempts and nonsuicidal self-injury in women with borderline personality disorder. *Journal of Abnormal Psychology*, 111(1), 198–202. doi:10.1037/0021-843X.111.1.198

Bryan, C. J. (2011). The clinical utility of a brief measure of perceived burdensomeness and thwarted belongingness for the detection of suicidal military personnel. *Journal of Clinical Psychology*, 67, 981–992.

Bryan, C. J., & Anestis, M. (2011). Re-experiencing symptoms and the interpersonal-psychological theory of suicidal behavior among deployed service members evaluated for traumatic brain injury. *Journal of Clinical Psychology*, 67(9), 856–865.

Bryan, C. J., Clemans, T. A., & Hernandez, A. M. (2012). Perceived burdensomeness, fearlessness of death, and suicidality among deployed military personnel. *Personality and Individual Differences*, 52, 374–379.

Bryan, C. J., & Cukrowicz, K. C. (2011). Associations between types of combat violence and the acquired capability for suicide. *Suicide and Life-Threatening Behavior*, 41(2), 126–136.

Bryan, C. J., Cukrowicz, K. C., West, C. L., & Morrow, C. E. (2010). Combat experience and the acquired capability for suicide. *Journal of Clinical Psychology*, 66(10), 1044–1056. doi:10.1002/jclp.20703

Bryan, C. J., Griffith, J., Pace, B. T., Hinkson, K., Bryan, A. O., Clemans, T. A., & Imel, Z. (2015). Combat exposure and risk for suicidal thoughts and behaviors among military personnel and veterans: A systematic review and meta-analysis. *Suicide and Life-Threatening Behavior*, 45(5), 63–649.

Bryan, C. J., Hernandez, A. M., Allison, S. & Clemans, T. (2013). Combat exposure and suicide risk in two samples of deployed military personnel. *Journal of Clinical Psychology*, 69(1), 64–77. doi:10.1002/jclp.21932.

Bryan, C. J., McNaughton-Cassill, M., Osman, A. & Hernandez, A. M. (2013). The associations of physical and sexual assault with suicide risk in nonclinical military and undergraduate samples. *Suicide and Life-Threatening Behavior*, 43(2), 223–234. doi:10.1111/sltb.12011

Bryan, C. J., Morrow, C. E., Anestis, M. A., & Joiner, T. E. (2009). A preliminary test of the interpersonal-psychological theory of suicidal behavior in a military sample. *Personality and Individual Differences*, 48, 347–350.

Bryan, C. J., & Rudd, M. D. (2006). Advances in the assessment of suicide risk. *Journal of Clinical Psychology, 62*(2), 185–200.

Bryan, C. J., & Rudd, M. D. (2012). Life stressors, emotional distress, and trauma-related thoughts occurring within 24 h of suicide attempts among active duty US Soldiers. *Journal of Psychiatric Research, 46*, 843–848.

Bryan, C. J., Rudd, M. D., & Wertenberger, E. (2013). Reasons for suicide attempts in a clinical sample of active duty soldiers. *Journal of Affctive Disorders, 144*(1–2), 148–152.

Butterfield, M. I., Stechuchak, K. M., Connor, K. M., Davidson, J. R. T., Wang, C., MacKuen, C. L., . . . Marx, C. E. (2005). Neuroactive steroids and suicidality in posttraumatic stress disorder. *American Journal of Psychiatry, 162*, 380–382.

Bryan, C. J., Stone, S. L., & Rudd, M. D. (2011). A practical, evidence-based approach for means-restriction counseling with suicidal patients. *Professional Psychology: Research and Practice, 42*(5), 339–346.

Centers for Disease Control and Prevention. (2012). *10 leading causes of death by age group, United States –2010.* Retrieved from http://www.cdc.gov/injury/wisqars/pdf/10LCID_All_Deaths_By_Age_Group_2010-a.pdf

Charney, D.S. (2004). Psychobiological mechanisms of resilience and vulnerability: Implications for successful adaptation to extreme stress. *The American Journal of Psychiatry, 161*, 195–216.

Clemans, T., Resick, P. A., Dondanville, K., Schuster, J., Peterson, A., Bryan, C. J. . . . STRONG STAR Consortium. (2010, June). *Impact of cognitive processing therapy on suicidal ideation among active duty military personnel.* Poster presented at the 2012 DoD/VA Suicide Prevention Conference, Washington DC.

Conner, K. R., Britton, P. C., Sworts, L. M., & Joiner, T. E. (2007). Suicide attempts among individuals with opiate dependence: The critical role of belonging. *Addictive Behaviors, 32*, 1395–1404.

Darke, S., & Ross, J. (2002). Suicide among heroin users: Rates, risk factors and methods. *Addiction, 97*, 1383–1394.

Department of the Army. (2012). *Army 2020: Generating health & discipline in the force ahead of the strategic reset: Report 2012.* Washington, DC: Author.

Department of Defense. (2010). *DoDSER: Department of Defense suicide event report, calendar year 2010 annual report.* Washington, DC: Author.

Drescher, K. D., Rosen, C. S., Burling, T. A., & Foy, D. W. (2003). Causes of death among male veterans who received residential treatment for PTSD. *Journal of Traumatic Stress, 16*, 535–543.

Farberow, N. L., Kang, H. K., & Bullman, T. A. (1990). Combat experience and postservice psychosocial status as predictors of suicide in Vietnam veterans. *The Journal of Nervous and Mental Disease, 178*, 32–37.

Fontana, A., Rosenheck, R., & Brett, E. (1992). War zone traumas and posttraumatic stress disorder symptomatology. *The Journal of Nervous and Mental Disease, 180*, 748–755.

Freeman, T. W., Roca, V., & Moore, W. M. (2000). A comparison of chronic combat-related posttraumatic stress disorder (PTSD) patients with and without a history of suicide attempts. *The Journal of Nervous and Mental Disease, 188*, 460–463.

Griffith, J. (2012). Suicide and war: The mediating effects of negative mood, traumatic stress disorder symptoms, and social support among Army National Guard Soldiers. *Suicide and Life-Threatening Behavior. 42*, 453–469. doi:10.1111/j.1943-278X.2012.00104.x

Heron, E. A., Bryan, C. J., Dougherty, C. A., & Chapman, W. G. (2013). Military mental health: the role of daily hassles while deployed. *The Journal of Nervous and Mental Disease, 201*, 1035–1039.

Holm-Denoma, J., Witte, T., Gordon, K., Herzog, D., Franko, D., Fichter, M., . . . Joiner, T. (2008). Case reports of anorexic women's deaths by suicide as arbiters between competing explanations of the anorexia-suicide link. *Journal of Affective Disorders, 107*, 231–236.

Joiner, T. E. Jr. (2005). *Why people die by suicide.* Cambridge, MA: Harvard University Press.

Joiner, T. E. Jr., Hollar, D., & Orden, K. V. (2006). On Buckeyes, Gators, Super Bowl Sunday, and the Miracle on Ice: "Pulling together" is associated with lower suicide rates. *Journal of Social and Clinical Psychology, 25*(2), 179–195.

Joiner, T. E. Jr., Van Orden, K. A., Witte, T. K., Selby, E. A., Ribeiro, J. D., Lewis, R., & Rudd, M. D. (2009). Main predictions of the interpersonal–psychological theory of suicidal behavior: Empirical tests in two samples of young adults. *Journal of Abnormal Psychology, 118*(3), 634–646.

Kang, H. K., & Bullman, T. A. (2008). Risk of suicide among US veterans returning from the Iraq or Afghanistan war zones. *JAMA: Journal of the American Medical Association, 300*(6), 652–653.

Knapik, J. J., Cuthie, J., Canham, M., Hewitson, W., & Laurin, M. J. (1997). *Injury incidence, injury risk factors, and physical fitness of US Army* basic trainees at Ft Jackson SC. Epidemiological Consultation Report. Report No. 29-HE-7513-98. Aberdeen Proving Ground, MD: US Army Center for Health Promotion and Preventive Medicine, 1998.

Krammer, T. L., Lindy, J. D., Green, B. L., Grace, M. C., & Leonard, A. C. (1994). The comorbidity of posttraumatic stress disorder and suicidality in Vietnam veterans. *Suicide and Life-Threatening Behavior, 24*, 58–67.

Lehmann, L., McCormick, R., & McCracken, L. (1995). Suicidal behavior among patients in the VA health care system. *Psychiatric Services, 46*, 1069–1071.

Maguen, S., Metzler, T. J., Bosch, J., Marmar, C. R., Knight, S. J., & Neylan, T. C. (2012). Killing in combat may be independently associated with suicidal ideation. *Depression and Anxiety, 29*(11), 918–923.

Maser, J. D., Akiskal, H. S., Schettler, P., Scheftner, W., Mueller, T., Endicott, J., . . . Clayton, P. (2002). Can temperament identify affectively ill patients who engage in lethal or near-lethal suicidal behavior? A 14-year prospective study. *Suicide and Life-Threatening Behavior, 32*, 10–32.

McLean, C. P., Lichner, T., Yadin, E., Peterson, A. L., Mintz, J., Evans, B. B., . . . STRONG STAR Consortium. (2012, June). *Trauma-related guilt and a predictor of suicidal ideation among active duty military personnel*. Poster presented at the annual meeting of the DoD/VA Suicide Prevention Conference, Washington, DC.

Nad, S., Marcinko, D., Vuksan-Eusa, B., Jakovljeviç, M., & Jakovljevic, G. (2008). Spiritual well-being, intrinsic religiosity, and suicidal behavior in predominantly Catholic Croatian war veterans with chronic posttraumatic stress disorder: A case control study. *The Journal of Nervous and Mental Disease, 196*, 79–83.

Nock, M. K., Borges, G., Bromet, E. J., Alonso, J., Angermeyer, M., Beautrais, A., & Bruffaerts, R. (2008). Cross-national prevalence and risk factors for suicidal ideation, plans, and attempts. *The British Journal of Psychiatry, 192*, 98–105.

Nock, M. K., Borges, G., Bromet, E. J., Cha, C. B., Kessler, R. C., & Lee, S. (2008). Suicide and suicidal behavior. *Epidemiologic Reviews, 30*, 133–154.

Nock, M. K., & Prinstein, M. J. (2004). A functional approach to the assessment of self-mutilative behavior. *Journal of Consulting and Clinical Psychology, 72*(5), 885–890. doi:10.1037/0022-006X.72.5.885

Nye, E. C., & Bell, J. B. (2007). Specific symptoms predict suicidal ideation in Vietnam combat veterans with chronic post-traumatic stress disorder. *Military Medicine, 172*, 1144–1147.

Ramchand, R., Acosta, J., Burns, R. M., Jaycox, L. H., & Pernin, C. G. (2011). *The war within: Suicide prevention in the US military*. Santa Monica: CA: RAND Corporation.

Rudd, M. D., Bryan, C. J., Wertenberger, E. G., Peterson, A. L., Young-McCaughan, S., Mintz, J., . . . Bruce, T. O. (2015). Brief cognitive behavioral therapy effects on post-treatment suicide attempts in a military sample: Results of a 2-year randomized clinical trial. *The American Journal of Psychiatry, 172*, 441–449.

Rudd, M. D., Goulding, J., & Bryan, C. J. (2011). Student veterans: A national survey exploring psychological symptoms and suicide risk. *Professional Psychology: Research and Practice, 42*, 354–360. doi:10.1037/a0025164

Rudd, M. D., Joiner, T., & Rajab, M. H. (1996). Relationships among suicide ideators, attempters, and multiple attempters in a young-adult sample. *Journal of Abnormal Psychology, 105*(4), 541–550. doi:10.1037/0021-843X.105.4.541

Sareen, J., Cox, B. J., Afifi, T. O., Stein, M. B., Belik, S., Meadows, G., & Asmundson, G. J. G. (2007). Combat and peacekeeping operations in relation to prevalence of mental disorders and perceived need for mental health care. *Archives of General Psychiatry, 64*, 843–852.

Scoville, S. L., Gardner, J. W., Magill, A. J., Potter, R. N., & Kark, J. A. (2004). Nontraumatic deaths during US Armed Forces basic training, 1977–2001. *American Journal of Preventative Medicine, 26*(3), 205–212.

Selby, E. A., Anestis, M. D., Bender, T. W., Ribeiro, J. D., Nock, M. K., Rudd, M. D., . . . Joiner, T. E. Jr. (2010). Overcoming the fear of lethal injury: Evaluating suicidal behavior in the military through the lens of the interpersonal-psychological theory of suicide. *Clinical Psychology Review, 30*(3), 298–307.

Selby, E. A., Anestis, M. D., & Joiner, T. E. (2007). Daydreaming about death: Violent daydreaming as a form of emotion dysregulation in suicidality. *Behavioral Modification, 31*, 867–879. doi:10.1177/0145445507300874

Shin, L.M., Rauch, S.L., & Pitman, R.K. (2006). Amygdala, medial prefrontal cortex, and hippocampal function in PTSD. *Annals of the New York Academy of Sciences, 1071*, 67–79.

Silverman, M. M., Berman, A. L., Sanddal, N. D., O'Carroll, P. W., & Joiner, T. E. (2007). Rebuilding the Tower of Babel: A revised nomenclature for the study of suicide and suicidal behaviors, part 2: Suicide-related ideations, communications, and behaviors. *Suicide and Life-Threatening Behavior, 37*, 264–277.

Thoresen, S., & Mehlum, L. (2008). Traumatic stress and suicidal ideation in Norwegian male peacekeepers. *The Journal of Nervous and Mental Disease, 196*, 814–821.

Tischler, L., Brand, S. R., Stavitsky, K., Labinsky, E., Newmark, R., Grossman. R., . . . Yehuda, R. (2006). The relationship between hippocampal volume and declarative memory in a population of combat veterans with and without PTSD. *Annals of the New York Academy of Sciences, 1071*, 405–9.

Van Orden, K. A., Witte, T. K., Cukrowicz, K. C., Braithwaite, S. R., Selby, E. A., & Joiner, T. E. Jr. (2010). The interpersonal theory of suicide. *Psychological Review, 117*(2), 575–600. doi:10.1037/a0018697

Van Orden, K. A., Witte, T. K., Gordon, K. H., Bender, T. W., & Joiner, T. E. Jr. (2008). Suicidal desire and the capability for suicide: Tests of the interpersonal-psychological theory of suicidal behavior among adults. *Journal of Consulting and Clinical Psychology, 76*(1), 72–83. doi:10.1037/0022-006X.76.1.72

Waller, S. J., Lyons, J. S., & Costantini-Ferrando, M. E. (1999). The impact of comorbid affective and alcohol use disorders on suicidal ideation and attempts. *Journal of Clinical Psychology, 55*, 585–595.

Whitlock, F. A., & Broadhurst, A. D. (1969). Attempted suicide and the experience of violence. *Journal of Biosocial Science, 1*, 353–368.

6

Combat-Related Killing and the Interpersonal-Psychological Theory of Suicide

Lindsey L. Monteith
Shira Maguen

Rates of suicide among veterans and military service members have recently reached unprecedented levels (Kuehn, 2009). This underscores the critical need to understand risk factors for suicide in these populations. Combat-related killing has been identified as a potentially relevant and understudied risk factor that may be particularly relevant to suicide in military personnel and veterans (Maguen et al., 2010; Maguen et al., 2012). In this chapter, we describe the interpersonal-psychological theory of suicide (IPTS; Joiner, 2005) and propose this framework as a theoretical model for understanding the relationship between combat-related killing and suicidal self-directed violence.

THE INTERPERSONAL-PSYCHOLOGICAL THEORY OF SUICIDE

The IPTS proposes that the desire for suicide is caused by perceived burdensomeness and thwarted belongingness (Joiner, 2005). *Perceived burdensomeness* involves self-hatred and is described as the perception that one's existence poses a liability to close others (Van Orden et al., 2010). *Thwarted belongingness* is defined as a sense of disconnection from others, consisting of loneliness and a dearth of relationships involving reciprocity and caring (Van Orden et al., 2010). Perceived burdensomeness and thwarted belongingness are constructs that can fluctuate over time. The IPTS posits perceived burdensomeness and thwarted belongingness as proximal risk factors for passive suicidal ideation while proposing that chronic and stable beliefs in both increase active suicidal desire (Van Orden et al., 2010).

Whereas the IPTS posits that suicidal desire (which encompasses suicidal ideation) is derived from burdensomeness and thwarted belongingness, it postulates that these constructs are not sufficient for suicide attempts to occur. The theory proposes that, because suicide is so contradictory to survival instincts, in order to enact suicidal self-directed violence, a person must repeatedly engage in painful and provocative experiences in which he or she habituates to the pain and fear of death (Joiner, 2005). Through these experiences, a person increasingly gains the confidence and competence to inflict suicidal self-directed violence. This acquired ability to enact lethal self-harm is called the *acquired capability for suicide* and is thought to influence suicidal self-directed violence, in addition to the lethality of such behavior (Joiner, 2005). The acquired capability for suicide is considered to be relatively stable over time and less amendable to intervention, compared to perceptions of burdensomeness and thwarted belongingness (Van Orden et al., 2010).

Further elaborations of the IPTS have proposed that acquired capability is comprised of two separate constructs: decreased fear of death and a high tolerance for physical pain (Van Orden et al., 2010). Multiple factors are theorized to influence one's tolerance for the pain of suicide, including expectations about pain, beliefs about one's pain tolerance, and whether habituation to pain has occurred (Van Orden et al., 2010). Proposed direct routes to acquiring the capability for suicide include nonfatal suicide attempts, nonsuicidal self-injury, and violence (Joiner, 2005). Indirect pathways have also been proposed and include engaging in physically painful behaviors, experiencing or observing physical pain, inflicting violence upon others, witnessing or experiencing violence, experiencing emotional pain, and substance use (Joiner, 2005).

Although the IPTS posits that the highest levels of suicidal desire occur when perceived burdensomeness and thwarted belongingness are simultaneously high (Van Orden et al., 2010), the acquired capacity for suicide is considered requisite for suicide attempt(s) to occur. Thus, according to the IPTS, the highest risk for suicide attempt occurs in the presence of high levels of perceived burdensomeness, thwarted belongingness, and the acquired capability for suicide (Van Orden et al., 2010). Whereas the acquired capacity for suicide is requisite for suicide attempts, it is not considered sufficient in and of itself to produce suicidal self-directed violence; rather, it must be accompanied by the constructs proposed to produce suicidal desire—burdensomeness and thwarted belongingness—for suicidal self-directed violence to occur (Van Orden, Merrill, & Joiner, 2005). Thus it is possible to have high levels of the acquired capability for suicide without high levels of perceived burdensomeness or thwarted belongingness. In this case, an individual would be expected to have a higher capability to engage in suicidal self-directed violence but would not be expected to have elevated levels of suicidal desire.

EMPIRICAL TESTS OF THE IPTS IN MILITARY AND VETERAN SAMPLES

The IPTS offers a potential explanation for suicide that has been tested in military personnel (Bryan, Clemans, & Hernandez, 2012; Bryan, Morrow, Anestis, & Joiner, 2010; Nademin et al., 2008) and veterans (Brenner et al., 2008; Gutierrez et al., 2013; Monteith, Menefee, Pettit, Leopoulos, & Vincent, 2013; Pfeiffer et al., 2014). Researchers have emphasized the potential roles of military training and combat in increasing the capability for suicide while proposing that perceived burdensomeness and thwarted belongingness may occur as a result of combat-related physical injuries, emotional distress, functional impairment, and difficulty transitioning from military to civilian life (Monteith, Green, Mathew, & Pettit, 2009; Selby et al., 2010).

Researchers testing the IPTS among military and veteran samples have typically utilized self-report measures to assess IPTS constructs. The Interpersonal Needs Questionnaire (INQ; Van Orden, Witte, Gordon, Bender, & Joiner, 2008) was developed to assess the extent to which participants feel like a burden (e.g., "These days the people in my life would be better off if I were gone") and feel disconnected from others (e.g., "These days, I feel disconnected from other people"). Participants rate each item on a 7-point Likert scale. The Acquired Capability for Suicide Scale (ACSS; Van Orden et al., 2008) was developed to assess the acquired capability for suicidal self-directed violence. Participants rate the extent to which various items describe them (e.g., "I am not at all afraid to die," "I can tolerate a lot more pain than most people"), using a 5-point Likert scale. Unless otherwise noted, the research described here used the INQ and ACSS to assess IPTS constructs.

Nademin and colleagues (2008) conducted the first test of the IPTS in a military sample. They examined interpersonal-psychological constructs as predictors of death by suicide among active duty Air Force personnel. Among living Air Force personnel, perceived burdensomeness, thwarted belongingness, and the acquired capacity for suicide were measured through the INQ and ACSS. However, for suicide decedents, psychological autopsy methods were utilized to measure IPTS constructs. Specifically, trained raters reviewed and coded information from medical records, suicide notes, and interviews with people who had known the deceased, using the Interpersonal-Psychological Survey (created by Nademin and colleagues) to rate each IPTS construct. A composite score consisting of burdensomeness, thwarted belongingness, and the acquired capability for suicide differentiated Air Force personnel who died by suicide. Differences were driven by higher levels of the acquired capability for suicide among military personnel who died by suicide compared to living active duty service members. Results supported the role of interpersonal-psychological constructs—and the acquired capability for suicide in particular—in differentiating who dies by suicide among military personnel.

Two subsequent studies found comparable support for the contribution of the acquired capability in predicting past suicidality in military personnel (Bryan, Morrow, et al., 2010; Bryan et al., 2012). Among Air Force personnel, the acquired capability for suicide and its interaction with perceived burdensomeness significantly predicted suicidal history (Bryan, Morrow, et al., 2010). Similarly, burdensomeness, acquired capability, and their interaction predicted suicidality among two separate samples of military personnel deployed to a combat zone (Bryan et al., 2010). Taken together, these results suggest an important association between perceived burdensomeness and the acquired capability for suicide with suicidality in military samples.

Recent studies have examined the IPTS among veterans in clinical settings within the Veterans Health Administration. Monteith and colleagues (2013) examined the IPTS in male combat veterans and female veterans in inpatient treatment. Perceived burdensomeness and its interaction with thwarted belongingness were significantly associated with past-week suicidal ideation, controlling for gender, depressive symptoms, and posttraumatic stress disorder (PTSD) symptoms. The authors also examined whether IPTS constructs statistically predicted lifetime history of suicide attempts. The interaction between burdensomeness and acquired capability, and the interaction between thwarted belongingness and acquired capability, were significant in differentiating veterans with no prior suicide attempts from those with multiple attempts; however, the hypothesized three-way interaction between perceived burdensomeness, thwarted belongingness, and acquired capability was not significantly associated with suicide attempt history. Pfeiffer and colleagues (2014) conducted an examination of the IPTS among veteran outpatients with depressive disorders using proxies to assess IPTS constructs. Perceived burdensomeness was associated with passive suicidal ideation both at baseline and at a three-month follow-up. In contrast, neither thwarted belongingness nor the interaction between burdensomeness and thwarted belongingness were significant in predicting suicidal ideation at baseline or at the three-month follow-up.

Quantitative research on the IPTS has been complemented by qualitative examinations of the IPTS. Brenner and colleagues (2008) conducted a qualitative study with veterans (primarily males) who served in Operation Enduring Freedom (OEF) and Operation Iraqi Freedom (OIF). Findings highlighted potential sources of perceived burdensomeness, thwarted belongingness, and the acquired capability for suicide in OEF/OIF veterans. Veterans reported multiple contexts for acquiring an increased tolerance for pain, including "repeated exposure to pain and danger" (p. 217) and decreased fear associated with combat. Sources of perceived burdensomeness included difficulty reintegrating into civilian life, trouble providing financially for family, and loss of sense of self. Veterans also reported feeling connected to fellow military service members but disconnected from civilians. Suicide was spontaneously mentioned in regard to coping with burdensomeness, disconnectedness, and intense pain. Gutierrez and colleagues (2013) employed a similar qualitative framework with female veterans formerly deployed to Iraq or Afghanistan. Participants described experiences during and after deployment leading to perceived burdensomeness, thwarted belongingness, and the acquired capability for suicide, suggesting that these constructs are also salient among female veterans. Additionally, a subsequent case series (Anestis, Bryan, Cornette, & Joiner, 2009) used IPTS constructs to conceptualize the emergence of suicidal ideation in two active duty OIF Air Force personnel, highlighting the utility of the IPTS in understanding suicidality among military service members.

In summary, quantitative research highlights the applicability of the IPTS to suicidality in military and veteran samples (Bryan et al., 2012; Bryan, Morrow, et al., 2010; Monteith et al., 2013; Nademin et al., 2008; Pfeiffer et al., 2014). Qualitative research findings have emphasized the relevance of interpersonal-psychological constructs to suicide in veterans (Brenner et al., 2008; Gutierrez et al., 2013) and military personnel (Anestis et al., 2009).

THE ACQUIRED CAPABILITY FOR SUICIDE AMONG MILITARY PERSONNEL AND VETERANS

Recent research aimed at applying the IPTS to understanding suicide among military personnel has focused on the acquired capability for suicide as a potential explanatory mechanism. Military service, training, and combat are inherently characterized by repeated opportunities to habituate to violence, pain, and death. As a result, researchers have proposed that the acquired capability for suicide is the most pertinent interpersonal-psychological construct for explaining suicide in military and veteran populations (e.g., Selby et al., 2010). Van Orden and colleagues (2010) noted that "combat exposure, which involves exposure to the fear of one's own possible death, as well as killing others, represents a relatively direct pathway" (p. 587) to the acquired capability for suicide.

Military personnel report elevated levels of the capacity for suicidal self-directed violence (Bryan, Morrow, et al., 2010). Active duty Air Force personnel who had recently completed military training reported a greater ability to inflict lethal self-harm compared to a nonmilitary clinical sample of adult outpatients, including a subsample of adults with a history of multiple suicide attempts (Bryan, Morrow, et al., 2010). This finding is noteworthy, considering that suicide attempts have been proposed to represent a direct pathway to acquiring the capacity for self-harm (Joiner, 2005; Van Orden et al., 2010). Moreover, military personnel in this sample had not yet deployed or experienced combat (Bryan, Morrow, et al., 2010). It is unknown whether the high levels of the capacity for suicide in the sample were due to habituation that occurred as a result of military training, prior life experiences, or an alternate explanation. Interestingly, military service members reported lower levels of burdensomeness than a civilian sample of undergraduate students. There were no differences in levels of thwarted belongingness. These findings suggest that, prior to deployment, active duty military personnel report an increased ability to engage in suicidal behaviors but do not report elevated levels of the constructs associated with suicidal ideation (Bryan, Morrow, et al., 2010).

Whereas military training is intended to increase habituation to pain and fear, combat inherently offers multiple opportunities to do so (Selby et al., 2010). Research with military personnel deployed in support of OIF supports the unique relationship between combat exposure and the capacity for suicidal self-directed violence (Bryan, Cukrowicz, West, & Morrow, 2010). Controlling for gender, past suicidality, and symptoms of depression and PTSD, combat exposure significantly predicted the acquired capability for suicide, suggesting that exposure to combat independently predicts the capability for suicide, above and beyond past suicidal behaviors (Bryan, Cukrowicz, et al., 2010). In contrast, combat experiences did not significantly predict perceptions of burdensomeness or thwarted belongingness. These findings suggest that combat may increase the ability to enact suicidal self-directed violence but does not directly increase the constructs theorized to increase suicidal desire.

Whereas combat is directly associated with the acquired capability for suicide (Bryan, Cukrowicz, et al., 2010), not all types of combat contribute equally to the ability to engage in suicidal behaviors. Bryan and Cukrowicz (2011) examined the types of combat that predict the acquired capability for suicide, including combat experiences involving injury and death (e.g., witnessing bodies or being exposed to body parts), mission duties (e.g., exposure to stressful or hostile situations, such as patrols), and aggressive and violent experiences (e.g., shooting at the enemy or being shot at). All three types of combat experiences significantly predicted high levels of the acquired capability; however, when examining these types of combat together, only combat involving aggressive and violent experiences significantly predicted the capability for suicide, highlighting the importance of considering specific combat experiences—namely, those characterized by violence and aggression—that are uniquely associated with the capacity for suicide.

COMBAT-RELATED KILLING

Killing has been identified as a potentially important and understudied risk factor that may hold particular relevance to understanding suicide among military personnel and veterans (Maguen et al., 2010; Maguen et al., 2012). Research indicates substantial rates of combat-related killing among military personnel and combat veterans (Hoge et al., 2004; Laufer, Gallops, & Frey-Wouters, 1984; Maguen et al., 2010; Maguen et al., 2009).

Using data from the National Vietnam Veterans Readjustment Study (NVVRS), 47% of a nationally representative sample of male combat veterans reported that they killed or thought they had killed someone during their military service (Maguen et al., 2009). Most of the killings involved an enemy combatant (47%), whereas a small percentage involved civilians (6%) or prisoners of war (4%). Comparable rates were obtained in a separate study of Vietnam combat veterans, approximately half of whom reported killing an enemy combatant (Laufer et al., 1984).

Military service members who served in OIF reported similar rates of combat-related killing. Among Army personnel returning from OIF, 40% of those who completed postdeployment screening reported that they engaged in, or were responsible for, killing while deployed (Maguen et al., 2010). Seventy-seven percent reported that they saw dead bodies, and 56% reported that they witnessed killing. In a separate study of Army and Marine Corps units who served in Iraq, up to 65% reported being responsible for the death of an enemy combatant, and up to 28% reported being responsible for the death of a noncombatant (Hoge et al., 2004). Lower rates of killing were reported in a study of Army personnel deployed in support of OIF: 8.8% reported direct responsibility for the death of an enemy combatant, and 3.2% reported direct responsibility for the death of a noncombatant during their most recent deployment (Killgore et al., 2008). Sixty-three percent reported seeing dead bodies, and many reported experiences with severe injury and death. Nearly half had deployed to Iraq two or more times; thus it is unknown if assessing rates of killing across deployments would have resulted in higher reported rates.

Lower rates of combat-related killing were reported by those deployed in support of the Gulf War and OEF. Eleven to 14% of Gulf War veterans reported that they killed another person while deployed (Carney et al., 2003; Maguen, Vogt, et al., 2011). Similarly, 12% of Army personnel who served in support roles during OEF reported being responsible for the death of an enemy combatant (Hoge et al., 2004).

Taken together, these findings suggest that combat-related killing represents a potentially significant component of deployment, particularly for service members who served in OIF (Hoge et al., 2004; Maguen et al., 2010) and the Vietnam War (Laufer et al., 1984; Maguen et al., 2009). Considering research highlighting the particular relevance of aggressive and violent experiences to the acquired capability for suicide (i.e., Bryan & Cukrowicz, 2011), understanding the relationship between killing and suicide is crucial.

COMBAT-RELATED KILLING AND SUICIDAL IDEATION

To our knowledge, only two studies have explicitly examined the relationship between combat-related killing and suicidal ideation. Using NVVRS data from a subsample of Vietnam veterans who had completed diagnostic interviews, Maguen and colleagues (2012) examined the relationship between killing and lifetime suicidal ideation. Killing was measured with four variables (killing the enemy; prisoners; civilians; or women, children, and the elderly). Suicidal ideation was analyzed as a dichotomous measure. As hypothesized, killing during combat independently predicted lifetime suicidal ideation, controlling for demographics and robust predictors of suicidal ideation (i.e., depression, combat, PTSD, and substance use disorders). Veterans who reported higher killing experiences were twice as likely to report lifetime suicidal ideation, relative to veterans whose killing experiences were low or nonexistent. Depression, substance use disorders, and PTSD were also significant predictors of lifetime suicidal ideation.

Whereas Maguen and colleagues (2012) found a direct relationship between killing and suicidal ideation among Vietnam veterans, research with OIF military personnel produced different results (Maguen, Luxton, et al., 2011). Maguen, Luxton, and colleagues (2011) examined predeployment and combat-related predictors of suicidal ideation in a large sample of military service members returning from an OIF deployment. Suicidal ideation was assessed during postdeployment screening, with a single item assessing thoughts of suicide in the past two weeks. Controlling for known predictors of suicidal ideation, killing did not significantly predict suicidal ideation or the desire for self-harm. Other variables related to military exposure (e.g., witnessing killing, being exposed to dead bodies, and injury) also were not significant. Whereas no support was obtained for a *direct* relationship between killing and recent suicidal ideation, mediation analyses indicated an *indirect* relationship between these constructs, such that symptoms of depression and PTSD mediated the relationship between killing and suicidal ideation, controlling for lifetime history of suicide attempt and psychiatric medications. Similarly, PTSD symptoms mediated the relationship between killing and current desire for self-harm.

It is unknown whether the differing results between these two studies is due to actual cohort differences (i.e., Vietnam veterans vs. OIF military personnel), examining lifetime versus recent suicidal ideation, the amount of time passed since the killing occurred, or differences in the covariates employed. However, taken together, these results suggest a significant relationship between killing and suicidal ideation, which may be direct (Maguen et al., 2012) or indirect (Maguen, Luxton, et al., 2011).

COMBAT-RELATED KILLING, PERCEIVED BURDENSOMENESS, AND THWARTED BELONGINGNESS

One possible explanation for the relationship between killing and suicidal ideation (e.g., Maguen et al., 2012) is that killing relates to suicidal ideation through its effects on perceived burdensomeness and thwarted belongingness. To our knowledge, no research has examined the impact of killing on perceptions of burdensomeness and belongingness. However, the limited research available on combat-related killing indicates that it is associated with interpersonal difficulties that could be considered proxies of, or risk factors for, the disconnectedness and loneliness that characterize thwarted belongingness. For example, combat-related killing is associated with marital and relationship problems (Maguen et al., 2010). Marital status is a significant predictor of suicide, with divorce and separation relating to significantly elevated rates of suicide (Hyman, Ireland, Frost, & Cottrell, 2012; Kposowa, 2000). A large-scale study with active duty military found that the magnitude of the relationship between divorce or separation with suicide was "roughly comparable to one deployment" (Hyman et al., 2012, p. S144).

Killing is associated with other outcomes that could impact the ability to feel connected to others and effective in daily life. There is a significant relationship between combat-related killing and violent behaviors toward others (Maguen et al., 2009), which could arguably disrupt relationships and produce feelings of disconnectedness. Maguen and colleagues examined the association between killing and functional impairment in various domains (e.g., finances, employment, family problems, health). Controlling for demographic variables and combat exposure, postwar functional impairment was predicted by killing noncombatants but not by killing enemy combatants. It is possible that functional impairment in relationships and employment could lead to thwarted belongingness, whereas impairment in finances could lead to perceptions of burdensomeness. Such findings would be consistent with results obtained by Brenner and colleagues (2008) with OEF/OIF veterans, who reported deriving a sense of burdensomeness from being unable to financially provide for their families. Thus it is possible that killing leads to suicidal desire through its effects on outcomes (e.g., violence toward others, functional impairment), which impact perceived burdensomeness and thwarted belongingness.

An alternate possibility—one that is compatible with the finding that killing indirectly relates to suicidal ideation through its effects on symptoms of PTSD and depression (Maguen, Luxton, et al., 2011)—involves the psychiatric sequelae of killing. Combat-related killing is associated with psychiatric outcomes that could theoretically impact perceptions of burdensomeness and thwarted belongingness—and thus suicidal ideation—over time. The most consistent finding regarding the effects of combat-related killing is its significant association with PTSD (Fontana & Rosenheck, 1999; Fontana, Rosenheck, & Brett, 1992; MacNair, 2002; Maguen et al., 2009; Maguen et al., 2010; Maguen et al., 2013; Maguen, Vogt, et al., 2011; Van Winkle & Safer, 2011). The relationship between killing and PTSD symptomatology remains significant controlling for a variety of covariates, including demographics, general combat exposure, exposure to death, perceptions of danger, and witnessing the killing of fellow military personnel (Maguen et al., 2010; Maguen et al., 2009; Maguen, Vogt, et al., 2011), suggesting a particularly robust association. In addition to its association with suicidal ideation and death by suicide (Jakupcak et al., 2009; Kang & Bullman, 2008; Pietrzak et al., 2010), PTSD is associated with a range of negative interpersonal outcomes, including interpersonal violence, family distress, relationship problems that are both severe and numerous, dissatisfaction in intimate relationships, and intimacy difficulties (Dekel & Monson, 2010; Galovski & Lyons, 2004; Hiley-Young, Blake, Abueg, & Rozynko, 1995; Monson, Taft, & Fredman, 2009). In this manner, PTSD symptoms could potentially produce feeling disconnected from loved ones and ineffective in relationships, thus indirectly relating to suicidal ideation through this pathway.

COMBAT-RELATED KILLING AND SUICIDE ATTEMPTS

Previously, we reviewed research on combat-related killing and suicidal ideation and proposed direct and indirect pathways from killing to thoughts of suicide. In this section, we review research on killing and suicide attempts.

Fontana and colleagues (1992) found a significant association between direct experiences with killing and suicide attempts. They examined the roles of behaviors that varied in the "degree of responsibility for the initiation of death and destruction" (p. 749) in a large sample of Vietnam veterans. Acting as an agent of killing and failing to prevent killing significantly predicted suicide attempts. In contrast, none of the other killing-related traumatic events—that is, witnessing, observing, or being the target of killing—significantly associated with suicide attempts. An explanation compatible with the IPTS is that acting as the agent of killing required more direct exposure to violence, which resulted in greater habituation to violence; in turn, this would be expected to increase the capacity for suicidal self-directed violence, thus increasing risk for suicide (Joiner, 2005). However, an alternate explanation is that higher levels of guilt associated with more direct involvement with killing resulted in subsequent suicide attempts (Hendin & Haas, 1991).

In contrast to findings by Fontana et al. (1992), Maguen and colleagues (2012) obtained different findings in a secondary analysis of predictors of lifetime suicide attempt among Vietnam veterans. Neither killing nor combat significantly predicted lifetime history of any suicide attempt. However, the exploratory nature of the analysis, in addition to the low number of suicide attempts in the sample ($n = 12$), suggests that further exploration of this relationship is warranted.

Research with active duty Canadian military personnel (Belik, Stein, Asmundson, & Sareen, 2009) also supports the relationship between combat-related killing and suicide attempts: intentionally injuring or killing another person was associated with increased odds of lifetime suicide attempt, after controlling for demographics, service type, lifetime psychiatric disorder, and psychiatric comorbidity. Similar results were obtained in a study of veterans in treatment for substance use disorders: killing significantly predicted history of multiple suicide attempts (Ilgen et al., 2010). Of note, however, there was no differentiation in regard to whether the killing occurred in (versus outside of) the context of combat.

Thus, with the exception of findings by Maguen and colleagues (2012), taken together these findings highlight a significant association between killing and attempted suicide (Belik et al., 2009; Fontana et al., 1992; Ilgen et al., 2010).

COMBAT-RELATED KILLING AND THE ACQUIRED CAPABILITY FOR SUICIDE

Whereas killing in combat could theoretically increase suicidal *ideation* through its influence on perceptions of burdensomeness and thwarted belongingness, killing could impact suicidal *attempts* through its impact on the acquired capability for suicide. Thus one possible explanation for the relationship between killing and suicide attempts (e.g., Belik et al., 2009; Fontana et al., 1992; Ilgen et al., 2010) is that the relationship between killing and suicidal self-directed violence is mediated by increases in the capability for suicide that occur as a result of killing. In this framework, the inherent violence associated with killing would cause a person to habituate to the violence associated with death, thus dampening fear of suicide and increasing the capacity for suicide.

To our knowledge, no research has tested the hypothesis that combat-related killing is associated with increased rates of suicide attempt through its effects on the acquired capability for suicide. However, research is consistent with the notion that combat and killing significantly relate to increased engagement in risky behaviors postdeployment (Killgore et al., 2008), particularly for military personnel with a prior history of risky behaviors (Thomsen, Stander, McWhorter, Rabenhorst, & Milner, 2011).

Killgore and colleagues (2008) examined whether specific types of combat experiences (measured within three days of returning from an OIF deployment) predicted risk-taking behaviors, assessed three months later. Controlling for demographics, greater exposure to violent combat significantly related to overall risk-taking propensity and to all risk-taking subscale scores, including danger-seeking and invincibility. Exposure to violent combat and killing (of both enemy combatants and noncombatants) predicted the propensity for risk-taking. Violent combat experiences and killing nonhostile forces predicted danger-seeking. Last, invincibility was predicted by violent combat exposure, killing enemy combatants, and killing nonhostile forces. Interestingly, of all the specific combat-related factors examined, killing predicted nearly all risk-taking behaviors, more so than violent combat exposure.

Although not directly testing the hypothesis that killing increases one's capacity for suicide, these findings are compatible with the notion that combat-related killing is associated with a range of postdeployment behaviors—such as danger-seeking, risk-taking, and invincibility (Killgore et al., 2008)—that could be considered proxies for the acquired capability for suicide. However, alternate explanations are also possible, and an important next step will be to examine whether these constructs relate to increases in suicidal behaviors over time.

Moreover, although we previously proposed a direct relationship between killing and the acquisition of the capability for suicide, an indirect relationship is also plausible. Killing is associated with a host of negative psychiatric outcomes, including alcohol abuse, PTSD, and dissociation (Fontana & Rosenheck, 1999; Fontana et al., 1992; Freeman, Keesee, Thornton, & Gillette, 1995; Ilgen et al., 2010; Maguen et al., 2010; Maguen et al., 2009; Maguen, Vogt, et al., 2011). Beyond their association with suicide attempts (Jakupcak et al., 2009; Kang & Bullman, 2008; Pietrzak et al., 2010; Petronis, Samuels, Moscicki, & Anthony, 1990; Rossow & Amundsen, 1995), such sequelae could also indirectly increase the capacity for suicidal self-directed violence.

For example, killing is related to problematic alcohol use, alcohol abuse, and an increased frequency and quantity of alcohol consumption (Maguen et al., 2010; Maguen, Vogt, et al., 2011). Disordered alcohol use could arguably represent a potential pathway to acquiring the capability for suicide through its influence on habituating to the pain and fear of self-harm. Alcohol may also lower inhibitions against suicide. Indeed, alcohol use disorders are associated with elevated rates of suicide attempts and suicide deaths (Black, Yates, Petty, & Noyes, 1986; Harris & Barraclough, 1997; Rossow & Amundsen, 1995; Waller, Lyons, & Costantini-Ferrando, 1999; Wilcox, Conner, & Caine, 2004).

In addition, the association between killing and PTSD (Fontana & Rosenheck, 1999; Fontana et al., 1992; MacNair, 2002; Maguen et al., 2009; Maguen et al., 2010; Maguen et al., 2013; Maguen, Vogt, et al., 2011; Van Winkle & Safer, 2011) is noteworthy, considering proposals that PTSD re-experiencing symptoms could facilitate habituation to the pain and fear associated with suicide (Monteith et al., 2009, Selby et al., 2010). Indeed, recent research has identified PTSD re-experiencing symptoms as a significant predictor of the acquired capability for suicide (Bryan & Anestis, 2011). Although no research has examined specific PTSD re-experiencing symptoms as predictors of the acquired capability for suicide, trauma-related nightmares represent one of four PTSD intrusive symptoms listed in the *Diagnostic and Statistical Manual of Mental Disorders* (5th ed., American Psychiatric Association, 2013). Repeated exposure to nightmares could theoretically represent an experience that could increase one's ability to inflict suicidal self-directed violence, via habituation to fear of death, although this has not been empirically tested. Nightmares are associated with suicidal ideation (Bernert & Joiner, 2007; Bernert, Joiner, Cukrowicz, Schmidt, & Krakow, 2005; Sjöström, Wærn, & Hetta, 2007), suicide attempts (Li, Lam, Yu, Zhang, & Wing, 2010; Sjöström, Hetta, & Waern, 2009), and suicide deaths (Tanskanen et al., 2001).

In sum, combat-related killing could theoretically increase risk for suicide attempts through its influence on the acquired capability for suicide. This could occur directly—by causing habituation to violence, pain, and fear—or indirectly, through the influence of combat-related killing on psychiatric symptoms that could also lead to increased levels of the capability for suicide.

CLINICAL IMPLICATIONS

These findings have several implications for clinicians working with military service members and veterans. Clinicians should assess the severity and intensity of combat exposure, with particular attention to violent and aggressive combat experiences, given their association with higher levels of the acquired capability for suicide (Bryan & Cukrowicz, 2011; Bryan, Cukrowicz, et al., 2010). Considering the prevalence of killing experiences in combat (e.g., Laufer et al., 1984; Maguen et al., 2009; Maguen et al., 2010), and their association with negative psychiatric and interpersonal outcomes that associate with risk for suicide (Freeman et al., 1995; Jakupcak et al., 2009; Kang & Bullman, 2008;

Petronis et al., 1990; Pietrzak et al., 2010; Rossow & Amundsen, 1995), it is particularly important that clinicians working with veterans and military personnel assess for a history of killing. Shame, guilt, and concerns about negative reactions may prevent veterans and military service members from disclosing killing experiences to providers. Thus it is essential that assessments of killing and combat occur with sensitivity, within a secure therapeutic alliance, and in a nonjudgmental context (Maguen et al., 2012). Creating an environment that is supportive, safe, and accepting is critical. One component of this may include communicating to clients how information will be kept confidential and used to assist with treatment. Clients may fear the potential repercussions (e.g., retribution or even criminal consequences) of the killing experiences they share. If this is the case, such issues should be addressed upfront. Addressing how the experiences they share will be documented may also help put clients at ease.

Considering the strong association between killing and PTSD (Fontana & Rosenheck, 1999; Fontana, Rosenheck, & Brett, 1992; MacNair, 2002; Maguen et al., 2010; Maguen et al., 2009; Maguen, Vogt, et al., 2011; Van Winkle & Safer, 2011), assessing for PTSD symptoms is important when working with clients who report a history of killing. If a clinician discovers that trauma secondary to killing is continuing to impact a veteran or service member, there are several possibilities for treatment. One option is utilizing cognitive-behavioral strategies to explore how the event(s) impacted the client's beliefs about him- or herself, others, and the world. Cognitive processing therapy (Resick, Monson, & Chard, 2007) is particularly well-suited for this process and has demonstrated positive outcomes for military-related PTSD symptoms (Alvarez et al., 2011; Monson et al., 2006). Another option—adaptive disclosure (Gray et al., 2012)—deals with the issue of moral injury within the context of a comprehensive trauma treatment. Additionally, a separate treatment, intended for use in conjunction with evidence-based treatments for PTSD, is currently being developed and targets killing-related cognitions, as well as issues of self-forgiveness. Finally, given the association between killing and alcohol abuse and other negative alcohol-related outcomes (Maguen et al., 2010; Maguen, Vogt, et al., 2011), therapies aimed at targeting comorbid PTSD symptoms and substance or alcohol abuse—such as Seeking Safety (Najavits, 2002)—may be of particular value to clients who report symptoms in both these areas.

Combat-related killing is associated with marital and relationship problems (Maguen et al., 2010), in addition to violent behaviors toward others (Maguen et al., 2009). Additionally, PTSD, a common sequelae of combat-related killing, is also related to numerous negative interpersonal outcomes (Dekel & Monson, 2010; Galovskia & Lyons, 2004; Hiley-Young, Blake, Abueg, & Rozynko, 1995; Monson, Taft, & Fredman, 2009). Thus for clients who report killing experiences, particularly those with significant PTSD symptoms, therapeutic modalities aimed at improving interpersonal relationships may be of particular benefit. To the extent that such processes can also decrease perceptions of burdensomeness and thwarted belongingness, according to the IPTS (Joiner, 2005), suicidal desire should also decrease (although this has not been empirically tested). Group therapy may represent a potential modality for deriving a sense of belongingness (e.g., by learning of others' experiences with killing and combat) and changing one's beliefs about burdensomeness. While it is often the case that some in the group will have killed and others may not share this experience, this is similar to other combat or life experiences that are shared and can be processed accordingly.

Considering the association of combat-related killing with suicidal ideation (Maguen et al., 2012) and suicide attempts (Belik et al., 2009; Fontana et al., 1992; Ilgen et al., 2010), suicide risk assessment will likely represent an important component of treatment for providers working with clients who report having killed in combat. Clinicians should assess for suicidality (e.g., ideation, intent, means, plan) and known risk factors for suicide (e.g., history of suicide attempt, depression, PTSD). For clients who are at risk for suicide, safety planning (Stanley & Brown, 2012; Stanley, Brown, Karlin, Kemp, & Von Bergen, 2008) and means restriction (Sarchiapone, Mandelli, Iosue, Andrisano, & Roy, 2011) are brief interventions aimed specifically at preventing suicide.

Another potentially important therapeutic goal for providers working with clients who report having killed in combat includes preventing increases in the capability for suicide. Active duty military personnel report high levels of the capacity for suicidal self-directed violence prior to deployment (Bryan, Morrow, et al., 2010). Furthermore, killing is associated with increased postdeployment risky behaviors (Killgore et al., 2008), in addition to alcohol abuse and consumption (Maguen et al., 2010; Maguen, Vogt, et al., 2011), which could theoretically further increase one's capability for suicide (Joiner, 2005). Thus clinicians may choose to work collaboratively with their clients to prevent behaviors that would further increase their acquired capability for suicide. Potential therapeutic targets in line with this goal might include developing coping techniques and distress tolerance strategies to prevent future suicide attempts and nonsuicidal self-injury, in addition to reducing the occurrence of risky behaviors and substance and alcohol abuse.

In sum, assessing combat-related killing and its impact on mental health and functioning should provide a more comprehensive assessment of suicide risk, in addition to a more thorough treatment planning process for military service members and veterans. Moreover, assessing killing experiences routinely (e.g., as part of an intake interview) communicates that killing is viewed as a normative experience in combat worth inquiring about, particularly in regard to how it impacts symptoms and functioning.

LIMITATIONS AND FUTURE DIRECTIONS

In this chapter, we reviewed research on combat-related killing and suicide. Although in its early stages, such research has made important contributions to the literature and highlights the importance of continued endeavors aimed at understanding the complex relationship between combat-related killing and suicide.

Research on the relationship between killing and suicidal ideation has produced mixed findings in regard to whether there is a direct (Maguen et al., 2012) or indirect (Maguen, Luxton, et al., 2011) relationship between killing and suicidal ideation. There may be a range of explanations for the discrepant findings (e.g., differences in samples, covariates, and time frames for suicidal ideation). Additional investigations are needed to further elucidate the relationship between combat-related killing and suicidal ideation.

Prior research on the relationship between combat-related killing and suicidal ideation has utilized single-item measures of suicidal ideation; research in this area would be strengthened by relying upon continuous measures of suicidal ideation consisting of multiple items (e.g., Beck Scale for Suicide Ideation [Beck & Steer, 1991]; Modified Scale for Suicidal Ideation [Miller, Norman, Bishop, & Dow, 1986]). This would provide a more psychometrically sound approach and would allow researchers to examine how killing relates to the severity and intensity of suicidal ideation. In addition, research investigating the impact of killing could potentially benefit from supplementing self-report measures of suicidal ideation with measures of attentional bias (e.g., Stroop task [Cha, Najmi, Park, Finn, & Nock, 2010]) or reaction time (e.g., the Death/Suicide Implicit Association Test [Nock et al., 2010]) specific to suicide-related stimuli. Such measures are considered to be less subject to social desirability (Nock et al., 2008) and have been shown to longitudinally predict suicidal behavior (Cha et al.; Nock et al., 2010).

Research on combat-related killing and suicide attempts has generally found killing to be a significant predictor of suicide attempt history (Belik et al., 2009; Fontana et al., 1992; Ilgen et al., 2010), though these results have not been obtained uniformly (i.e., Maguen et al., 2012). Furthermore, although these studies have yielded important findings, the majority have utilized a single item based on self-report to assess suicide attempts. Future research aimed at understanding the relationship between combat-related killing and attempted suicide would be strengthened by utilizing interviewer-rated measures of suicide attempt history, such as the Lifetime Suicide Attempt Self Injury Interview (Linehan & Comtois, 1996), which allows differentiation between nonsuicidal self-injury and suicide attempts.

Additionally, prior research on killing and attempted suicide has typically focused on *lifetime* history of suicide attempts, with no differentiation between suicide attempts that occurred prior to and following deployments; thus, from such studies, it is not possible to control for prior suicide attempts (or to rule out the possibility that predeployment suicidal self-directed violence led to an increased propensity for engaging in combat-related killing). Future examinations of the relationship between killing and suicide attempts would be strengthened by distinguishing between suicide attempts that precede and follow killing that occurred in combat, thus enabling an examination of the impact of combat-relating killing on subsequent suicide attempts, controlling for prior history of attempted suicide. Controlling for other forms of violence that occurred prior to and following combat will also be important.

Similarly, based on the cross-sectional nature of prior research on killing and suicidality, it is not possible to determine the temporal nature of this relationship. Although research on combat-related killing and suicidal behaviors has understandably been cross-sectional, conducting longitudinal research in this area would provide further clarification regarding the impact of killing on suicidal outcomes over time. Some of the research reviewed in this chapter involved interviewing military personnel following their return from deployment and assessing their killing-related experiences. Interviewing military service members again following this initial interview and assessing suicidal ideation, suicide attempts, and potential covariates (e.g., psychiatric outcomes) would provide important information regarding how killing in combat relates to suicidal ideation and attempts over time, particularly following reintegration into civilian life.

Along similar lines, research on the impact of killing on suicidality has focused exclusively on suicidal ideation and suicide attempts, to the exclusion of death by suicide as an outcome. Although past suicide attempts are important predictors of eventual death by suicide, the relationship between suicidal ideation and suicide is substantially less robust (Large, Sharma, Cannon, Ryan, & Nielssen, 2011; Large, Smith, Sharma, Nielssen, & Singh, 2011). Thus an important next step will be to examine whether combat-related killing associates with increased odds of suicide. Doing so will enable a more comprehensive understanding of whether killing in combat represents a risk factor for suicide among military personnel and veterans.

Finally, we proposed that combat-related killing impacts suicidal ideation through direct and indirect effects on perceptions of burdensomeness and thwarted belongingness. Similarly, we proposed that killing in combat impacts suicide attempts by directly and indirectly increasing the acquired capability for suicide. No research has tested these hypotheses. Consequently, future research examining these interpersonal-psychological constructs (Joiner, 2005) as potential mediators of the relationships between combat-related killing and suicide-related outcomes will be an important next step—one with the potential to clarify possible mechanisms by which killing relates to suicidality, thus informing effective suicide risk assessment and prevention among our nation's military personnel and veterans.

AUTHORS' NOTE

This material is based upon work supported in part by the Department of Veterans Affairs and the Rocky Mountain MIRECC. The views expressed are those of the authors and do not necessarily represent the views or policy of the Department of Veterans Affairs or the U.S. Government.

REFERENCES

Alvarez, J., McLean, C., Harris, A. S., Rosen, C. S., Ruzek, J. I., & Kimerling, R. (2011). The comparative effectiveness of cognitive processing therapy for male veterans treated in a VHA posttraumatic stress disorder residential rehabilitation program. *Journal of Consulting and Clinical Psychology*, 79(5), 590–599. doi:10.1037/a0024466

American Psychiatric Association. (2013). *Diagnostic and statistical manual of mental disorders* (5th ed.). Washington, DC: Author.

Anestis, M. D., Bryan, C. J., Cornette, M. M., & Joiner, T. E. (2009). Understanding suicidal behavior in the military: An evaluation of Joiner's interpersonal-psychological theory of suicidal behavior in two case studies of active duty post-deployers. *Journal of Mental Health Counseling*, 31(1), 60–75.

Beck, A. T., & Steer, R. A. (1991). *Beck Scale for Suicide Ideation: Manual*. San Antonio, TX: Psychological Corporation.

Belik, S., Stein, M. B., Asmundson, G. G., & Sareen, J. (2009). Relation between traumatic events and suicide attempts in Canadian military personnel. *The Canadian Journal of Psychiatry/La Revue Canadienne De Psychiatrie*, 54(2), 93–104.

Bernert, R. A., & Joiner, T. E. (2007). Sleep disturbances and suicide risk: A review of the literature. *Neuropsychiatric Disease and Treatment*, 3(6), 735–743.

Bernert, R. A., Joiner, T. E., Cukrowicz, K. C., Schmidt, N. B., & Krakow, B. (2005). Suicidality and sleep disturbances. *Sleep*, 28(9), 1135–1141.

Black, D. W., Yates, W., Petty, F., & Noyes, R. (1986). Suicidal behavior in alcoholic males.

Comprehensive Psychiatry, 27(3), 227–233. doi:10.1016/0010-440X(86)90046-5

Brenner, L. A., Gutierrez, P. M., Cornette, M. M., Betthauser, L. M., Bahraini, N., & Staves, P. J. (2008). A qualitative study of potential suicide risk factors in returning combat veterans. *Journal of Mental Health Counseling*, 30(3), 211–225.

Bryan, C. J., & Anestis, M. D. (2011). Reexperiencing symptoms and the interpersonal-psychological theory of suicidal behavior among deployed service members evaluated for traumatic brain injury. *Journal of Clinical Psychology*, 67(9), 856–865. doi:10.1002/jclp.20808

Bryan, C. J., Clemans, T. A., & Hernandez, A. (2012). Perceived burdensomeness, fearlessness of death, and suicidality among deployed military personnel. *Personality and Individual Differences*, 52(3), 374–379. doi:10.1016/j.paid.2011.10.045

Bryan, C. J., & Cukrowicz, K. C. (2011). Associations between types of combat violence and the acquired capability for suicide. *Suicide and Life-Threatening Behavior*, 41(2), 126–129. doi:10.1111/j.1943-278X.2011.00023.x

Bryan, C. J., Cukrowicz, K. C., West, C. L., & Morrow, C. E. (2010). Combat experience and the acquired capability for suicide. *Journal of Clinical Psychology*, 66(10), 1044–1056.

Bryan, C. J., Morrow, C. E., Anestis, M. D., & Joiner, T. E. (2010). A preliminary test of the interpersonal-psychological theory of suicidal behavior in a military sample. *Personality and Individual Differences*, 48(3), 347–350. doi:10.1016/j.paid.2009.10.023

Carney, C. P., Sampson, T. R., Voelker, M., Woolson, R., Thorne, P., & Doebbeling, B. N. (2003). Women in the Gulf War: Combat experience, exposures, and subsequent health care use. *Military Medicine*, 168(8), 654–661.

Cha, C. B., Najmi, S., Park, J. M., Finn, C. T., & Nock, M. K. (2010). Attentional bias toward suicide-related stimuli predicts suicidal behavior. *Journal of Abnormal Psychology*, 119(3), 616–622. doi:10.1037/a0019710

Dekel, R., & Monson, C. M. (2010). Military-related post-traumatic stress disorder and family relations: Current knowledge and future directions. *Aggression and Violent Behavior*, 15(4), 303–309. doi:10.1016/j.avb.2010.03.001

Fontana, A., & Rosenheck, R. (1999). A model of war zone stressors and posttraumatic stress disorder. *Journal of Traumatic Stress*, 12(1), 111–126. doi:10.1023/A:1024750417154

Fontana, A., Rosenheck, R., & Brett, E. (1992). War zone traumas and posttraumatic stress disorder symptomatology. *Journal of Nervous and Mental Disease*, 180(12), 748–755. doi:10.1097/00005053-199212000-00002

Freeman, T. W., Keesee, N., Thornton, C., & Gillette, G. (1995). Dissociative symptoms in posttraumatic stress disorder subjects with a history of suicide attempts. *The Journal of Nervous and Mental Disease*, 183(10), 664–666. doi:10.1097/00005053-199510000-00010

Galovski, T., & Lyons, J. A. (2004). Psychological sequelae of combat violence: A review of the impact of PTSD on the veteran's family and possible interventions. *Aggression and Violent Behavior*, 9(5), 477–501. doi:10.1016/S1359-1789(03)00045-4

Gray, M. J., Schorr, Y., Nash, W., Lebowitz, L., Amidon, A., Lansing, A., . . . Litz, B. T. (2012). Adaptive disclosure: An open trial of a novel exposure-based intervention for service members with combat-related psychological stress injuries. *Behavior Therapy*, 43(2), 407–415. doi:10.1016/j.beth.2011.09.001

Gutierrez, P. M., Brenner, L. A., Rings, J. A., Devore, M. D., Kelly, P. J., Staves, P. J., . . . Kaplan, M. S. (2013). A qualitative description of female veterans' deployment-related experiences and potential suicide risk factors. *Journal of Clinical Psychology*, 69(9), 923–935. doi:10.1002/jclp.21997

Harris, E., & Barraclough, B. (1997). Suicide as an outcome for mental disorders: A meta-analysis. *The British Journal of Psychiatry*, 170(3), 205–228. doi:10.1192/bjp.170.3.205

Hendin, H., & Haas, A. P. (1991). Suicide and guilt as manifestations of PTSD in Vietnam combat veterans. *The American Journal of Psychiatry*, 148(5), 586–591.

Hiley-Young, B., Blake, D., Abueg, F. R., & Rozynko, V. (1995). Warzone violence in Vietnam: An examination of premilitary, military, and postmilitary factors in PTSD in-patients. *Journal of Traumatic Stress*, 8(1), 125–141. doi:10.1002/jts.2490080109

Hoge, C. W., Castro, C. A., Messer, S. C., McGurk, D., Cotting, D. I., & Koffman, R. L. (2004). Combat duty in Iraq and Afghanistan, mental health problems, and barriers to care. *The New England Journal of Medicine*, 351(1), 13–22. doi:10.1056/NEJMoa040603

Hyman, J., Ireland, R., Frost, L., & Cottrell, L. (2012). Suicide incidence and risk factors in an active duty US military population. *American Journal of Public Health*, 102(Suppl. 1), S138–S146.

Ilgen, M. A., Burnette, M. L., Conner, K. R., Czyz, E., Murray, R., & Chermack, S. (2010). The association between violence and lifetime suicidal thoughts and behaviors in individuals treated for substance use disorders. *Addictive Behaviors*, 35(2), 111–115. doi:10.1016/j.addbeh.2009.09.010

Jakupcak, M., Cook, J., Imel, Z., Fontana, A., Rosenheck, R., & McFall, M. (2009). Posttraumatic stress disorder as a risk factor for suicidal ideation in Iraq and Afghanistan war veterans. *Journal of Traumatic Stress*, 22(4), 303–306. doi:10.1002/jts.20423

Joiner, T. E. (2005). *Why people die by suicide*. Cambridge, MA: Harvard University Press.

Kang, H. K., & Bullman, T. A. (2008). Risk of suicide among US veterans after returning from the Iraq or Afghanistan war zones. *JAMA: Journal of the American Medical Association*, 300(6), 652–653. doi:10.1001/jama.300.6.652

Killgore, W. S., Cotting, D. I., Thomas, J. L., Cox, A. L., McGurk, D., Vo, A. H., . . . Hoge, C. W. (2008). Post-combat invincibility: Violent combat experiences are associated with increased risk-taking propensity following deployment. *Journal of Psychiatric Research*, 42(13), 1112–1121. doi:10.1016/j.jpsychires.2008.01.001

Kposowa, A. J. (2000). Marital status and suicide in the National Longitudinal Mortality Study. *Journal of Epidemiology and Community Health*, 54(4), 254–261. doi:10.1136/jech.54.4.254

Kuehn, B. M. (2009). Soldier suicide rates continue to rise: Military, scientists work to stem the tide. *JAMA: Journal of the American Medical Association*, 301(11), 1111.

Large, M., Sharma, S., Cannon, E., Ryan, C., & Nielssen, O. (2011). Risk factors for suicide within a year of discharge from psychiatric hospital: A systematic meta-analysis. *Australian & New Zealand Journal of Psychiatry*, 45(8), 619–628. doi:10.3109/00048674.2011.590465

Large, M. M., Smith, G. G., Sharma, S. S., Nielssen, O. O., & Singh, S. P. (2011). Systematic review and meta-analysis of the clinical factors associated with the suicide of psychiatric in-patients. *Acta Psychiatrica Scandinavica*, 124(1), 18–19. doi:10.1111/j.1600-0447.2010.01672.x

Laufer, R. S., Gallops, M. S., & Frey-Wouters, E. (1984). War stress and trauma: The Vietnam veteran experience. *Journal of Health and Social Behavior*, 25(1), 65–85. doi:10.2307/2136705

Li, S. X., Lam, S. P., Yu, M. M., Zhang, J., & Wing, Y. K. (2010). Nocturnal sleep disturbances as a predictor of suicide attempts among psychiatric outpatients: A clinical, epidemiologic, prospective study. *Journal of Clinical Psychiatry*, 71(11), 1440–1446. doi:10.4088/JCP.09m05661gry

Linehan, M. M. & Comtois, K. (1996). *Lifetime parasuicide history*. Unpublished manuscript. University of Washington, Seattle.

MacNair, R. M. (2002). Perpetration-induced traumatic stress in combat veterans. *Peace and Conflict: Journal of Peace Psychology*, 8(1), 63–72. doi:10.1207/S15327949PAC0801_6

Maguen, S., Lucenko, B. A., Reger, M. A., Gahm, G. A., Litz, B. T., Seal, K. H., . . . Marmar, C. R. (2010). The impact of reported direct and indirect killing on mental health symptoms in Iraq War veterans. *Journal of Traumatic Stress*, 23(1), 86–90.

Maguen, S., Luxton, D. D., Skopp, N. A., Gahm, G. A., Reger, M. A., Metzler, T. J., & Marmar, C. R. (2011). Killing in combat, mental health symptoms, and suicidal ideation in Iraq war veterans. *Journal of Anxiety Disorders*, 25(4), 563–567. doi:10.1016/j.janxdis.2011.01.003

Maguen, S., Madden, E., Bosch, J., Galatzer-Levy, I., Knight, S. J., Litz, B. T., . . . McCaslin, S. E. (2013). Killing and latent classes of PTSD symptoms in Iraq and Afghanistan veterans. *Journal of Affective Disorders*, 145(3), 344–348. doi:10.1016/j.jad.2012.08.021

Maguen S., Metzler T. J., Bosch, J., Marmar, C. R., Knight, S. J., & Neylan, T. C. (2012). Killing in combat may be independently associated with suicidal ideation. *Depression and Anxiety*, 29(11), 918–923. doi:10.1002/da.21954.

Maguen, S., Metzler, T. J., Litz, B. T., Seal, K. H., Knight, S. J., & Marmar, C. R. (2009). The impact of killing in war on mental health symptoms and related functioning. *Journal of Traumatic Stress*, 22(5), 435–443. doi:10.1002/jts.20451

Maguen, S., Vogt, D. S., King, L. A., King, D. W., Litz, B. T., Knight, S. J., & Marmar, C. R. (2011). The impact of killing on mental health symptoms in Gulf War veterans. *Psychological Trauma: Theory, Research, Practice, and Policy*, 3(1), 21–26. doi:10.1037/a0019897

Miller, I. W., Norman, W. H., Bishop, S. B., & Dow, M. G. (1986). The Modified Scale for Suicidal Ideation: Reliability and validity. *Journal of Consulting and Clinical Psychology*, 54(5), 724–725. doi:10.1037/0022-006X.54.5.724

Monson, C. M., Schnurr, P. P., Resick, P. A., Friedman, M. J., Young-Xu, Y., & Stevens, S. P. (2006). Cognitive processing therapy for veterans with military-related posttraumatic stress disorder. *Journal of Consulting and Clinical Psychology*, 74(5), 898–907.

Monson, C. M., Taft, C. T., & Fredman, S. J. (2009). Military-related PTSD and intimate relationships: From description to theory-driven research and intervention development. *Clinical Psychology Review*, 29(8), 707–714. doi:10.1016/j.cpr.2009.09.002

Monteith, L. L., Green, K. L., Mathew, A. R., & Pettit, J. W. (2009). The interpersonal-psychological theory of suicidal behaviors as an explanation of suicide among war veterans. In L. Sher & A. Vilens (Eds.), *War and suicide* (pp. 249–264). Hauppauge, NY: Nova Publishers.

Monteith, L. L., Menefee, D. S., Pettit, J. W., Leopoulos, W. L., & Vincent, J. P. (2013), Examining the interpersonal–psychological theory of suicide in an inpatient veteran sample. *Suicide and Life-Threatening Behavior, 43*, 418–428. doi:10.1111/sltb.12027

Nademin, E., Jobes, D. A., Pflanz, S. E., Jacoby, A. M., Ghahramanlou-Holloway, M., Campise, R., . . . Johnson L. (2008). An investigation of interpersonal-psychological variables in Air Force suicides: A controlled-comparison study. *Archives of Suicide Research, 12*(4), 309–326.

Najavits, L. M. (2002). *Seeking safety: A treatment manual for PTSD and substance abuse.* New York: Guilford Press.

Nock, M. K., Borges, G., Bromet, E. J., Cha, C. B., Kessler, R. C., & Lee, S. (2008). Suicide and suicidal behavior. *Epidemiologic Reviews, 30*, 133–154. doi:10.1093/epirev/mxn002

Nock, M. K., Park, J. M., Finn, C. T., Deliberto, T. L., Dour, H. J., & Banaji, M. R. (2010). Measuring the suicidal mind: Implicit cognition predicts suicidal behavior. *Psychological Science, 21*(4), 511–517. doi:10.1177/0956797610364762

Petronis, K. R., Samuels, J. F., Moscicki, E. K., & Anthony, J. C. (1990). An epidemiologic investigation of potential risk factors for suicide attempts. *Social Psychiatry and Psychiatric Epidemiology, 25*(4), 193–199.

Pfeiffer, P. N., Brandfon, S., Garcia, E., Duffy, S., Ganoczy, D., Kim, H. M., & Valenstein, M. (2014). Predictors of suicidal ideation among depressed veterans and the interpersonal theory of suicide. *Journal of Affective Disorders, 152*, 277–281. doi:10.1016/j.jad.2013.09.025

Pietrzak, R. H., Goldstein, M. B., Malley, J. C., Rivers, A. J., Johnson, D. C., & Southwick, S. M. (2010). Risk and protective factors associated with suicidal ideation in veterans of Operations Enduring Freedom and Iraqi Freedom. *Journal of Affective Disorders, 123*(1-3), 102–107. doi:10.1016/j.jad.2009.08.001

Resick, P. A., Monson, C. M., & Chard, K. M. (2007). *Cognitive processing therapy treatment manual: Veteran/military version.* Boston: Veterans Administration.

Rossow, I., & Amundsen, A. (1995). Alcohol abuse and suicide: A 40-year prospective study of Norwegian conscripts. *Addiction, 90*(5), 685–691. doi:10.1111/j.1360-0443.1995.tb02206.x

Sarchiapone, M., Mandelli, L., Iosue, M., Andrisano, C., & Roy, A. (2011). Controlling access to suicide means. *International Journal of Environmental Research and Public Health, 8*(12), 4550–4562.

Selby, E. A., Anestis, M. D., Bender, T. W., Ribeiro, J. D., Nock, M. K., Rudd, M., . . . Joiner, T. E. (2010). Overcoming the fear of lethal injury: Evaluating suicidal behavior in the military through the lens of the interpersonal–psychological theory of suicide. *Clinical Psychology Review, 30*(3), 298–307. doi:10.1016/j.cpr.2009.12.004

Sjöström, N., Hetta, J., & Waern, M. (2009). Persistent nightmares are associated with repeat suicide attempt: A prospective study. *Psychiatry Research, 170*(2–3), 208–211. doi:10.1016/j.psychres.2008.09.006

Sjöström, N., Wærn, M., & Hetta, J. (2007). Nightmares and sleep disturbances in relation to suicidality in suicide attempters. *Sleep: Journal of Sleep and Sleep Disorders Research, 30*(1), 91–95.

Stanley, B., & Brown, G. K. (2012). Safety planning intervention: A brief intervention to mitigate suicide risk. *Cognitive and Behavioral Practice, 19*(2), 256–264.

Stanley, B., Brown, G. K., Karlin, B., Kemp, J. E., & Von Bergen, H. A. (2008). *Safety plan treatment manual to reduce suicide risk: Veteran version.* New York: New York Suicide Prevention Center, Department of Psychiatry, Columbia University and New York State Psychiatric Institute.

Tanskanen, A., Tuomilehto, J., Viinamäki, H., Vartiainen, E., Lehtonen, J., & Puska, P. (2001) Nightmares as predictors of suicide. *Sleep: Journal of Sleep and Sleep Disorders Research, 24*(7), 844–847.

Thomsen, C. J., Stander, V. A., McWhorter, S. K., Rabenhorst, M. M., & Milner, J. S. (2011). Effects of combat deployment on risky and self-destructive behavior among active duty military personnel. *Journal of Psychiatric Research, 45*(10), 1321–1331. doi:10.1016/j.jpsychires.2011.04.003

Van Orden, K., Merrill, K., & Joiner, T. R. (2005). Interpersonal-psychological precursors to suicidal behavior: A theory of attempted and completed suicide. *Current Psychiatry Reviews, 1*(2), 187–196.

Van Orden, K. A., Witte, T. K., Cukrowicz, K. C., Braithwaite, S. R., Selby, E. A., & Joiner, T. E. (2010). The interpersonal theory of suicide. *Psychological Review, 117*(2), 575–600. doi:10.1037/a0018697

Van Orden, K. A., Witte, T. K., Gordon, K. H., Bender, T. W., & Joiner, T. R. (2008). Suicidal desire and the capability for suicide: Tests of the interpersonal-psychological theory of suicidal behavior among adults. *Journal of Consulting and Clinical Psychology, 76*(1), 72–83. doi:10.1037/0022-006X.76.1.72

Van Winkle, E. P., & Safer, M. A. (2011). Killing versus witnessing in combat trauma and reports of PTSD symptoms and domestic violence. *Journal of Traumatic Stress, 24*(1), 107–110.

Waller, S. J., Lyons, J. S., & Costantini-Ferrando, M. F. (1999). Impact of comorbid affective and alcohol use disorders on suicidal ideation and attempts. *Journal of Clinical Psychology, 55*(5), 585–595.

Wilcox, H. C., Conner, K. R., & Caine, E. D. (2004). Association of alcohol and drug use disorders and completed suicide: An empirical review of cohort studies. *Drug and Alcohol Dependence, 76*(Suppl. 7), S11–S19. doi:10.1016/j.drugalcdep.2004.08.003

7

Suicide Risk Assessment with Combat Veterans—Part I

Contextual Factors

Christopher G. AhnAllen
Abby Adler
Phillip M. Kleespies

This chapter provides a discussion of the contextual factors that need to be understood in order to engage in suicide risk assessment and management with combat veterans. It is intended to assist the clinician in attaining a more complete understanding of the combat veteran when attempting to assess the veteran's risk of suicidal or self-injurious behavior. In this sense, it is a companion chapter with chapter 8, "Suicide Risk Assessment with Combat Veterans—Part II: Assessment and Management."

MILITARY CULTURE

Understanding the relationship between military veterans and suicidality necessitates an awareness and appreciation of the military as a unique cultural context. The military is a large-scale operational system that has been described by many as a cultural framework that is worthy of distinction (Moore, 2011; Reger, Etherage, Reger, & Gahm, 2008). Understanding military culture allows for a clearer understanding of the experiences that military service members and veterans endure. Military organizational culture is typically more collectivist, respectful of an internal hierarchy, and less influenced by salary than civilian working cultures (Soeters, Poponete, & Page, 2006).

In addition, the military emphasizes the importance of its own language(s), customs, and traditions. While it is understood that there are aspects of each branch of the military that are different, the commonalities across the Army, Navy, Air Force, Marine Corps, and Coast Guard have been noted to be relevant in understanding the military experience (Christian, Stivers, & Sammons, 2009; Hall, 2008).

As described in this section, military culture is a complex system that affects all service members. While culturally competent practice with veterans requires an appreciation and understanding of military culture, military culture is expected to impact each member individually, and it is not the case that all service members and veterans experience the impact of military culture in the same way. Nonetheless, having an understanding of the military culture will enable clinicians to pursue culturally competent assessment and practice with veterans.

Developmental Context of Joining the Military

Military service often begins and occurs during late adolescence and young adulthood as it can represent a pathway to adulthood for a variety of reasons. In fact, nearly 50% of military personnel are between the ages of 17 and 24 years old (Kelty, Kleykamp, &

Segal, 2010). The military therefore serves a purpose for training young people to become adults with specific traits and expertise. The opportunity to gain knowledge that is transferable to occupational eligibility in the civilian world is often a reason young people join the military. Some have heralded the military service period as a "moratorium" within the developmental transition into adulthood given that there may be deferment of adult-type roles and responsibilities (Elder, 1987). In fact, the military promotes development of personal responsibility, health, physical fitness and training, codes of conduct, and community and civic engagement (Kelty et al., 2010). Relatedly, enlistees are provided much of their basic needs during this time period including wages, medical care, housing, and educational and/or training opportunities.

The decision to join the military as a volunteer is one that provides insight into the customs and values of one's family as well as one's individualism. Although this decision is likely a personal one for the enlistee, there are similar thematic reasons for why young adults may join the military across the different branches (Moore, 2011). First, service members may join the military to improve their quality of life. Many service members are raised in homes of working- or lower-middle-class parents from rural and small towns where opportunities are limited (Lutz, 2008).

Second, service members may have joined the military because service is a family tradition. This reason may develop as a personal goal of the enlistee or it may have been an explicit expectation during childhood. In fact, service members who have a family history of military service are the norm, with fewer service members from nonmilitary families (Gegax & Thomas, 2005).

Third, young adults may enlist in a branch of the military because they seek to develop a sense of belonging with others. The military may attract those who seek to become part of a tight-knit culture that promotes camaraderie, connectedness, and cohesion (Coll, Weiss, & Yarvis, 2011). These may include young adults who place a high degree of value on these experiences as well as those who may not have fit in well in prior social groups.

Finally, young adults may pursue the military as they see a need to demonstrate their patriotism and become part of the response to terrorist attacks on Americans. Since 9/11, the rates of enlistment in the military have increased, and the rate of enlistment has consistently met the goals set by each branch. In fact, in April of 2011, the Army reduced its maximum enlistment age from 42 back to 35 given such high rates of enlistment in this branch.

Military Cultural Traditions and Values

As general of the Army, Douglas MacArthur stated in 1962 when accepting the Sylvanus Thayer Award that three important ideals for the American soldier include "duty, honor and country" (Department of Defense, 1964). These three words were expressed as a way to inspire soldiers to achieve all that is personally possible through instilling courage, hope, and faith. In order to understand the military culture, an appreciation for the importance of traditions and values held within the military is needed. Across the branches of service, a set of core military values are a central component of the military experience (Hsu, 2010). These include honor, courage, loyalty, integrity, and commitment. While there are unique values for each branch of the military, those that are shared provide a framework from which clinicians can understand the military experience.

Service members conform to certain elements, norms, and traditions within the military experience (Moore, 2011). For example, there is a high degree of importance placed upon discipline within the hierarchical class system of the military. Service members also prioritize group-based goals and tasks compared with individualistic ones. This mentality supports the accomplishment of goals that necessitate the cooperation of large groups of people in an effective and efficient manner. Collectivism enhances morale and cohesion within military units and, therefore, promotes inclusion and effective task orientation (Segal & Segal, 1983).

Understanding the language of the military is also an important cultural competence goal for clinicians. The military, as a government entity, promotes the use of colloquial terms, acronyms, and abbreviations that serve a need of efficient communication. Not understanding such cultural language can be problematic when attempting to understand a veteran's or service member's experience (Reger et al., 2008). References are available to assist in understanding military language and other aspects of military culture (see deploymentpsych.org and www.dtic.mil/doctrine/dod_dictionary/).

Other relevant cultural elements of the military include the mindset of missions and solution-focused problem-solving skills. Service members are trained to identify a solution to known problems to be able to devise a strategy of approach and accomplishment. Therefore, cultural expectations call upon service members to be able to problem-solve using the resources available to them in the field. This is a highly transferable skill into the civilian world and can be a helpful framework for clinicians to engage with veterans seeking mental health care. On the contrary, military culture also promotes an emphasis on physical and emotional strength that is devoid of weakness or vulnerability. While this may promote action and mission-oriented goals within the military, it may undermine mental health engagement and treatment if it is not conceptualized as a relevant mission for problem-solving.

Personnel

The face of the typical military service member has changed over time as the population of the United States has changed and as increasing opportunities have been given to diverse Americans. Increasingly over recent years, the personnel demographics within the military are young Hispanic men and women (Kelty et al., 2010). As noted, nearly 50% of the military is between 17 and 24 years of age with women constituting only 15% of this subgroup. The military has historically been a male-dominated culture; however, since it became all volunteer after the Vietnam War and there have been other legal reforms, the presence of women has grown from 1.6% in 1973 to 15% in 2005 (Manning, 2005). Women's roles within the military vary according to the branch, with the Air Force containing the largest group (20%) of female enlisted personnel and officers in 2008. Of note, women increasingly depart the military earlier than men, which has been linked to poorer job satisfaction, childbearing plans, or negative reactions of others to such plans (Pierce, 2006).

Racial and ethnic minority engagement in the military has also changed significantly over time. According to Harris and Jones (2007), nearly 36% of active duty service members identify as a member of at least one minority group (African American, Hispanic American, Native American, Alaskan Native, Asian American, Pacific Islander, or multiracial). The role of service members of color has varied over time with movement toward an integrated military. In fact, African Americans served in a segregated service prior to the Korean War. More recently, since 2001, African American engagement in the military has declined to below 20%, while Hispanic American participation has risen sharply to 13% in 2006. Ethnic minority veterans have also historically been subject to increased levels of trauma exposure and unique trauma due to their racial and ethnic identities. Asian American veterans experienced race-related stressors while in the Vietnam War such as racial stigmatization and exclusion, bicultural conflict, racial/cultural identification with the enemy, and being mistaken for the enemy by other service members (Loo, Lim, Koff, Morton, & Kiang, 2007). Native American veterans have sought out military service as a means of increasing their opportunities for employment outside their local communities and reservations. While these service members have also endured racism and discrimination in the service, many Native Americans report positive overall experiences of their military service and note the reverence with which military service has been treated within their native communities (Harada, Villa, Reifel, & Bayhylle, 2005). There are also increasing roles in the military for immigrants to the United States who are allowed to serve in the military. A total of 65,000 noncitizens and naturalized citizens have been identified within recent surveys of military personnel (Batolova, 2008).

Inclusion of sexual orientation minorities (gay, lesbian, bisexual) within the military has also changed significantly over time (Herek & Belkin, 2006). While there has been documentation of military service by gay men since the Revolutionary War, they were prohibited from serving in the military from 1950 to January 1993 prior to the Don't Ask, Don't Tell, Don't Pursue policy (Kelty et al., 2010). There has been increasing support within and outside the military culture for gay men and lesbian women to serve openly within the military. In 2012, there was a repeal of the Don't Ask, Don't Tell, Don't Pursue policy allowing for such open expression of minority sexual orientation.

The military cultural experience also transcends service members to those within a service member's family, including spouses, partners, and children. While young single men historically postponed marriage and parenthood until after the service, increasing numbers of servicemen and women are marrying within the service, and dual-service unions are increasingly common (Harris & Jones, 2007; Kelty et al., 2010). The military may serve to preserve marriages for active duty service members who have been deployed, though following discharge the rates of divorce are much greater than the rate for those not in the military (Kelty et al., 2010). Stressors within families in the military include financial, spousal employment, separation from networks of support and deployed partners, relocations around the world, and unpredictable duty hours. While some assignments within the military may allow service members to bring their families, this may come with stress of integrating into a new culture, country, or lifestyle, including increased stress for families of color who may reside in locales without significant diversity (Harris & Jones, 2007). Regarding children, most service members (nearly 75%) have dependent children, and women service members tend to have children slightly earlier than civilian women.

UNDERSTANDING THE STRESSES OF BEING DEPLOYED, SERVING IN A COMBAT ZONE, AND RETURNING HOME

In evaluating a combat veteran who has posttraumatic stress disorder (PTSD) and who may be at risk for suicide, it is clearly important to have an awareness of the many stresses that may have had an emotional impact on him or her through his or her deployment to, service in, and return from a combat zone. In this section, we discuss these stresses in order to enhance understanding of the combat veteran's experience.

Deployment to combat environments is a disruptive and stressful experience emotionally, physically, and psychologically. The experience of being deployed for an active duty service person or activated reservist includes many of the challenges associated with leaving home, being in a combat zone, and finally readjusting to home life. The experience of war for a combat veteran includes some commonalities across service members but is ultimately a unique experience that may include myriad exposures to stress of varying degrees depending upon the combat mission(s), military duties, and many other factors. Understanding all of these factors is critical in conceptualizing the impact of deployment on an individual veteran.

Home Departure Stress

Military deployments to combat-related missions are associated with a number of psychological challenges requiring a period of adjustment for the service member (Vasterling et al., 2011). While the military provides training for service members on the duties required for combat-related missions and the cultural expectations of the combat environment, the transition to the war zone overseas is nonetheless a difficult process. For some, the call to deploy to a combat environment can come with little advance notice. Military personnel leave behind both the comforts and the social supports of their home environment. The separation from established social supports, while anticipated in many service members, may be difficult given the nature of the planned missions and the loss of established methods of communication. During the recent wars in Iraq and Afghanistan, service members often employed technology (e.g., Skype) to communicate back home. This direct communication could involve bidirectional challenges in negotiating what information to share. Both sides of the exchange could feel a need to convey only positive news and to keep the burden of their own challenges to themselves. Service members might minimize the danger and rigors of their environment when communicating back home. This could lead, in turn, to family members sharing bad news that they might otherwise have held back. Increased access to direct methods of communication could potentially increase the stress level of service members in combat zones.

Military families must often manage increased burdens when a service member is deployed (Eaton et al., 2008). Evidence suggests that there are many psychological consequences for the children and families that remain at home (Lincoln, Swift, & Shorteno-Fraser, 2008). The stressful experiences for these families may include undetermined return dates, frequent deployment extensions, and multiple deployments. Children with a deployed service member parent demonstrate increased rates of depression and anxiety (Kelly, 1994). In fact, there are nearly 700,000 children who have had at least one parent who was deployed overseas in the recent conflicts in Iraq and Afghanistan (Johnson et al., 2007). For remaining spouses, the departure of a military partner overseas to combat operations also means that changes in roles and responsibilities within the home are inevitable, with increased duties for the remaining spouse (Eaton et al., 2008). It is unknown how these changes within the family system as well as issues related to the children's mental health are communicated to the combat-deployed service member. Greater access to faster methods of communication, however, suggests that these home-front factors may affect the stress level of service members in combat zones. Therefore, the transition for the service member to a combat environment can include multiple stresses both in the combat environment itself and at home.

Combat Environment Stress

Unit cohesion. Upon arrival in the combat zone, service members must depend upon those within their unit for emotional, physical, and psychological support. Evidence suggests that service members who report being understood and accepted by their peers and superiors and whose unit has a high degree of cohesiveness report fewer negative consequences from exposure to stressful situations (Iverson et al., 2008). In their study of 4,700 UK armed forces personnel, Iverson and colleagues (2008) reported that social support, feeling well informed, having a sense of belonging, and effective leadership were associated with reduced risk of PTSD. On the other hand, those personnel who reported experiencing low morale during deployment were more likely to exhibit PTSD symptoms. Other researchers have identified how social connectedness with fellow service members may result in formation of a social identity and allow for increased communication among service personnel (Cobb, 1976; Cohen & Willis, 1985).

Cultural changes and adaptations. The transition into the combat environment often includes an introduction to a new culture in a foreign land. The language spoken is usually an unfamiliar one. There can be great difficulty in communicating with the local people. Language differences may result in increased stress for the service member who needs to negotiate interactions with local people and perform assessments of high-risk environments. Personal cultural beliefs may also be challenged, including one's spiritual or religious beliefs (Nash, 2007). For example, war might cause service members to question their faith in God or other higher power. Others may find their beliefs strengthened, which could also be destabilizing.

Environmental characteristics. War zones are physically demanding, and service members are exposed to elements of nature for prolonged periods of time with limited protection (Nash, 2007). Environmental exposure, for example, can include extreme temperatures of heat during the day and cold during the night. These conditions are only exacerbated by protective gear that must be worn or by physically uncomfortable travel in armored vehicles. Other challenges include maintaining hydration, guarding against excessive wetness, sleep deprivation, malnutrition, and noises associated with war (e.g., sniping, firefights, falling mortars).

Potential for physical injury. War zones are inherently unsafe environments. Service members are at risk of witnessing, learning about, or personally experiencing physically, psychologically, and emotionally traumatic events. These events may result in physical damage or injury requiring medical attention. As a result of improved medical treatment, service members in the Operation Enduring Freedom (OEF) and Operation Iraqi Freedom (OIF) wars are surviving serious physical injuries more frequently than in previous combat environments (Pryce, Pryce, & Shackelford, 2012).

In some cases, service members are exposed to the deaths of fellow service members, enemy combatants, and/or civilians. The service member may also be involved in caring for those who are seriously injured or in collecting human remains. They may need to seriously hurt or kill others. Some war zones involve the threat of chemical, biological, and radiological warfare. Service members' concerns about such exposure may have implications for their psychological and physical health.

Military sexual trauma. Military sexual trauma (MST) is defined as a psychological trauma resulting from a physical or verbal assault of a sexual nature, including unwanted sexual touching or grabbing, threatening or offensive sexual remarks about the body, and threatening or unwanted sexual advances (Street & Stafford, 2004). These offenses are typically perpetrated by peers or supervisors and may occur either while deployed or stateside. Such victimization may impact the service member's mental health and sense of unit cohesion, something that is particularly important for survival in a combat zone.

Given the close working environment of the war zone, service members who are victims of MST can be put into positions where they will continue to be exposed to the risk of future assault. Many victims of MST are hesitant to report experiences of unwanted sexual encounters or experiences with other service members or have had their reports invalidated and dismissed. The combat environment may also be associated with increased rates of MST relative to peacetime military operations (Wolfe et al., 1998). Ultimately, service members who are exposed to MST experience poorer psychological and physical well-being and are less satisfied with their employment and health (Street & Stafford, 2004). Other evidence suggests that military sexual assault results in poorer mental health functioning than sexual trauma that is experienced outside of military operations (Himmelfarb, Yaeger, & Mintz, 2006; Street et al., 2011).

Race and ethnicity-related issues. Military service members of color who identify as minorities or are seen by others as minorities may also encounter unwanted racist comments and actions from fellow service members (Litz & Orsillo, 2004). This may be particularly salient for minority service members who identify or are viewed by others as being from the same racial and/or ethnic group(s) as those of the enemy combatants. It is possible that minority service members may experience conflict between their self-identity as a person of color and the race/ethnicity of the enemy combatants. Of concern is that units may not always include other service members of color in military operations, an experience that would be expected to increase the risk of identity conflict as well as prejudice, racism, and maltreatment by fellow service members.

Emotional reactions. Emotional reactions from exposure to traumatic events are often complex. A portion of service members who are exposed to such traumatic experiences manage their emotional distress in a healthy manner, while others have extended emotional reactions. For some, the experience of injury and/or death to fellow service members may lead to grief for extended periods of time (Pivar & Field, 2004). Veterans returning from the recent wars in Iraq and Afghanistan are experiencing mental health problems including PTSD, major depressive disorder (MDD) or depressive symptoms, and traumatic brain injury (Tanielian & Jaycox, 2008). A government study indicates that as the number of deployments increases, the risk of developing PTSD, MDD, or anxiety increases (Government Accountability Office, 2008). Service members may need to cope emotionally with the death of or injury to comrades, which can include reactions of shock, disbelief, guilt, shame, anger, and longing (Nash, 2007). Given the demands associated with being in a combat environment, there is often little opportunity to express emotions and/or grief given the need to perform subsequent missions. Intense emotional experiences, such as helplessness, horror, and fear, are distressing emotions that can be associated with reactions to a multitude of stressful experiences within a combat zone. Some service members return from war zones with feelings of guilt regarding their own survival.

Home Arrival Stress

The arrival home for service members serving abroad in combat is often conceptualized as a positive event and experience for the soldier, family, and friends. While this may be true for some returning veterans, the transition back to a civilian lifestyle may hold many challenges and stresses. Sloan and Friedman (2008) suggest that the readjustment to home life may be even more difficult than adjusting to life in a combat war zone.

Service members returning from combat experience a process and period of readjustment to civilian life that varies in length for each soldier and is understood not to occur "overnight" (Munroe, 2012). The return of the service member from the combat environment is likely to be anticipated with relief and with a motivation to return to one's predeployment lifestyle. The homecoming is often thought of as a "honeymoon" period whereby couples reunite, welcome home banners are displayed, and parties are planned, but this phase usually ends as problems develop (Sloan & Friedman, 2008). Veterans may not initially understand that their return home will require a period of readjustment, although this often the case. Psychological distress during this process is not unusual and can include irritability, disrupted sleeping patterns, and poor concentration (Vasterling et al., 2012). More significant levels of psychopathology may also develop both early in the home arrival period and after periods of months or years. These include symptoms of posttraumatic stress, depression, and anxiety, as well as substance use problems (Bliese, Wright, Adler, Thomas, & Hoge, 2007; Hoge, Auchterlonie, & Milliken, 2006; Kessler, Sonnega, Bromet, Hughes, & Nelson, 1995; Lapierre,

Schwegler, & LaBauve, 2007, Shipherd, Stafford, & Tanner, 2005).

Returning combat veterans arrive back to a civilian lifestyle having developed highly specialized and critically important battlefield skills. A challenging aspect about the transition out of a war zone environment is the need to transition away from a reliance upon combat survival skills that are no longer necessary outside of a combat arena. Essentially, readjustment to home life is dependent upon each veteran's ability to transition out of a war-zone mindset. Munroe (2012) identified 14 unique skills of combat veterans that must be modified: (1) being vigilant to increase safety in an inherently unsafe war zone; (2) learning not to trust people who may harm you; (3) using anger and aggression as defenses in war; (4) learning that routines are highly susceptible to vulnerabilities of attack by the enemy; (5) keeping "tight-lipped" about one's thoughts, intent, or actions; (6) having an intense focus and concentration on one primary task; (7) having a preference for immediate and effective decision-making; (8) adopting an "act first, think later" mentality; (9) regarding everyone as a potential enemy; (10) numbing one's emotions; (11) following a chain of command; (12) having an intense closeness to comrades versus avoiding closeness to deal with the loss of comrades; (13) having a "move on" mentality in regard to loss; and (14) avoiding talk about war. Whereas the survival of the combat veteran depended upon the development and execution of these specialized skills, all of them now require reshaping for the service member to be able to adapt successfully back into his or her home environment. This is not a simple matter and identifies the importance of social supports (e.g., family, friends, partners) in recognizing a process of readjustment as an ongoing need for recently returned veterans.

In addition, there can be "homecoming stress" involving the civilian population's cultural and political views of war and their view of the military and servicemen and women who have fought in that war. Varying experiences of homecoming stress have existed for veterans depending upon whether they returned home from the Vietnam War or the more recent conflicts such as those in Iraq and Afghanistan.

Family dynamics and social support arrangements are again subject to accommodation given the return home of the veteran. Family members who remained stateside often grow accustomed to managing functions that the deployed family member had previously performed. The necessary reassessment of these roles upon the veteran's return can be stressful and challenging for all members of the family. These functions can include management of finances, childcare duties, home repairs, and even routine household duties such as cooking, cleaning, and waste disposal. Negotiations about who should perform these roles and responsibilities now that the family is reunited can be challenging, particularly if the veteran is experiencing mental health problems. Relationships with romantic partners require effort to reconnect emotionally, physically, and sexually. There are some veterans who return home to disrupted romantic relationships, including separation or divorce. Some veterans return to greet infants or toddlers whom they had not previously met or return to children who have aged and may have difficulty connecting with a parent who has been at war for a period of time (Sloan & Friedman, 2008).

Returning veterans need to re-establish connections with employers and other social supports such as friends and acquaintances. While service members may re-secure their predeployment jobs, some veterans return to significant changes in work duties. With regard to friendships, returning veterans often experience challenges in reconnecting with their peers who were not deployed and may not understand the experience of being in combat in a war zone (Sloan & Friedman, 2008). The veteran may end up distancing himself or herself from his or her predeployment social support network. Such developments can be a concern, particularly if these veterans are having trouble re-engaging in other parts of their life. Recent studies point to increasingly more difficult experiences for activated reservists returning home compared with regular active duty soldiers. Thus there have been increases in posttraumatic stress, particularly if the veterans reside in rural areas with limited mental health services (Sloan & Friedman, 2008; Vasterling et al., 2010).

MILITARY CULTURE AS A POTENTIAL BARRIER TO MENTAL HEALTH CARE

Although it is clear that suicidal combat veterans pose a significant risk to themselves, many do not receive the mental health treatment that could be beneficial to them. In fact, many veterans with likely mental health diagnoses (whether suicidal or not) do not receive treatment. Schell and Marshall (2008) described a study conducted by the RAND Corporation that included interview data from 1,965 OEF/OIF veterans. They found that only 52.7% of those having probable major depression or probable PTSD met with a physician or mental health provider at least once in the previous 12 months. Furthermore, only 30% received minimally adequate treatment, which was defined in the study as at least eight psychotherapy sessions or four psychopharmacology visits. There are a number of reasons why veterans do not seek the mental health care they need, including factors related to the military culture, beliefs about treatment, and logistical barriers.

One significant barrier to mental health treatment related to military culture was described earlier in this chapter as "battlemind" or the war-zone mindset. Some of the skills identified by Munroe (2012) that are unique to combat veterans may interfere with help-seeking behavior postdeployment. In particular, learning not to trust people who may harm you; staying tight-lipped about one's thoughts, intent, or actions; regarding everyone as a potential enemy; numbing of emotions; the "move on" mentality in the face of loss; and avoiding talking about war may make it difficult for a veteran to approach a mental health provider about his or her suffering and emotional concerns. Additionally, the opinion of peers and superiors in such a hierarchical system may also influence treatment-seeking behavior. In support of this issue, Hoge and colleagues (2004) compared barriers to mental health treatment between OEF/OIF combat infantry service members postdeployment who met screening criteria for a mental disorder and those who did not. The authors reported that those who met screening criteria for a mental disorder endorsed at a higher rate the following items: being perceived as weak, being treated differently by leadership, being blamed by leadership for the problem, and others having less confidence in them. These beliefs about the impact of seeking treatment on one's role in the military can interfere with getting the needed help.

In addition to factors related to military culture, beliefs about treatment may also decrease utilization among those in need. Pietrzak and colleagues (2009) assessed whether beliefs about mental health care were associated with barriers to care and treatment utilization among 272 OEF/OIF veterans. Barriers to care were assessed based on five items from the Perceived Stigma and Barriers to Care for Psychological Problems scale (Britt et al., 2008; Hoge et al., 2004). Pietrzak and colleagues found that specific beliefs, "Therapy is not effective for most people" and "Therapy is a sign of weakness," were positively associated with these barriers to care. Additionally, they found that negative beliefs about mental health care predicted attendance at fewer counseling or medication sessions. From this sample of 272 OEF/OIF veterans, 34 respondents endorsed contemplating suicide within two weeks prior to completing the survey. This subsample of suicide contemplators scored higher on this measure of perceived barriers to care compared to nonsuicide contemplators (Pietrzak et al. 2010). Beyond negative beliefs about treatment, logistical concerns, such as transportation, financial, and scheduling problems, were also identified as significant barriers to care among veterans with mental disorders (Pietrzak et al., 2009, 2010).

CONCLUSION

Understanding suicide risk assessment in combat veterans requires cultural competence in working with military veterans. Military experiences and cultural bases for behavior in veterans, while diverse depending upon the service of each veteran, includes some common factors associated with why individuals join the service, the traditions and values inherent in military service, and the policies and procedures within this system. Military veterans who have deployed to combat undergo significant stress associated with leaving home, entering new and unfamiliar environments that are in a state of war, and then require reintegration into home life. Aspects of the military culture, such as battlemind and negative beliefs about mental health treatment, may serve as barriers to accessing mental health care for veterans. In conclusion, this chapter highlighted the value of understanding suicide risk assessment in military veterans through developing an awareness of military culture and practices.

REFERENCES

Batolova, J. (2008). *Immigrants in the U.S. Armed Forces*. Retrieved from www.migrationinformation.org/feature/display.cfm?ID=683

Bliese, P. D., Wright, K. M., Adler, A. B., Thomas, J. L., & Hoge, C. W. (2007). Timing of postcombat mental health assessments. *Psychological Services, 4*, 141–148.

Britt, T. W., Greene-Shortridge, T. M., Brink, S., Nguyen, Q. B., Rath, J., Cox, A. L., Hoge, C. W., & Castro C. A. (2008). Perceived stigma and barriers to care for psychological treatment: Implications for reactions to stressors in different contexts. *Journal of Social and Clinical Psychology*, 27, 317–335

Christian, J., Stivers, J., & Sammons, M. (2009). Training to the warrior ethos: Implications for clinicians treating military members and their families. In S. Morgillo-Freeman, B. A. Moore, & A. Freeman (Eds.), *Living and surviving in harm's way: A psychological treatment handbook for pre- and post-deployment of military personnel* (pp. 27–49). New York: Routledge/Taylor & Francis.

Cobb, S. (1976). Social support as a moderator of life stress. *Psychosomatic Medicine*, 38, 300–314.

Cohen, S., & Willis, T. A. (1985). Stress, social support, and the buffering hypothesis. *Psychological Bulletin*, 98, 310–357.

Coll, J. E., Weiss, E. L., & Yarvis, J. S. (2011). No one leaves unchanged: Insights for civilian mental health care professionals into the military experience and culture. *Social Work in Health Care*, 50, 487–500.

Department of Defense. (1964). *Address to the Cadets of the U.S. Military Academy, 12 May, 1962*. Pamphlet DoD PAM GEN-1A. Washington, DC: Author.

Eaton, K., Hoge, C., Messer, S., Whitt, A., Cabrera, O., McGurk, D., . . . Castro, C. (2008). Prevalence of mental health problems, treatment need, and barriers to care among primary care-seeking spouses of military service members involved in Iraq and Afghanistan deployments. *Military Medicine*, 173, 1051–1056.

Elder, G. H. (1987). War mobilization and the life course: A cohort of World War II veterans. *Sociological Forum*, 3, 449–472.

Gegax, T. T., & Thomas, E. (2005, June 20). The family business. *Newsweek*, 145, 24–31.

Government Accountability Office (GAO). (May 30, 2008). *DOD health care: Mental health and traumatic brain injury screening efforts implemented, but consistent pre-deployment medical record review policies needed.* (GAO-08-615).

Hall, L. (2008). *Counseling military families: What mental health professionals need to know*. New York: Routledge/Taylor & Francis.

Harada, N. D., Villa, V. M., Reifel, N., & Bayhylle, R. (2005). Exploring veteran identity and health services use among Native American veterans. *Military Medicine*, 170, 782–786.

Harris, J. J., & Jones, N. G. (2007). African-American military service members and their families: A different environment. In L. A. See (Ed.), *Human behavior in the social environment from an African-American perspective* (pp. 133–152). Binghamton, NY: Hawthorn Press.

Herek, G. M., & Belkin, A. (2006). Sexual orientation and military service: Prospects for organizational and individual change in the United States. In T. W. Britt, A. B. Adler, & C. A. Castro (Eds.), *Military life: The psychology of serving in peace and combat* (Vol. 4, pp. 119–142) London: Praeger Security International.

Himmelfarb, N., Yaeger, D., & Mintz, J. (2006). Posttraumatic stress disorder in female veterans with military and civilian sexual trauma. *Journal of Traumatic Stress*, 19, 837–846.

Hoge, C. W., Auchterlonie, J. L., & Milliken, C. S. (2006). Mental health problems, use of mental health services, and attrition from military service after returning from deployment in Iraq or Afghanistan. *JAMA: Journal of the American Medical Association*, 295, 1023–1032.

Hoge, C. W., Castro, C. A., Messer, S. C., McGurk, D., Cotting, D. L., & Koffman, R. L. (2004). Combat duty in Iraq and Afghanistan, mental health problems, and barriers to care. *The New England Journal of Medicine*, 351, 13–22.

Hsu, J. (2010, September). *Overview of military culture*. Retrieved from http://www.apa.org/about/gr/issues/military/military-culture.pdf

Iverson, A. C., Fear, N. T., Ehler, A., Hacker Hughes, J., Hull, L., Earnshaw, M., . . . Hotopf, M. (2008). Risk factors for post-traumatic stress disorder among UK Armed Forces personnel. *Psychological Medicine*, 38, 511–522.

Johnson, S. J., Sherman, M. D., Hoffman, J. S., James, J. C., Johnson, P. L., Lochman, J. E., . . . Stepney, B. (2007). *The psychological needs of U.S. military service members and their families: A preliminary report*. Presidential Task Force on Military Deployment Services for Youth, Families and Service Members. Washington, DC: American Psychological Association.

Kelly, M. L. (1994). Military-induced separation in relation to maternal adjustment and children's behavior. *Military Psychology*, 6, 163–176.

Kelty, R., Kleykamp, M., & Segal, D. R. (2010). The military and the transition to adulthood. *Future of Our Children*, 20, 181–207.

Kessler, R., Sonnega, A., Bromet, E., Hughes, M., & Nelson, C. (1995). Posttraumatic stress disorder in the National Comorbidity Survey. *Archives of General Psychiatry*, 52, 1048–1060.

Lapierre, C., Schwegler, A., & LaBauve, B. (2007). Posttraumatic stress and depression symptoms in soldiers returning from combat operations in Iraq and Afghanistan. *Journal of Traumatic Stress, 20,* 933–943.

Lincoln, A., Swift, E., & Shorteno-Fraser, M. (2008). Psychological adjustment and treatment of children and families with parents deployed in military combat. *Journal of Clinical Psychology, 64,* 984–992.

Litz, B., & Orsillo, S. (2004). The returning veteran of the Iraq war: Background issues and assessment guidelines. Iraq War Clinician Guide. National Center for PTSD, Department of Veterans Affairs.

Loo, C. M., Lim, B. R., Koff, G., Morton, R. K., & Kiang, P. N. C. (2007). Ethnic-related stressors in the war zone: Case studies of Asian American Vietnam veterans. *Military Medicine, 172,* 968–971.

Lutz, A. (2008). Who joins the military? A look at race, class, and immigration status. *Journal of Political and Military Sociology, 36*(2), 167–188.

Manning, L. (2005). *Women in the military: Where they stand* (5th ed.). Washington DC: Women's Research and Education Institute.

Moore, B. (2011). Understanding and working within military culture. In B. Moore & W. Penk (Eds.), *Treating PTSD in military personnel: A clinical handbook.* (pp. 9–22). New York: Guilford Press.

Munroe, J.F. (2012). *Transitioning war zone skills: Information for veterans and those who care.* Unpublished manuscript.

Nash, W. (2007). The stressors of war. In C. R. Figley & W. P. Nash (Eds.), *Combat stress injury: Theory, research and management* (pp. 33–63). New York: Routledge.

Pierce, P. F. (2006). The role of women in the military. In T. W. Britt, A. B. Adler, & C. A. Castro (Eds.), *Military life: The psychology of serving in peace and combat* (Vol. 4, pp. 97–118). London: Praeger Security International.

Pietrzak, R. H., Johnson, D. C., Goldstein, M. B., Malley, J. C., & Southwick, S. M. (2009). Perceived stigma and barriers to mental health care utilization among OEF-OIF veterans. *Psychiatric Services, 60,* 1118–1122.

Pivar, I., & Field, N. (2004). Unresolved grief in combat veterans with PTSD. *Journal of Anxiety Disorders, 18,* 745–755.

Pryce, J. G., Pryce, D. H., & Shackelford, K. K. (2012). *The costs of courage: Combat stress, warriors and family survival.* Chicago: Lyceum Books.

Reger, M., Etherage, J., Reger, G., & Gahm, G. (2008). Civilian psychologists in an Army culture: The ethical challenge of cultural competence. *Military Psychology, 20,* 21–35.

Schell, T. L., & Marshall, G. N. (2008). Survey of individuals previously deployed for OEF/OIF. In T. Tanielian & L. H. Jacvox (Eds.), *Invisible wounds of war: Psychological and cognitive injuries, their consequences, and services to assist recovery* (pp. 87–115). Santa Monica, CA: RAND Corporation.

Segal, D. R., & Segal, M. W. (1983). Change in military organization. *Annual Review of Sociology, 9,* 151–170.

Shipherd, J.C., Stafford, J., & Tanner, L.R. (2005). Predicting alcohol and drug abuse in Persian Gulf War veterans: What role do PTSD symptoms play? *Addictive Behaviors, 30,* 595–599.

Sloan, L.B., & Friedman, M.J. (2008). *After the war zone: A practical guide for returning veterans.* Philadelphia, PA: De Capo Press.

Soeters, J. L., Poponete, C., & Page, J. T. (2006). Culture's consequences in the military. In T. W. Britt, A. B. Adler, & C. A. Castro (Eds.), *Military life: The psychology of serving in peace and combat* (Vol. 4, pp. 13–34). London: Praeger Security International.

Street, A. E., Kimerling, R., Bell, M. E., & Pavao, J. (2011). Sexual harassment and sexual assault during military service. In J. Ruzek, P. Schnurr, J. Vasterling, & M. Friedman (Eds.), *Posttraumatic stress reactions: Caring for the veterans of the global war on terror* (pp. 131–150). Washington D.C.: American Psychological Association Books.

Street, A., & Stafford, J. (2004). Military sexual trauma issues in caring for Veterans. In National Center for PTSD (Ed.), *Iraq War Clinician's Guide.* Boston, MA: Department of Veterans Affairs.

Tanielian, T., & Jaycox, L. H. (2008). *Invisible wounds of war: Psychological and cognitive injuries, their consequences, and services to assist recovery.* Santa Monica, CA: RAND Corporation.

Vasterling, J. J., Daly, E. S., & Friedman, M. J. (2011). Posttraumatic Stress Reactions Over Time: The Battlefield, Homecoming, and Long-term Course. In J. I. Ruzek, P. P. Schnurr, J. J. Vasterling, & M. J. Friedman (Eds.), *Caring for Veterans with Deployment-related Stress Disorders: Iraq, Afghanistan and Beyond.* Washington D.C.: American Psychological Association.

Vasterling, J. J., Proctor, S. P., Friedman, M. J., Hoge, C. W., Heeren, T., King, L. A., & King, D. W. (2010). PTSD symptom increases in Iraq-deployed soldiers: comparison with nondeployed soldiers and associations with baseline symptoms, deployment experiences and postdeployment stress. *J Trauma Stress, 23,* 41–51.

Wolfe, J., Sharkansky, E., Read, J., Dawson, R., Martin, J., & Ouimette, P. (1998). Sexual harassment and assault as predictors of PTSD symptomatology among U.S. female Persian Gulf War military personnel. *Journal of Interpersonal Violence, 13,* 40–57.

8

Suicide Risk Assessment with Combat Veterans—Part II

Assessment and Management

Phillip M. Kleespies

Abby Adler

Christopher G. AhnAllen

Jeffrey Lucey, a 23-year-old lance corporal in the Marine Reserves and an Iraq War combat veteran from Massachusetts, hung himself with a garden hose in the basement of his family's home after seeking care at a Department of Veterans Affairs medical center (VAMC). At the VA, he was reportedly told that he needed to attain sobriety before he could be treated for his posttraumatic stress disorder (PTSD; Sege, 2005). Jonathan Schulze, a 25-year-old Marine Corps veteran of the Iraq War and the recipient of two Purple Hearts, hung himself in a basement with an electrical cord in January 2007. His death followed an agonizing effort by family and friends to have him seek mental health care at his local VAMC in Minnesota. Although he had apparently said that he felt suicidal, he had been told in a phone contact that he was 26th on a treatment waiting list for a 12-bed PTSD unit (Sennott, 2007). Joshua Lee Omvig, a 22-year-old Army specialist and Iraq War combat veteran, shot and killed himself in front of his distraught mother in Iowa in 2005. He had recently learned that his battle buddy in Iraq had been killed and stated that he felt that he should have been there to protect him. Although showing clear signs of PTSD, he had reportedly not sought treatment because he had felt that it would interfere with his ability to advance in the Army Reserve.

The suicides of these and a number of other Iraq War and Afghanistan War veterans received a great deal of coverage in the national media. At this same time, there were reports of a possible rise in suicides among active duty military personnel in Iraq, particularly in the Marine Corps and Army (Nelson, 2004), and the concern about suicide among those on active duty fueled concerns about an increase in suicides as soldiers transitioned to veteran status. These concerns were brought to the attention of members of the U.S. Congress, and one of the veterans noted previously, Joshua Omvig, became the human face for the Joshua Omvig Veterans Suicide Prevention Act of 2007 (Cvetanovich & Reynolds, 2008). This legislation directed the VA to develop and implement a comprehensive program that included establishing a VA Crisis Hotline, appointing suicide prevention coordinators at every VAMC in the nation, increasing efforts to heighten the awareness of VA staff to suicide risk among the veteran population, and requiring that the VA provide education and training for staff on suicide risk detection and suicide prevention.

The actual rate of suicide among veterans, however, has not been so easy to determine. Large community-based studies of veteran suicide have had inconsistent results in terms of whether or not veteran status per se (i.e., simply having served in the armed forces but not necessarily in combat) confers an elevated risk of suicide (Gibbons, Brown, & Hur, 2012; Kaplan, McFarland, Huguet, & Newsom, 2012; Kaplan, Huguet, McFarland, & Newsom, 2007; Miller, Azrael, Barber, Mukamal, & Lawler, 2012; Miller et al., 2009). A Blue Ribbon Work Group on Suicide Prevention in the Veteran Population (2008; appointed by James B. Peake, MD, the Secretary of Veterans Affairs) came to this same conclusion and as its first recommendation stated that the Veterans Health Administration should establish an analysis and research plan to resolve conflicting study results and ensure that there is a consistent approach to describing the rates of suicide and suicide attempts in veterans.

In a large register linkage study that focused only on veterans who used the VA health care system, an elevated standardized *mortality* ratio for suicide (SMR = 1.66, 95% confidence interval [CI]: 1.58–1.75) was found for veterans in the VA system when compared with age-matched individuals in the general population (McCarthy et al., 2009). Further, in a study that focused specifically on Operation Iraqi Freedom/Operation Enduring Freedom (OIF/OEF) veterans, Kang and Bullman (2008) linked 490,346 veterans who served between 2001 and 2005 in OIF/OEF with cause of death data available through the National Death Index. They found that OIF/OEF veterans who had returned from the Iraq and Afghanistan war zones had a standardized mortality ratio of 1.33 (95% CI: 1.03–1.69) for death by suicide. This finding clearly seemed to indicate an increased rate of suicide with returning OIF/OEF combat veterans, albeit not the epidemic described in the media (CBS Evening News, 2007). In a still more recent study, Ilgen and colleagues, (2012) linked all OEF/OIF veterans who had a health care encounter in the VA health care system in fiscal years 2007 and 2008 to the National Death Index. They found a significant interaction between psychiatric conditions and OEF/OIF status. More specifically, having a diagnosed mental health condition was associated with a greater risk of suicide among OEF/OIF veterans who sought care in the VA (hazard ratio = 4.41; 95% CI: 2.57–7.55, $p < .01$).

The publication of the study by Kang and Bullman in 2008 put a focus on combat experience as a potential risk factor for suicide. Subsequent information, however, has suggested a situation more complicated than one in which combat experience is taken as a sole or direct risk factor for suicide. A task force of the U.S. Department of Defense (2010) has reported that 40 percent of those who died by suicide in the military did so prior to ever having been deployed to a combat area. As Rudd (2013) has pointed out, such data raises the question of whether there may be different paths to suicide in military and veteran populations. Recently published findings from the Army Study to Assess Risk and Resilience in Servicemembers (Army STARRS) has given further impetus to this question. While the study found that the increase in Army suicide rates was associated with being currently or previously deployed between the years 2004 and 2009, it also found that soldiers who were never deployed during that period also exhibited an increased suicide rate during that time (Schoenbaum et al., 2014).

In the previous chapter (i.e., "Suicide Risk Assessment with Combat Veterans—Part I: Contextual Factors"), the stresses of the combat veteran's experience as well as the military culture that can form part of the backdrop to the potential development of suicide risk are discussed. The barriers that veterans may face in attempting to request or obtain mental health services when feeling hopeless and despairing are also presented. The focus of the current chapter, however, is on a review of the evidence indicating whether combat veterans, and particularly those veterans with combat-related PTSD, have an elevated risk of suicide and suicidal behavior. Following this review, the process of assessing and managing suicide risk in combat veterans is discussed.

IS COMBAT EXPERIENCE A RISK FACTOR FOR SUICIDE?

Interest in combat experience per se as a risk factor for suicide has received considerable attention from researchers seeking to validate the interpersonal-psychological theory of suicidal behavior proposed by Joiner (2005). In his theory, Joiner contends that three distinct variables must be present for an individual to commit suicide: perceived burdensomeness, thwarted belongingness, and acquired capability for suicide. This latter variable, acquired capability for suicide, refers to the degree to which an individual has become capable of engaging in a lethal suicide attempt. Since potentially lethal suicide attempts can be very frightening and can involve pain and discomfort, Joiner further suggests that a prerequisite for serious suicidal behavior is habituation to the fear and pain involved. Repeated exposure to aversive and painful events such as those experienced in military training and in actual combat can presumably lead to such habituation. So, for example, basic military training involves exposure to painful and fear-inducing events while combat experience itself often involves experience with increasingly violent behaviors.

The notion that military, and combat experience in particular, can lead to an acquired capability for suicide has inspired a number of studies investigating whether there is evidence to support this contention (e.g., Bryan & Anestis, 2011; Bryan, Clemans, & Hernandez, 2012; Bryan & Cukrowicz, 2011; Bryan, Cukrowicz, West, & Morrow, 2010). These studies have focused on the recruitment of military personnel engaged in the Iraq and Afghanistan Wars as participants, and the Acquired Capability for Suicide Scale (ACSS; Van Orden, Witte, Gordon, Bender, & Joiner, 2008) has been used as a measure of such acquired ability. The ACSS is a five-item self-report questionnaire and the participants respond to each item (e.g., "I am not at all afraid to die") using a 7-point Likert scale. Findings have indicated that a range of combat experiences predicted acquired capability for suicide on the ACSS, with combat characterized by violence and high levels of injury and death strongly associated with such capability. The acquired capability for suicide, however, is not sufficient to make an individual suicidal. In Joiner's theory, there must also be a desire to commit suicide that is driven by perceived burdensomeness and thwarted belongingness.

Possibly because military suicide rates have historically been found to be lower than matched civilian rates (Eaton, Messer, Garvey Wilson, & Hoge, 2006), there have been very few studies that have investigated whether combat experience per se is a risk factor for suicide. It is only recently that U.S. Army suicide rates for soldiers engaged in the Iraq and Afghanistan Wars have been found to be greater than matched civilian rates (Suicide Risk Management & Surveillance Office, 2008) and even greater than the rate of deaths due to combat (Bell, Harford, Amoroso, Hollander, & Kay, 2010). Of course, these findings have heightened interest in whether combat experience, or perhaps how frequently one is deployed to combat, has been a factor in driving an increase in the suicide rate.

As noted, Kang and Bullman (2008), in a large register linkage study, found that former active duty OIF/OEF veterans had a significantly elevated rate of suicide (SMR = 1.33; 95% CI: 1.03–1.69). Moreover, in an earlier study (described later), Bullman and Kang (1994) found that Vietnam combat veterans on the Agent Orange Registry who had no clinical diagnosis nonetheless had a two-fold risk of suicide when compared to the general U.S. population. In addition, in the study by Schoenbaum and colleagues (2014), cited earlier, OIF/OEF Army soldiers who were currently or previously deployed had an increased rate of suicide while the rate of deaths by accident in this population were reported to be inconsistent. Further, in another of the STARRS studies, Nock and colleagues (2014), in a cross-sectional survey of 5,428 active duty soldiers, found that the odds ratios for suicidal ideation, plan, and attempt were consistently elevated among soldiers who were ever deployed versus those who were never deployed. The odds ratios were highest for those soldiers who had three or more deployments. These studies taken together seem to suggest that combat experience might be a risk factor for suicide and/or suicidal behavior, but it is unclear how strong of a risk factor it may be. There is a need for further research specifically focused on this question.

COMBAT-RELATED PTSD AS A RISK FACTOR FOR SUICIDE

While studies supportive of combat experience per se as a risk factor for suicide have been relatively scarce, there have been reports that certain groups of combat veterans (e.g., those with combat-related PTSD) may be more vulnerable to suicide than others. Krysinska and Lester (2010), however, recently completed a meta-analysis of 50 studies that examined the association between PTSD and suicidal behavior and concluded that, although there was clear evidence of a relationship between PTSD and suicidal behavior, there was no evidence for an increased risk of completed suicide in individuals with PTSD. There were four studies of veteran suicide in their analysis, two that found an elevated rate of suicide for veterans with PTSD (Bullman & Kang, 1994; Drescher, Rosen, Burling, & Foy, 2003) and two that did not (Desai, Dausey, & Rosenheck, 2005; Zivin et al., 2007). The authors seem to have concluded that these studies nullified each other and that therefore there was no evidence of an elevated rate of suicide associated with PTSD. Their findings, however, are in stark contrast to the more extensive review by Panagioti, Gooding, and Tarrier (2009) who concluded that "there is a strong relationship between PTSD as a consequence of combat trauma and subsequent suicidal behavior" (p. 474). In addition, Panagioti and colleagues noted that the relationship was confirmed across studies examining completed suicides, suicide attempts, and suicidal ideation.

If we, in fact, look more closely at the studies reviewed by Krysinska and Lester (2010), we see, for example, that, in the study by Desai et al. (2005), the investigators sampled all veterans discharged from VA inpatient psychiatric programs in the years 1994 to 1998 ($N = 121,933$). They linked this sample to the National Death Index to determine deaths by suicide. They then compared those who committed suicide and had a diagnosis of PTSD with the remainder of the sample (which included patients with major depression, bipolar depression, schizophrenia, and substance abuse). They found that those with a diagnosis of PTSD had a significantly lower suicide rate relative to those patients with these other psychiatric diagnoses. Since the comparison group contained patients with high suicide risk diagnoses, some of which might well have higher associated suicide rates than PTSD (e.g., major depression, bipolar depression), this study hardly seems to be a test of whether or not there is an elevated risk of suicide associated with PTSD.

In a second study with negative findings, Zivin et al., (2007) conducted a very large record linkage study in which they collected data from the VA's National Registry for Depression and linked it to the National Death Index. The investigators found that the suicide rate in the VA treatment population with depression (88.25 per 100,000) was seven to eight times higher than the suicide rate for those in the general U.S. population in the same age group. Since depression is the diagnosis most frequently associated with suicide, this finding does not seem surprising. They also found that those veterans who had comorbid diagnoses of depression and substance abuse had a suicide rate that was significantly higher than those veterans who only had a diagnosis of depression. This finding is also consistent with previous findings (Cornelius et al., 1995; Murphy, 1992). Surprisingly, however, they found that veterans with co-morbid diagnoses of depression and PTSD had a significantly lower rate of suicide than those veterans who were depressed but did not have a diagnosis of PTSD.

This finding is curious and hard to explain, but again it does not speak directly to the question of whether combat-related PTSD in itself has an increased risk for suicide. More to the point, Bullman & Kang (1994), in a register linkage study of suicide rates among Vietnam combat veterans with PTSD, used the Agent Orange Registry (AOR) to follow 4,247 male veterans who had a diagnosis of PTSD and 12,010 randomly selected male veterans on the AOR who had no clinical diagnosis. They checked vital status by linking these two groups with the VA-maintained Beneficiary Identification and Record Locator Subsystem (BIRLS). Over a 4-5 year follow-up period, they found that those veterans with PTSD relative to those with no clinical diagnosis had a Risk Ratio (RR) of 3.97 (95% CI: 2.20-7.03) for suicide and a RR of 2.89 (95% CI: 1.03-8.12) for death by accidental poisoning.

When compared to the U.S. population rate of suicide, the veterans with PTSD in the Bullman & Kang (1994) study had an almost seven-fold increased risk for suicide (SMR = 6.74, 95% CI: 4.40-9.87) while, as noted earlier, those combat veterans with no clinical diagnosis had an almost two-fold increased risk (SMR = 1.67, 95% CI: 1.05-2.53). The investigators also studied a subgroup of 1001 veterans with PTSD who had co-morbid diagnoses. Fifty-six percent had co-morbid alcohol and drug dependency disorders, while 11% had co-morbid neurotic disorder and 10% had co-morbid depressive disorder. Among this group with co-morbid disorders, there was almost a 10-fold increased risk for suicide (SMR = 9.81, 95% CI: 4.48-18.63). The study authors interpreted their findings as indicating that, for Vietnam veterans on the AOR, PTSD was associated with a significantly increased risk for death by suicide and accidental poisoning.

In a second register linkage study, Bullman & Kang (1996) investigated the suicide rate among Vietnam combat veterans who had received a non-lethal wound or wounds during the time period of 1969-73. The 34,534 subjects were randomly selected from the Casualty Information System Tape maintained by the U.S. Army, and their vital status was determined by linkage with the BIRLS registry as well as with the Social Security Administration file of deaths. Relative to the U.S. male population adjusted for age, race, and calendar year, those veterans who had been wounded two or more times had a significantly elevated risk of suicide (SMR = 1.58, 95% CI: 1.06-2.26) as did those who were wounded more than once and had to be hospitalized for their wound(s) on at least one occasion (SMR = 1.73. CI: 1.10-2.60). The relative risk of suicide for those wounded two or more times compared to those wounded once was 1.50 (95% CI = 1.01, 2.24). The study authors felt that there was a trend of increasing risk of suicide with increasing occurrence of combat-related physical (and likely emotional) trauma. A limitation of the study was that the authors did not have data on the subjects' psychiatric diagnoses or their psychological characteristics. Thus, it was not known to what degree PTSD or other mental and emotional disorders that might have been associated with being wounded were factors contributing to the outcome of suicide. Nonetheless, the investigators felt that their findings indicated that being wounded in combat in Vietnam should be considered a risk factor for post-service suicide.

Beyond these studies of Vietnam veterans by Bullman and Kang, Ilgen and colleagues (2010) recently published a study in which they obtained diagnostic information from the VA National Patient Care Database on all individuals who used VA services in fiscal year 1999. They followed these patients for 7 years through 2006 and linked their database to the National Death Index. They reported that slightly less than half (46.8%) of those who died by suicide had at least one psychiatric condition at entry into the study. Among the psychiatric disorders, bipolar disorder was found to have the strongest association with suicide followed by depression, but several other disorders, including PTSD, were also found to have significant associations with completed suicide. The association of PTSD with suicide was not as robust as that of a number of other psychiatric conditions (such as bipolar disorder or depression), but it was nonetheless significant.

Finally, in a study that directly targeted OEF/OIF combat veterans, Rudd (in press) explored the relationship between severity of combat exposure, psychological symptoms, and suicide risk using an electronic survey that included measures such as the Combat Exposure Scale (Keane et al., 1989), the PTSD Checklist (military version) (PCL-M; Weathers et al., 1993), and the Suicide Behaviors Questionnaire-Revised (SBQ-R; Osman et al., 2001). He found that those veterans who had heavy combat exposure (compared to all other categories of lesser combat exposure) had greater symptomatology across the board, including general anxiety, depression, posttrauma symptoms, sleep disturbance, and suicide risk as assessed by the SBQ-R. Those with heavy combat exposure rated themselves as more likely to attempt suicide in the future. Ninety percent of those reporting that a suicide attempt was likely in the future scored above the PCL-M cutoff for a diagnosis of PTSD. In addition, it was found that combat exposure per se did not provide any additional power in predicting suicidality or ratings of the likelihood of future suicide attempts once the variance attributed to posttraumatic symptoms was allocated. Rudd (2013) concluded that these results suggest very high rates of posttraumatic symptoms and potential PTSD diagnoses among those with heavy combat experience. Moreover, Rudd has reported that a clinical trajectory for suicide risk is readily recognizable for OEF/OIF veterans with heavy combat experience and that posttraumatic symptoms appear to be prominent factors in that trajectory.

In summary, there is a clear need for further research on the issue of whether combat experience per se is a risk factor for suicide. On the other hand, although the meta-analytic review by Krysinska and Lester (2010) led those authors to conclude that there was no evidence for an increased risk of completed suicide in individuals with PTSD, recent studies, in combination with earlier studies, suggest to us that the preponderance of the evidence now favors the conclusion that those with combat-related PTSD are at an elevated risk of suicide.

CONDITIONS COMORBID WITH PTSD AS RISK FACTORS FOR SUICIDE

As may have been evident in the previous section, there is a high rate of co-occurring mental and emotional disorders among veterans with PTSD. For example, higher rates of current major depression, bipolar disorder, panic disorder, and social phobia were reported among veterans with war-zone related PTSD compared with veterans without PTSD who also served in a war zone (Orsillo et al., 1996). Major depressive disorder (MDD) is the most commonly diagnosed comorbid disorder among veterans with PTSD at 55% (Orsillo et al.). The reverse is also true; that is, high rates of PTSD (36%) have been reported among veterans diagnosed with MDD (Campbell et al., 2007).

Substance use disorders (SUDs) are another class of psychiatric disorders that often co-occur with PTSD and depression. This co-occurrence may constitute a coping strategy in response to PTSD or depressive symptoms. Such a coping strategy is commonly referred to as self-medicating (e.g., Harris & Edlund, 2005). In support of this model, Kruse Steffen, Kimbrel, and Gulliver (2011) in their chapter on co-occurring PTSD and SUDs highlighted findings that show "the onset of PTSD typically precedes the development of SUDs (e.g., Kessler, 2000) ... [and] reductions in PTSD severity (in particular, reductions in hyperarousal symptoms; Back, Brady, Sonne, & Verduin, 2006) are strongly associated with improvements in SUDs" (p. 220). In addition to evidence of high comorbidity with PTSD, research also shows a clear association between depression and SUDs and increased suicidal behavior and suicide in veterans (Merrill, Milner, Owens, & Vale, 1992; Zivin et al., 2007).

The association between PTSD with comorbid conditions and completed suicide, however, seems uncertain. Bullman and Kang (1994), in a study (cited earlier) of Vietnam veterans on the Agent Orange Registry with and without PTSD, found that those with PTSD had an increased rate of suicide while those with PTSD and an additional diagnosis (most frequently [56%] substance abuse) had a 10-fold excess of suicides. As described previously in this chapter, however, Zivin et al. (2007) reported that, among depressed veterans, an elevated rate of suicide was found among those with comorbid substance abuse but a lower rate was found among those with comorbid PTSD. Such variable findings seem to leave the question of whether diagnoses comorbid with PTSD confer an increased risk of completed suicide as yet unsettled. They seem to raise the question of whether some comorbidities may interact in an additive or even synergistic way in terms of suicide risk while others may have a suppressing effect.

Beyond depression and SUDs, another potential comorbid risk factor for suicide is traumatic brain injury (TBI). There has been little research, however, on TBI as a specific risk factor for suicide in combat veterans. Many service members returning from the OEF/OIF conflicts have suffered a TBI from exposure to improvised explosive device blasts. Moreover, research suggests high rates of comorbidity between mild TBI and PTSD (Hoge et al., 2008). Yet we could identify only three articles in the literature that mentioned TBI in combat veterans in relation to suicide risk (i.e., Brenner et al., 2011; Posey, 2009; Reeves & Panguluri, 2011). The recent study by Brenner and colleagues examined rates of suicide attempts among veterans, though not specifically combat veterans. These authors found that the odds of a suicide attempt in veterans with both PTSD and a TBI was 3.3 times the odds of an attempt for those with a TBI alone.

Further research in this area seems strongly warranted given these findings and that an association between suicide and TBI has been found in nonveteran populations (e.g., Silver, Kramer, Greenwald, & Weissman, 2001; Simpson & Tate, 2002; Teasdale & Engberg, 2001). Teasdale and Engberg, for example, reported an increased incidence of completed suicide among persons with a TBI compared to the general population in Denmark, and the incidence increased with the severity of the TBI. The authors suggested that the association between TBI and suicide risk is also mediated by psychiatric conditions that co-occur with TBI, such as depression, PTSD, and substance abuse.

ASSESSING SUICIDE RISK WITH COMBAT VETERANS

In an initial evaluation of suicide risk with a combat veteran, the clinician is well advised to be sensitive to the potential stressors and barriers noted in the preceding chapter and to the associated pain, suffering, and stress that the veteran may have recently experienced. This sensitivity or awareness can be crucial in communicating to the veteran that the clinician has an understanding of the veteran's experience, which can become the basis of a working relationship. As has been learned through clinical experience, war-related survival skills and a battle mindset are often not easily set aside upon reentry into civilian life (Munroe, 2012). Moreover, combat veterans have often been exposed to events in a war zone that they felt were beyond their control or threatened to become so. Such circumstances can engender a deep sense of vulnerability and may be the breeding ground for the development of PTSD (Keane & Barlow, 2002). As a result, combat veterans often feel a particular need to be in control of their circumstances while being fearful that, if they say they are at risk, control may be taken from them. In this regard, the collaborative approach to suicide risk assessment and management proposed by Jobes (2006) may be very appropriate in evaluating the combat veteran with PTSD. In this approach, the clinician does not engage in a struggle with the patient over whether or not he or she can commit suicide. Rather, the focus is on working with the veteran to find viable alternatives to deal with the pain and suffering that is leading the individual to consider suicide.

Munroe (2012) has noted that combat veterans, especially those with PTSD symptoms, can be hypervigilant, distrusting, quick to anger, aggressive, and avoidant of talking about emotions and war experiences. As noted in the preceding chapter, they have often been inculcated with the warrior culture in which one does not show emotion since it can be equated with weakness. The clinician needs to recognize the source of these attitudes and reactions and not simply take them as signs of resistance.

As Claassen and Knox (2011) have pointed out, there is evidence that suggests that active duty OEF/OIF soldiers who report suicidal ideation are more likely to die by suicide than their civilian counterparts who report suicidal ideation. Moreover, Nock and colleagues (2015) in the New Soldier Study, a component of the Army STARRS, found that among new recruits in basic training, depression was the strongest predictor of suicidal ideation, but disorders characterized by anxiety, irritability, and impulsive/aggressive behavior predicted movement from suicidal ideation to suicide attempts. It is not clear that this heightened vulnerability is present to the same degree when active duty soldiers transition to veteran status. Nonetheless, as mentioned in the preceding chapter, the clinician needs to remain aware that the period of transitioning back to civilian life can be very stressful. Often problems with PTSD, depression, anxiety, and/or substance abuse emerge at this time. These diagnoses have all been associated with an elevated risk for suicide (Kleespies & Hill, 2011).

There is no one accepted or validated way of evaluating suicide risk either with veteran or civilian populations. When working with a veteran for whom suicidality is a question, clinicians typically inquire about evidence-based risk factors and protective factors and estimate whether the degree of risk for suicide in a particular case is high, moderate, or low. Risk factors can be chronic (associated with elevated lifetime risk) or acute (associated with near-term risk). Many chronic risk factors are static or unchangeable (e.g., a history of a suicide attempt or a history of violence), but others may be modifiable or dynamic (e.g., a mental disorder that can be treated effectively or limitations in coping ability that can be improved with intervention). Acute risk factors are most often dynamic and changeable.

Protective factors (e.g., children under 18 years of age in the home) may provide a counterweight to risk factors, but how risk factors and protective factors may interact is not always clear. No evidence-based system exists that informs clinicians about how much weight to assign any given risk or protective factor or any given combination of risk or protective factors. The co-occurrence of certain risk factors may be additive (i.e., equal to the sum of the risk associated with each factor), subadditive (i.e., equal to a risk greater than that with one factor but not equal to the sum of the two), or synergistic (i.e., the combined risk may be greater than the simple sum of the risk associated with each factor).

It should be noted that, although the clinician should make all efforts to arrive at an evidence-based or evidence-informed risk formulation, the clinical formulation of suicide risk is ultimately a clinical judgment. Moreover, in assessing for acute or short-term risk for suicide, chronic or static risk factors may provide a foundation for the risk evaluation, but particular attention should be given to acute risk factors and protective factors. In the violence risk assessment literature, research has suggested that acute and dynamic risk factors contribute appreciably to assessments of short-term risk (McNiel, Gregory, Lam, Sullivan, & Binder, 2002). It seems plausible that the case may be similar with the assessment of short-term risk for suicide. In fact, a working group of the American Association of Suicidology took the position that risk factors (and particularly those that are longstanding risk factors) are insufficient to assess immediate suicide risk (Rudd et al., 2006). They argued that the current state of the individual must also be taken into account. Thus they attempted to contrast the concept of suicide risk with that of a suicide crisis (i.e., a time-limited emotional state in which there is imminent risk of suicide). They also posited that there are warning signs in the suicide crisis state that suggest near-term risk (i.e., risk within the next few hours to days) of suicidal behavior.

By expert consensus, Rudd and colleagues (2006) suggested a two-tier model of warning signs for suicide. The first tier consists of three warning signs that should direct the individual to call 911 or seek immediate professional help: (a) someone threatening to hurt or kill him- or herself, (b) someone looking for ways to kill him- or herself, and (c) someone talking or writing about death, dying, or suicide. The second tier consists of nine warning signs that should direct the individual to seek help without specifying the need to get immediate assistance. This second tier includes the following factors: (a) hopelessness; (b) rage, anger, seeking revenge; (c) acting reckless or engaging in risky activities seemingly without thinking; (d) feeling trapped—like there's no way out; (e) increasing alcohol or drug use; (f) withdrawing from friends, family, or society; (g) anxiety, agitation, unable to sleep, or sleeping all the time; (h) dramatic changes in mood; and (i) feeling like there is no reason for living or no sense of purpose in life.

SUICIDE RISK FACTORS ASSOCIATED WITH COMBAT-RELATED PTSD

These factors are general risk factors for suicide. Clinicians should certainly be aware of them when doing an evaluation for suicide risk, but awareness of additional factors specific to combat-related PTSD is also needed. For example, based on two studies discussed previously, severity of trauma and the intensity of combat experience should be considered when determining the level of risk in combat veterans. Higher severity of trauma (i.e., being wounded more than once and/or being hospitalized for a wound; Bullman & Kang, 1996) as well as heavier combat experience (Rudd, 2013) were shown to increase risk for suicide.

Potentially more important for the evaluation of acute suicide risk in combat veterans is the present psychological state of veterans with PTSD. Hyer, McCranie, Woods, and Boudewyns (1990) found that Vietnam veterans who reported higher levels of guilt, emotional lability, and lower levels of psychological adjustment were most likely to engage in suicidal behavior. Similarly, Hendin and Haas (1991) found increased levels of guilt, depression, and anxiety to be associated with elevated risk of suicide attempts in veterans with PTSD. Both of these studies identified guilt as an important factor in discriminating veterans at risk of suicidal behavior. Hendin and Haas further described the guilt that veterans reported as either related to combat actions (e.g., killing or harming noncombatants) or to surviving when others died or were seriously injured. They indicated that most individuals who reported guilt related to combat actions also reported survival guilt and that the combination of the two was most strongly associated with suicidal behavior.

Witnessing or participating in atrocities (e.g., torturing and killing prisoners of war, killing noncombatants, mutilating the bodies of enemies by cutting off ears or putting heads on sticks, and so forth) may also be a risk factor for suicide in combat veterans. In studies with Vietnam veterans who had chronic PTSD, witnessing or participating in atrocities was associated with heightened levels of PTSD symptom severity, particularly re-experiencing or criterion B symptoms (Beckham, Feldman, & Kirby, 1998; Yehuda, Southwick, & Giller, 1992). Hiley-Young, Blake, Abueg, Rozynko, and Gusman (1995), in a study of Vietnam veterans with PTSD who had engaged in war-zone violence, found that participation in the mutilation of bodies predicted postmilitary suicide attempts.

Finally, Bell and Nye (2007) assessed Vietnam combat veterans for suicidal ideation and PTSD. Using *Diagnostic and Statistical Manual of Mental Disorders* (fourth ed.) criteria for PTSD, they found that re-experiencing symptoms (criterion B; e.g., nightmares, intrusive thoughts, flashbacks) were significantly associated with suicide ideation while avoidance/numbing symptoms (criterion C) and arousal symptoms (criterion D) were not. Thus evaluation of specific PTSD symptoms seems necessary when determining the level of risk for combat veterans.

These studies, of course, do not allow us to conclude that there is a causative link between participating in or witnessing atrocities, guilt, re-experiencing symptoms, and increased risk of suicidality. They do, however, raise questions of such linkage that can only be determined by further research. Clinicians, on the other hand, might do well to assess for these risk factors and observe caution about suicide risk, particularly with veterans who have experienced heavy combat or severe trauma, who have intense re-experiencing symptoms, who may have participated in atrocities, and who have guilt about their combat actions and/or survivor guilt.

MANAGING SUICIDE RISK WITH COMBAT VETERANS

As noted by Kleespies and Hill (2011), there is no absolute rule for when a suicidal patient can be managed and treated on an outpatient basis or when one must make an emergency intervention and hospitalize the patient. The clinician needs to be guided by a carefully considered estimate of the level of risk as described previously, and, with combat veterans, he or she must be particularly attuned to those risk factors that may have special importance for them. The next sections offer guidance for making the at times difficult decision to proceed on an outpatient basis or to pursue hospitalization.

When and How Can the Suicidal Patient Be Managed on an Outpatient Basis?

Clinicians may be inclined to hospitalize patients with suicidal ideation because they feel it is safer and because they have a high index of concern about liability issues. There is little evidence, however, that hospitalization per se ultimately prevents suicide. It may help to reduce immediate risk, but the clinician should bear in mind that the two- to four-month period following hospital discharge is known to be a high risk period for suicide (Kleespies et al., 2011; Morgan & Stanton, 1997).

Many patients with suicidal ideation can be treated successfully on an outpatient basis, and, as suggested, the estimated level of risk is the key in making the decision to manage the patient as an outpatient or to hospitalize. Generally, outpatient management for patients assessed at mild or moderate risk has been found to be feasible and safe (Rudd, Joiner, & Rajab, 2001; Sullivan & Bongar, 2009). An example of a combat veteran at moderate risk might be a young ex-Marine with a few distal risk factors (e.g., chronic, moderate PTSD, and chronic guilt about combat actions), several acute, proximal risk factors (e.g., combat-related nightmares, anger control issues, and episodic suicidal ideation), and some protective factors that are beginning to weaken (e.g., a supportive spouse whose patience is being tried and a five-year-old son who is beginning to withdraw in the face of his father's anger).

For patients at a mild or moderate level of risk who are to be treated on an outpatient basis, Stanley and Brown (2008, 2012) have recommended that their management include a Safety Planning Intervention (SPI). The SPI consists of a prioritized written list of coping strategies and sources of support, developed collaboratively by the clinician and the patient, for use by the patient preceding or during a state of heightened suicidality. Stanley and Brown have developed a veteran version of the SPI as well as a more generic version for use with any suicidal patient (Stanley & Brown, 2012).

The SPI consists of six steps, the first of which involves helping patients to identify and pay attention to the warning signs that occur when they begin to think about suicide. These warning signs, which may be thoughts, behaviors, or moods, are listed in the safety plan in the patient's own words. Second, patients are asked to identify internal coping strategies that they might use to take their mind off their problems. These strategies might include such things as going for a walk, exercising, or reading. Third, if the internal coping strategies are ineffective in reducing suicidal ideation or intent, the patient's safety plan should include a list of social contacts that might serve as a distraction. Such contacts might include coffee shops or places of religion. Veterans as well as others should be advised to avoid environments where alcohol or illicit drugs may be present or served as these can exacerbate rather than dampen suicide ideation. Fourth, veterans are asked if they could designate family members and/or friends whom they feel they could talk with and inform that they are having thoughts of suicide. This step differs from the third step in that patients are to explicitly tell the family member or friend that they are having such thoughts. Although it is not considered mandatory, the clinician could work collaboratively with the patient to see if he or she would feel comfortable actually sharing the safety plan with someone he or she trusts. Fifth, in the event that the previous coping strategies are ineffective, patients generate a list of clinicians or professional agencies that they could contact. This part of the plan should include contacts that can be reached during nonbusiness hours. For veterans, one such contact is the VA 24-Hour Crisis Hotline (1-800-273-TALK [8255]). Since some patients may fear being hospitalized or rescued in a way that they do not want, potential obstacles to rescue efforts should be discussed when making the safety plan. Finally, the clinician should discuss with patients what means they might use in a suicide attempt and then work collaboratively with them on eliminating or limiting access to lethal means. It is particularly important to ask veterans about firearms since they are known to have high rates of gun ownership and they have been found to be more likely than nonveterans to end their lives by gunshot (Kaplan et al., 2007). Bryan, Stone, and Rudd (2011) have offered a useful and evidence-based approach to what they refer to as means-restriction counseling with suicidal patients.

While Stanley and Brown (2008, 2012) have focused on the management of suicidal states by helping the patient to develop coping strategies, Rudd and Joiner (1998) have suggested certain contingencies that the clinician who is working with a suicidal outpatient might wish to consider. For the patient who is at moderate risk, they have suggested that the clinician consider the following: (a) an increase in outpatient visits and/or telephone contacts, (b) frequent assessment of suicide risk, (c) recurrent evaluation for hospitalization while the risk continues, (d) 24-hour availability or coverage, (e) re-evaluation of the treatment plan as needed, (f) consideration of a medication evaluation or change in regimen, and (g) use of professional consultation as warranted. We recommend that clinicians who work with suicidal combat veterans utilize the SPI proposed by Stanley and Brown while keeping in mind the contingencies recommended by Rudd and Joiner.

When Is Emergency Intervention Needed?

When, in the clinician's considered judgment, the suicide risk is high, emergency intervention to protect the patient is needed. In many cases, the patient presents in an emotional crisis and the clinician is involved in attempting to achieve a resolution that will enable the patient to continue in outpatient treatment. At times, the clinician and the patient, working collaboratively, can arrive at such a safe resolution. As Comstock (1992) has pointed out, however, hospitalization is indicated when it is not possible to establish or reinstate a treatment alliance, when crisis intervention techniques fail, and when the patient continues to voice suicidal intent in the near future. Although, as noted earlier, there is little evidence that hospitalization prevents suicide in the long run, it does provide a relatively safer environment during a period of heightened suicide risk. Typically, one or two hours with a patient who continues to seem to be at imminent suicide risk are sufficient to convince clinicians to hospitalize.

Most suicidal patients who are engaged with their therapist in an evaluation for heightened suicide risk have some ambivalence about such a final action as suicide and agree to voluntary hospital admission.

When such patients refuse to be hospitalized, however, the clinician is faced with a decision about temporary involuntary commitment. This decision can be difficult because we know that the estimation of suicide risk is not always reliable, and, in addition, involuntary hospitalization can damage the therapeutic relationship. Should the process of hospitalization require the use of restraints, it can be difficult for combat veterans who have been traumatized and who may be particularly concerned about being rendered defenseless. The clinical staff involved should remain sensitive to such issues and attempt to use the least restrictive means in trying to work safely with the veteran. In the final analysis, the decision to hospitalize involuntarily must be based on sound, evidence-based judgment that considers the estimated seriousness of the suicide risk and the risk-benefit ratio of hospitalization versus outpatient management. For the clinician, it can be helpful to keep in mind that, once hospitalization has occurred, resistant patients often begin to perceive the caring nature of the clinician's actions and re-establish a treatment alliance.

CONCLUSION

Combat veterans with PTSD, and particularly those with severe PTSD, are at increased risk of suicide. As noted in the preceding chapter, having an understanding of the influence of military culture on veterans as well as an understanding of the coping strategies that they developed in the context of war can be crucial in working with them to assess and manage risk. If suicide prevention efforts can succeed, there are now evidence-based treatments for combat-related PTSD that offer hope of relief from the inner turmoil and suffering that attends this disorder. The strongest evidence for effective treatment supports the use of cognitive-behavioral therapies, particularly prolonged exposure therapy and cognitive processing therapy, as well as the use of pharmacotherapy, especially some of the selective serotonin reuptake inhibitors (Rothbaum, Gerardi, Bradley, & Friedman, 2011).

REFERENCES

Beckham, J., Feldman, M., & Kirby, A. (1998). Atrocities exposure in Vietnam combat veterans with chronic posttraumatic stress disorder: Relationship to combat exposure, symptom severity, guilt, and interpersonal violence. *Journal of Traumatic Stress, 11,* 777–785.

Bell, J., & Nye, E. (2007). Specific symptoms predict suicidal ideation in Vietnam combat veterans with chronic post-traumatic stress disorder. *Military Medicine, 172,* 1144–1147.

Bell, N., Harford, T., Amoroso, P., Hollander, I., & Kay, A. (2010). Prior health care utilization patterns and suicide among U.S. Army soldiers. *Suicide and Life-Threatening Behavior, 40,* 407–415.

Blue Ribbon Work Group on Suicide Prevention in the Veteran Population. (2008). *Report to James B. Peake, MD, Secretary of Veterans Affairs.* Unpublished manuscript, Department of Veterans Affairs, Washington, DC.

Brenner, L., Betthauser, L., Homaifar, B., Villarreal, E., Harwood, J., Staves, P., & Huggins, J. (2011). Posttraumatic stress disorder, traumatic brain injury, and suicide attempt history among veterans receiving mental health services. *Suicide and Life-Threatening Behavior, 41,* 416–423.

Bryan, C., & Anestis, M. (2011).Reexperiencing symptoms and the interpersonal-psychological theory of suicidal behavior among deployed service members evaluated for traumatic brain injury. *Journal of Clinical Psychology, 67,* 856–865.

Bryan, C., Clemans, T., & Hernandez, A. (2012). Perceived burdensomeness, fearlessness of death, and suicidality among deployed military personnel. *Personality and Individual Differences, 52,* 374–379.

Bryan, C., & Cukrowicz, K. (2011). Associations between types of combat violence and acquired capability for suicide. *Suicide and Life-Threatening Behavior, 41,* 126–136.

Bryan, C., Cukrowicz, K., West, C., & Morrow, C. (2010). Combat experience and the acquired capability for suicide. *Journal of Clinical Psychology, 66,* 1044–1056.

Bryan, C., Stone, S., & Rudd, M.D. (2011). A practical, evidence-based approach for means-restriction counseling with suicidal patients. *Professional Psychology: Research and Practice, 42,* 339–346.

Bullman, T., & Kang, H. (1994). Posttraumatic stress disorder and the risk of traumatic deaths among Vietnam veterans. *The Journal of Nervous and Mental Disease, 182,* 604–610.

Bullman, T., & Kang, H. (1996). The risk of suicide among wounded Vietnam veterans. *American Journal of Public Health, 86,* 662–667.

Campbell, D. G., Felker, B. L., Liu, C. F., Yano, E. M., Kirchner, J. E., Chan, D., . . . Chaney, E. F. (2007). Prevalence of depression-PTSD comorbidity: Implications for clinical practice guidelines and primary care-based interventions. *Society of General Internal Medicine, 22,* 711–718.

CBS Evening News. (2007, November 13). The veteran suicide "epidemic": A CBS News investigation uncovers a suicide rate for veterans twice that of other Americans. Retrieved from http://www.cbsnews.com/stories/2007/11/13/cbsnews_investigates/main3496470.shtml?s.

Claassen, C., & Knox, K. (2011). Assessment and management of high-risk suicidal states in postdeployment Operation Enduring Freedom and Operation Iraqi Freedom military personnel. In J. Ruzek, P. Schnurr, J. Vasterling, & M. Freidman (Eds.), *Caring for veterans with deployment-related stress disorders: Iraq, Afghanistan, and beyond* (pp. 109–127). Washington, DC: APA Books.

Comstock, B. (1992). Decision to hospitalize and alternatives to hospitalization. In B. Bongar (Ed.), *Suicide: Guidelines for assessment, management, and treatment* (pp. 204–217). New York: Oxford University Press.

Cornelius, J., Salloum, I., Mezzich, J., Cornelius, M., Fabrega, H., Ehler, J., . . . Mann, J. (1995). Disproportionate suicidality in patients with co-morbid major depression and alcoholism. *The American Journal of Psychiatry, 152,* 358–364.

Cvetanovich, B., & Reynolds, L. (2008). Joshua Omvig Veterans Suicide Prevention Act of 2007. *Harvard Journal on Legislation, 45,* 619–640.

Desai, R., Dausey, D., & Rosenheck, R. (2005). Mental health service delivery and suicide risk: The role of individual patient and facility factors. *The American Journal of Psychiatry, 162,* 311–318.

Drescher, K., Rosen, C., Burling, T., & Foy, D. (2003). Causes of death among male veterans who received residential treatment for PTSD. *Journal of Traumatic Stress, 16,* 535–543.

Eaton, K., Messer, S., Garvey Wilson, A., & Hoge, C. (2006). Strengthening the validity of population-based suicide rate comparisons: An illustration using U.S, military and civilian data. *Suicide and Life-Threatening Behavior, 36,* 182–191.

Gibbons, R., Brown, C. H., & Hur, K. (2012). Is the rate of suicide among veterans elevated? *American Journal of Public Health, 102*(Suppl. 1), S17–S19.

Harris, K., M., & Edlund, M. J. (2005). Self-medication of mental health problems: New evidence from a national survey. *Health Services Research, 40,* 117–134.

Hendin, H., & Haas, A. (1991). Suicide and guilt as manifestations of PTSD in Vietnam combat veterans. *The American Journal of Psychiatry, 148,* 586–591.

Hiley-Young, B., Blake, D., Abueg, F., Rozynko, V., & Gusman, F. (1995). Warzone violence in Vietnam: An examination of premilitary, military, and postmilitary factors in PTSD in-patients. *Journal of Traumatic Stress*, 8, 125–141.

Hoge, C., McGurk, D., Thomas, J., Cox, A., Engel, C., & Castro, C. (2008). Mils traumatic brain injury in U. S. soldiers returning from Iraq. *New England Journal of Medicine*, 358, 453–463.

Hyer, L., McCranie, E., Woods, M., & Boudewyns, P. (1990). Suicidal behavior among chronic Vietnam theatre veterans with PTSD. *Journal of Clinical Psychology*, 46, 713–721.

Ilgen, M., Bohnert, A., Ignacio, R., McCarthy, J., Valenstein, M., Myra Kim, H., & Blow, F. (2010). Psychiatric diagnoses and risk of suicide in veterans. *Archives of General Psychiatry*, 67, 1152–1158.

Ilgen, M., McCarthy, J., Katz, I., Ignacio, R., Bohnert, A., Valenstein, M., & Blow, F. (2012). Psychopathology, Iraq and Afghanistan service, and suicide among Veterans Health Administration patients. *Journal of Consulting and Clinical Psychology*, 80, 323–330.

Jobes, D. (2006). *Managing suicidal risk: A collaborative approach*. New York: Guilford Press.

Joiner, T. (2005). *Why people die by suicide*. Cambridge, MA: Harvard University Press.

Kang, H., & Bullman, T. (2008). Risk of suicide among U.S. veterans after returning from the Iraq or Afghanistan war zones. *JAMA: Journal of the American Medical Association*, 300, 652–653.

Kaplan, M., Huguet, N., McFarland, B., & Newsom, J. (2007). Suicide among male veterans: A prospective population-based study. *Journal of Epidemiology and Community Health*, 61, 619–624.

Kaplan, M., McFarland, B., Huguet, N., & Newsom, J. (2012). Estimating the risk of suicide among US veterans: How should we proceed from here? *American Journal of Public Health*, 102(Suppl. 1), S21, S23.

Keane, T., & Barlow, D. (2002). Posttraumatic stress disorder. In D. Barlow (Ed.), *Anxiety and its disorders* (pp. 418–453). New York: Guilford Press.

Keane, T., Fairbank, J., Caddell, J., Zimering, R., Taylor, K., & Mora, C. (1989). Clinical evaluation of a measure to assess combat exposure. *Psychological Assessment: A Journal of Consulting and Clinical Psychology*, 1, 53–55.

Kessler, R. (2000). Posttraumatic stress disorder: The burden to the individual and to society. *Journal of Clinical Psychiatry*, 61(Suppl. 5), 4–14.

Kleespies, P., AhnAllen, C., Knight, J., Presskreischer, B., Barrs, K., Boyd, B., & Dennis, J. (2011). A study of self-injurious and suicidal behavior in a veteran population. *Psychological Services*, 8, 236–250.

Kleespies, P., & Hill, J. (2011). Behavioral emergencies and crises. In D. Barlow (Ed.), *The Oxford handbook of clinical psychology* (pp. 739–751). New York: Oxford University Press.

Kruse, M. I., Steffen, L. E., Kimbrel, N. A., & Gulliver, S. B. (2011). Co-occurring substance use disorders. In B. A. Moore & W. E. Penk (Eds.), *Treating PTSD in military personnel* (pp. 217–238). New York: Guilford Press.

Krysinska, K., & Lester, D. (2010). Post-traumatic stress disorder and suicide risk: A systematic review. *Archives of Suicide Research*, 14, 1–23.

McCarthy, J., Valenstein, M., Myra Kim, H., Ilgen, M., Zivin, K., & Blow, F. (2009). Suicide mortality among patients receiving care in the Veterans Health Administration Health System. *American Journal of Epidemiology*, 169, 1033–1038.

McNiel, D., Gregory, A., Lam, J., Sullivan, G., & Binder, R. (2002), Utility of decision support tools for assessing acute risk of violence. *Journal of Consulting and Clinical Psychology*, 71, 945–953.

Merrill, J., Milner, G., Owens, J., & Vale, A. (1992). Alcohol and attempted suicide. *British Journal of Addiction*, 87, 83–89.

Miller, M., Azrael, D., Barber, C., Mukamal, K., & Lawler, E. (2012). A call to link data to answer pressing questions about suicide risk among veterans. *American Journal of Public Health*, 102(Suppl. 1), S20, S22.

Miller, M., Barber, C., Azrael, D., Calle, E., Lawler, E., & Mukamal, K. (2009). Suicide among U.S. veterans: A prospective study of 500,000 middle aged and elderly men. *American Journal of Epidemiology*, 170, 494–500.

Morgan, H., & Stanton, R. (1997). Suicide among psychiatric inpatients in a changing clinical scene. *The British Journal of Psychiatry*, 171, 561–563.

Munroe, J. (2012). *Transitioning war zone skills: Information for veterans and those who care*. Unpublished manuscript. Psychology Service, VA Boston Healthcare System, Boston, MA.

Murphy, G. (1992). *Suicide in alcoholism*. New York: Oxford University Press.

Nelson, R. (2004). Suicide rates rise among soldiers in Iraq. *The Lancet*, 363, 300.

Nock, M., Stein, M., Heeringa, S., Ursano, R., Colpe, L., Fullerton, C., . . . Kessler, R. (2014). Prevalence and correlates of suicidal behavior among soldiers: Results from the Army Study to Assess Risk and Resilience in Servicemembers (Army STARRS). *JAMA Psychiatry*, 71(5), 514–522.

Nock, M., Ursano, R., Heeringa, S., Stein, M., Jain, S., Raman, R., . . . Army STARRS collaborators (2015). Mental disorders, comorbidity, and pre-enlistment suicidal behavior among new soldiers in the U. S. Army: Results from the Army Study to Assess Risk and Resilience in Servicemembers (Army STARRS). *Suicide and Life-Threatening Behavior, 45,* 588–599.

Orsillo, S. M., Weathers, F. W., Litz, B., Steinberg, H. R., Huska, J. A., & Keane, T. M. (1996). Current and lifetime psychiatric disorders among veterans with war zone-related posttraumatic stress disorder. *Journal of Nervous and Mental Disease, 184,* 307–313.

Osman, A., Bagge, C., Gutierrez, P., Konick, L., Kopper, B., & Barrios, F. (2001). The Suicidal Behavior Questionnaire—Revised (SBQ-R): Validation with clinical and nonclinical samples. *Assessment, 8,* 443–454.

Panagioti, M., Gooding, P., & Tarrier, N. (2009). Post-traumatic stress disorder and suicidal behavior: A narrative review. *Clinical Psychology Review, 29,* 471–482.

Posey, S. (2009). Veterans and suicide: A review of potential increased risk. *Smith College Studies in Social Work, 79,* 368–374.

Reeves, R. R., & Panguluri, R. J. (2011). Neuropsychiatric complications of traumatic brain injury. *Journal of Psychosocial Nursing, 49,* 42–50.

Rothbaum, B., Gerardi, M., Bradley, B., & Friedman, M. (2011). Evidence-based treatments for posttraumatic stress disorder in Operation Enduring Freedom and Operation Iraqi Freedom military personnel. In J. Ruzek, P. Schnurr, J. Vasterling, & M. Freidman (Eds.), *Caring for veterans with deployment-related stress disorders: Iraq, Afghanistan, and beyond* (pp. 215–239). Washington, DC: APA Books.

Rudd, M. D. (2013). *Severity of combat exposure, psychological symptoms, social support, and suicide risk in OEF/OIF veterans.* Unpublished manuscript. National Center for Veterans Studies, University of Utah, Salt Lake City, UT.

Rudd, M. D., & Joiner, T. (1998). The assessment, management, and treatment of suicidality: Toward clinically informed and balanced standards of care. *Clinical Psychology: Science and Practice, 5,* 135–150.

Rudd, M. D., Berman, A., Joiner, T., Nock, M., Silverman, M., Mandrusiak, M., . . . Witte, T. (2006). Warning signs for suicide: Theory, research, and clinical applications. *Suicide and Life-Threatening Behavior, 36,* 255–262.

Rudd, M. D., Joiner, T., & Rajab. M. (2001). *Treating suicidal behavior: An effective, time-limited approach.* New York: Guilford Press.

Schoenbaum, M., Kessler, R., Gilman, S., Colpe, L., Heeringa, S., Stein, M., . . . Cox, K. (2014). Predictors of suicide and accident death in the Army Study to Assess Risk and Resilience in Servicemembers (Army STARRS): Results from the Army Study to Assess Risk and Resilience in Servicemembers (Army STARRS). *JAMA Psychiatry, 71*(5), 493–503.

Sege, I. (2005, March 1). "Something happened to Jeff": Jeff Lucey returned from Iraq a changed man. Then he killed himself. *The Boston Globe.* Retrieved from http://www.boston.com/yourlife/health/mental/articles/2005/03/01/je

Sennott, C. (2007, February 11). Told to wait, a Marine dies: VA care in spotlight after Iraq War veteran's suicide. *The Boston Globe,* pp. A1, A10.

Silver, J. M., Kramer, R., Greenwald, S., & Weissman, M. (2001). The association between head injuries and psychiatric disorders: Findings from the New Haven NIMH epidemiologic catchment area study. *Brain Injury, 15,* 935–945.

Simpson, G., & Tate, R. (2002). Suicidality after traumatic brain injury: Demographic, injury and clinical correlates. *Psychological Medicine, 32,* 687–697.

Stanley, B., & Brown, G. (2008). *Safety plan treatment manual to reduce suicide risk: Veteran version.* Washington, DC: US Department of Veterans Affairs.

Stanley, B., & Brown, G. (2012). Safety planning intervention: A brief intervention to mitigate suicide risk. *Cognitive and Behavioral Practice, 19*(2), 256–264.

Suicide Risk Management & Surveillance Office. (2008). *Army suicide event report (ASER) calendar year 2007.* Tacoma, WA: Office of the Surgeon General of the Army.

Sullivan, G., & Bongar, B. (2009). Assessing suicidal risk in the adult patient. In P. Kleespies (Ed.), *Behavioral emergencies: An evidence-based resource for evaluating and managing risk of suicide, violence, and victimization* (pp. 59–78). Washington, DC: APA Books.

Teasdale, T. W., & Engberg, A. W. (2001). Suicide after traumatic brain injury: A population study. *Journal of Neurology, Neurosurgery, & Psychiatry, 71,* 436–440.

US Department of Defense, Task Force on the Prevention of Suicide by Members of the Armed Forces. (2010). *The challenge and the promise: Strengthening the force, preventing suicide, and saving lives: Final report of the Department of Defense task force on the prevention of suicide by members of the armed forces.* Washington, DC: Department of Defense.

Van Orden, K., Witte, T., Gordon, K., Bender, T., & Joiner, T. (2008). Suicidal desire and the capability for suicide: Tests of the interpersonal-psychological theory of suicidal behavior among adults. *Journal of Consulting and Clinical Psychology, 76,* 72–83.

Weathers, F., Litz, B., Herman, D., Huska, J., & Keane, T. (1993). The PTSD Checklist (PCL): Reliability, validity, and diagnostic utility. In *Proceedings of the annual convention of the International Society for Traumatic Stress Studies.* San Antonio, TX.

Yehuda, R., Southwick, S., & Giller, E. Jr. (1992). Exposure to atrocities and severity of chronic post-traumatic stress disorder in Vietnam combat veterans. *The American Journal of Psychiatry, 149,* 333–336.

Zivin, K., Kim, M., McCarthy, J., Austin, K., Hoggatt, K., Walters, H., & Valenstein, M. (2007). Suicide mortality among individuals receiving treatment for depression in the Veterans Affairs Health System: Associations with patient and treatment setting characteristics. *American Journal of Public Health, 97*(12), 1–6.

9

Driving Themselves to Death
Covert and Subintentioned Suicide among Veterans

Glenn Sullivan
Phillip C. Kroke
Timothy B. Hostler

Postdeployment mortality is significantly higher for recent veterans of war zones than for their military peers. During the first five years after deployment, Vietnam veterans were 1.93 times more likely to die in a motor-vehicle accident (MVA) than veterans of the same era who served in the United States, West Germany, or South Korea (Boyle et al., 1987). These excess MVA deaths remained significant even after controlling for presence of a passenger, driving conditions (day or night), and driver's blood alcohol level. Military personnel who had been deployed to Vietnam were also 1.72 times more likely to die by suicide and 1.52 times more likely to die by homicide during the five-year readjustment period. Six years or more after returning from Vietnam, these veterans were less likely than the nondeployed comparison group to die by suicide or homicide but were still 1.16 times more likely to die in an MVA.

Selection effects appear to be the source of these discrepant outcomes. However, at the time of their entry into the Army, "no important differences were apparent in background characteristics between Vietnam and non-Vietnam veterans" (Boyle et al., 1987, p. 57). In fact, the non-Vietnam group had higher rates of dishonorable discharges, AWOLs, and confinement time, all of which presumably would be associated with higher levels of risky driving. Severity of combat exposure was also not a factor: Vietnam veterans who served in noncombat roles (e.g., shipping clerk, aircraft mechanic) were just as likely as combat infantrymen to exhibit excessive mortality after returning home. The authors noted that excess mortality rates also had been found for veterans of World War II and the Korean War, relative to the civilian male population.

So what is it about serving in a war zone that increases postdeployment mortality risk? Boyle and colleagues (1987) suggested that young Vietnam veterans may have used risky driving as a way to manage the stress, aggression, hostility, or frustration that could have been caused by their experiences overseas. At the same time, their overseas experience might have generated a need for sensation-seeking that they attempted to satisfy through risky driving. Finally, the authors suggested that service in Vietnam may have increased these veterans' tolerance for risk in everyday life: That which had been perceived as intolerably dangerous prior to deployment may have been viewed as significantly less hazardous in light of their war-zone experiences.

An excess of vehicle-related accidental deaths was also found among Desert Storm veterans (Kang & Bullman, 1996). Interestingly, the MVA mortality rate ratios were greater for female than for male veterans. Female Persian Gulf War veterans were 1.81 times more likely to suffer an MVA death than female service members who had not been deployed. Male Persian Gulf War veterans, however, were 1.27 times more likely to suffer an MVA death than male service members who had not been deployed. Further, female Persian Gulf War veterans were 1.47 times more likely to die by suicide than nondeployed females. In contrast, deployment to the Persian Gulf seems to have had a protective effect for male service members, if only for suicide and not for MVA deaths; they produced a suicide rate only 88% that of their nondeployed male counterparts.

Findings such as these have been replicated with other samples of veterans. A meta-analysis encompassing 20 studies of postdeployment mortality among Vietnam or Persian Gulf War veterans from the United States, United Kingdom, and Australia found that those serving in conflict zones exhibited higher rates of injury-related mortality and that much of that excess mortality was accounted for by MVAs (Knapik, Marin, Grier, & Jones, 2009). One of the leading hypotheses to account for this consistent finding is "altered risk perception." This explanation suggests that veterans of war zones return home and wear seatbelts and motorcycle helmets less regularly. They pass on the right and drive at high speeds. They became habituated to risks that they would have found unacceptable prior to their deployments. It is important to note that this excess mortality risk seems to diminish over time, but perhaps in part because those veterans at highest risk die early as a consequence of their altered behaviors.

Another contributing factor to excessive motor-vehicle deaths among veterans could be trauma exposure and its sequelae. Two studies comparing female Vietnam veterans (almost all of whom served as nurses) and female veterans who had not served in Vietnam also found excess mortality for postdeployment MVAs. Interestingly, while there was no excess risk for death by suicide (Risk Ratio [RR] = 0.96) and no excess risk for all-cause mortality (RR = .93), the excess risk of death from MVA ranged from 2.6 times (Cypel & Kang, 2008) to 3.2 times greater (Thomas, Kang, & Dalager, 1991). These increased MVA mortality rates were even greater than those found among male Vietnam veterans. This very well might be related to the extreme levels of trauma exposure experienced by Vietnam nurses. Similarly, Bullman and Kang (1994) found that Vietnam veterans with posttraumatic stress disorder (PTSD) were 3.97 times more likely to die by suicide and 1.85 times more likely to die in a MVA than Vietnam veterans without a PTSD diagnosis.

There is both a theoretical and empirical relationship between combat exposure and increased risk for completed suicide. Joiner (2005) has proposed that repeated exposure to the dangers of combat results in a habituation to fear that could facilitate risky postdeployment behaviors, including suicide. Bryan, Cukrowicz, West, and Morrow (2010) presented evidence that combat exposure increases veterans' *acquired capability* for suicide, that is, a reduction in the fear associated with self-harm. However, there was no association between combat exposure and the *desire to die* (which is composed of thwarted belongingness and perceived burdensomeness). Nevertheless, because acquired capability is a primary risk factor for suicide, combat veterans who engage in risky postdeployment behaviors are acting along a continuum of suicidal behavior.

RISKY DRIVING BY VETERANS

The nature of the Iraq War was different in many ways from previous American wars. For many veterans of Iraq, the iconic image of their war is not a dogfight over the trenches, an amphibious landing on a Pacific beach, or a helicopter assault into a jungle clearing—it is looking out of dust-covered windshield, down a seemingly endless highway, surrounded by empty desert terrain, and nervously anticipating the detonation of an improvised explosive device (IED) that will destroy your vehicle, amputate your limbs, and kill your friends. IED attacks on vehicle convoys accounted for a greater percentage of casualties in the Iraq War than in any other American war (Stevenson, 2009). To combat the threat of IEDs and vehicle-borne IEDs (or suicide-bomb vehicles), American soldiers and Marines adopted what could be called "combat survival driving." They drove fast and made sudden, unpredictable lane changes. Some units intoned the mantra "Speed is security." They "rode the zipper" (i.e., drove down the middle of the road) and kept other, non-American vehicles at a distance, ramming them or driving them off the road if necessary. Many deployed service members did not wear seatbelts, despite regulations requiring them to do so, because they believed that seatbelts would make it more difficult to exit the vehicle in case of an emergency and that they would make it more difficult to use their weapons (Okpala, Ward, & Bhullar, 2007). They constantly scanned their environment for any sign of an IED or other threat. They calculated that it was best to consider civilian bystanders as a lethal threat until proven otherwise.

After a year or longer in such an environment, it is unsurprising that these veterans' postdeployment driving behaviors have been frequently characterized as "risky" (Sullivan & Kimsey, 2008). They continue their disuse of seatbelts, they drive at excessive speeds; they do not allow others to merge into their lane. They think about the war or "just space out" while driving. They use their horn more than they used to; they yell at or make obscene gestures toward other drivers; they tailgate or follow other cars to "teach them a lesson." They drive when they have not had enough sleep, when they are under the influence of alcohol or other drugs, or just to escape.

Sullivan and Kimsey (2008) found that, compared to a control group of male ROTC cadets, veterans of Iraq or Afghanistan were more likely to report feeling "super alert or on guard" while driving, driving when angry or upset, driving without adequate sleep, feeling angry or irritated while driving, and that passengers have told them to calm down. The greater the severity of PTSD symptoms (as measured by the Minnesota Multiphasic Personality Inventory-2 [MMPI-2] PK scale), the more likely these veterans were to go for a drive just to get away from people, cross double yellow lines to see if they can pass a slow moving vehicle, drive drunk, go for a drive just to do something exciting (relieve boredom) and have felt scared, keyed up, tense, or panicky for no clear reason while driving or suddenly realize while driving that they could not remember what happened during all or part of the trip.

Both veterans and ROTC controls in this study produced MMPI-2 4-9/9-4 profiles below the clinical threshold. However, the veterans produced significantly higher scores on Clinical Scales 4 (Psychopathic Deviancy), 6 (Paranoia), and 8 (Schizophrenia). Using the evolutionary adaptive model of MMPI interpretation (Caldwell, 2001), these veterans could be described as being in the process of "ceasing to care" about others and themselves, acutely vigilant to (and protective against) valid cues of hostility in others, and behaving in a manner that encourages others to leave them alone. In this study, 25% of the veterans (versus 7% of the controls) endorsed as "false" the MMPI-2 item *I have no enemies who really wish to harm me*. Only 12% of the MMPI-2 norming sample endorsed that item in the deviant direction, in contrast to the first author's clinical database of Iraq and Afghanistan veterans, who endorsed that item 75% of the time. Rather than interpret these responses as reflective of clinical paranoia, it is perhaps more prudent to consider that many veterans return home with new knowledge—about the world, other people, and themselves—that many of us would prefer not to know.

Combat deployment has been linked to an increase in speeding, not wearing a seatbelt, and other risky driving behaviors (Fear et al., 2008). Sheppard and Earleywine (2013) employed an anonymous, unmatched count technique to estimate the base rates of risky driving behaviors among veterans of Iraq and Afghanistan. More than half of the veterans participating in that study (51%) reported carrying a gun while driving, and nearly 78% reported driving drunk. In a study of 474 male veterans attending a 60-day residential PTSD treatment program, 11.7% reported having intentionally driven their vehicle into another object (e.g., car, tree, etc.; Kuhn, Drescher, Ruzek, & Rosen, 2010). Twenty percent of the Iraq or Afghanistan veterans in this study reported using their seatbelts less than "sometimes." Strom and colleagues (2012) found that 59% of the Iraq or Afghanistan veterans in their sample of 395 patients receiving treatment at a VA Medical Center reported "sacrificing speed for safety when driving a car." Further, 54% reported "thinking about suicide" and nearly 6% reported "intentionally driving your vehicle into another object (tree, other car)."

Thomsen and colleagues (2010) found that deployment increased the prevalence of high-risk recreational behavior ("doing things for fun that were so dangerous you were likely to be injured or killed") by 255% but only among those service members who had previously engaged in such behaviors prior to their deployment or in their civilian lives: "the strongest predictor of current engagement in each specific type of risky behavior was having engaged in that type of behavior previously" (p. 1325). After deployment, the overall prevalence rate for risky behaviors increased from just over 30% to nearly 50%. Interestingly, Thomsen and colleagues found higher rates of high-risk recreation during civilian life than during military service, either pre- or postdeployment. This might be related to a maturation effect or to the influence of the structured and disciplined military environment.

In their study of 1,252 U.S. Army soldiers recently returned from Iraq, Killgore and colleagues (2008) reported associations between specific combat experiences and risk-taking propensity. The strongest association was between violent combat exposure and postdeployment danger seeking. The combat experience of having killed the enemy was positively associated with both danger seeking and a heightened sense of "invincibility." It should be noted that postcombat "invincibility" is measured by three items on a 24-item questionnaire. Endorsement of these items in the risk-taking direction suggests that the respondent (a) feels that he is always right and never wrong, (b) has a propensity to accelerate through yellow traffic signals, and (c) would likely "proceed immediately" down an unfamiliar stairwell even after the lights have gone out. Given that the first item is likely to be associated with hostile confrontations and that the remaining two items are clearly associated with increased risk of physical injury or death in exchange for only minimal gain, it would not seem inappropriate to label this subscale "Death Seeking" and not "Invincibility."

Earlier, Grigsby (1991) observed that many combat veterans reported what he described as *combat rush*: "When experiencing this state of arousal, not only are soldiers trying to kill other people, but they are also thoroughly engrossed in and enjoying their activity.... While the combat rush is in certain respects frightening, many patients nevertheless try to reproduce this experience of intense excitement by participating in highly stimulating civilian pursuits. Fighting, fast driving, and heavy drug or alcohol use are common means of seeking such stimulation" (p. 358).

Nevertheless, postdeployment danger-seeking behaviors are not merely attempts to simulate the adrenaline rush of combat. For some, these behaviors might be a way to tame the death anxiety triggered by combat exposure. (Although, Killgore et al. [2008] found no association between having "survived a close call" and risk taking.) Risky postdeployment behaviors are also a means of courting death. Dying in a postdeployment "accident" enables one to avoid the problem of having to manage the difficult transition between combat and civilian life. It could facilitate a fantasized reunion with dead comrades. It could expiate the guilt of actions done (or not done) in the war zone. In sum, we should view the thrill-seeking veteran not merely as an adrenaline junkie but as a person who is deeply ambivalent about being alive.

DRIVING-RELATED SUICIDAL IDEATION

In a sample of 41 patients reporting suicidal ideation, Hamburger (1969) found that nearly 15% of these were "contemplating death using a motor vehicle" (p. 441). Avoidance of the stigma associated with suicide was cited by most of these patients as a reason for considering that means of death. Others expressed a desire not to jeopardize life insurance policies for their survivors. One male, aged 52, "decided against an obvious suicidal attempt, due to religious aspects. He felt that he could not be buried on consecrated grounds, and thought of the social stigma to his children. In view of these considerations, he had decided to use his automobile and drive it off a cliff in the nearby mountains" (p. 442). In this sample, half of the cases involving driving-related suicidal ideation also involved alcohol abuse.

Murray and de Leo (2007) contacted 1,196 residents of Australia's Gold Coast who were known to have a history of suicidal ideation. Nearly 35% of these individuals had also made "suicide plans or arrangements," and about 19% had made suicide attempts. The researchers found that nearly 15% of the suicide planners had thought to kill themselves by means of a motor vehicle "accident." They also found that 8.3% of the attempters had previously attempted suicide by means of a motor vehicle collision. In this large, community-dwelling sample of past suicidal ideators, the lifetime rate of vehicular planning was 5%, and the rate of past vehicular attempts was 1.6%.

In a sample of 146 Social Security disability claimants, nearly 35% of the 29 claimants reporting current suicidal ideation also indicated thoughts involving vehicular suicide (Sullivan & Hostler, unpublished data collected by the authors). This yields an overall rate for recent driving-related suicidal ideation of nearly 7%. Some responses involved passive thoughts about "just letting go of the steering wheel." These thoughts were sometimes followed by imagining "the sound of the crash," "the sound of glass breaking," or, "what it would be like to hit a tree going full speed." Many respondents had specific locations in mind for their imagined vehicular suicide: "the road home from work," "at the railroad crossing," or "that place on [the interstate] that curves real sharp." The most frequently imagined scenario was a head-on collision with a tractor-trailer truck. No respondents acknowledged having had any thoughts as to the potential impact, either physical or psychological, on the other driver.

Critically, the degree to which the vehicular mode of death was perceived to be swift, certain, and painless seemed to correspond to the intensity of these suicidal thoughts. Claimants who reported thoughts of intentionally driving in front of an oncoming tractor-trailer truck were more likely to describe that action as "quick" or "easy." One claimant observed, "There wouldn't be much left. I probably wouldn't even know what happened." Most ideators offered some opinion as to the nonsurvivability of the act. Conversely, one female claimant noted, "With my luck, the air bag will save me." Claimants who expressed less certainty regarding the lethality of the method appeared less attached to it. One claimant noted that "I would probably steer away [from the trees] at the last second"; another observed, "That's just what I need, to end up in a wheelchair after a car wreck."

Kroke and Sullivan (2013) examined a sample of 96 ROTC cadets (86% male; 79% White) and found that nearly 17% reported having thought about "crashing your vehicle into oncoming traffic," "crashing your vehicle on purpose," or "killing yourself by crashing your vehicle" at some point during the previous 24 months. The driving-related ideators reported fewer "reasons for living" but greater "fear of suicide" on the Reasons for Living Inventory (Linehan, Goodstein, Nielsen, & Chiles, 1983). No differences were found between ideators and nonideators on the Beck Hopelessness Scale (Beck & Steer, 1988). This study was perhaps the first to provide an estimated base rate of driving-specific suicidal ideation in a nonclinical sample.

SUICIDE BY MOTOR VEHICLE CRASH

Mental health clinicians are expected to conduct, as a matter of routine, thorough suicide risk assessments with all of their patients. In the event that suicidal ideation is present, clinicians are usually encouraged to assess and reduce the lethality of the patient's environment (e.g., Bongar & Sullivan, 2013). Lethality-reducing interventions are usually focused on firearms, because self-inflicted gunshot is the most common means of completed suicide in the United States. Bryan, Stone, and Rudd (2011) have presented a useful approach to means-restriction counseling that involves collaboration among the clinician, patient, and supportive others with the primary goal of reducing a suicidal patient's access to firearms.

It must be noted, however, that to a suicidal patient the means to one's own self-annihilation are potentially legion. A patient who has allowed his family to divest him of his firearms may begin to hoard his psychiatric medication, or eye the kitchen knives, or study the many YouTube videos on "how to tie a noose." It is not at all clear that simply because a patient has indicated a preference for a specific suicide method during a risk assessment, eliminating access to that one method will reduce risk sufficiently. Given that serious, and quite often fatal, suicide attempts are often made with acetaminophen, plastic bags used for self-suffocation, or broken glass, it must be recognized that a patient's environment is never "suicide-proof." The fact that suicides often occur even within the confines of psychiatric hospitals is instructive in this case.

While access to firearms is a striking risk factor for completed suicide, access to a motor vehicle is far less salient. This may be because the percentage of driver fatalities that are intentional is presumed to be very low. In an early study of MVAs in Baltimore over a six-year period, it was found that only 2% of the fatal accidents were suicides and that 1% of the nonfatal accidents were suicide attempts (Schmidt, Shaffer, Zlotowitz, & Fisher, 1977). Moreover, among all of the MVA victims, only 8% of the fatally injured drivers and 6% of the nonfatally injured drivers had experienced suicidal thoughts or impulses at some point during the two years prior to their crashes. These base rates for suicidal ideation are no greater than those typically found among the general population.

Of the three fatal cases determined to have been suicides, all had driven their vehicles at high speed into a fixed object, with no sign of braking, skidding, or swerving. The results of this study seemed to suggest that suicide by motor vehicle was not a widespread phenomenon. However, it should be noted that a large percentage of the accident victims in this study were either African American or female and that both of these groups represent low suicide risk relative to White males. The fact that the study was conducted in an urban center should also be considered, as vehicular suicide could be more likely on interstate highways than on city streets.

Evidence of the suicide-by-automobile phenomenon has also been offered in case reports. Peck and Warner (1995) presented six cases of vehicular suicide and noted that "death by automobile offers a unique opportunity for concealment of suicide intent" (p. 463). The cases presented were (a) a 38-year-old male facing criminal charges who drove his car at high speed into the rear end of a truck; (b) a 39-year-old female anticipating a divorce who parked at the bottom of a hill on an interstate and drove into the underside of a tractor-trailer truck as it passed her; (c) a 23-year-old female with schizophrenia and polysubstance abuse who intentionally crashed her car into another vehicle, killing two other people; (d) a 47-year-old male who, sober, stepped out into traffic on a busy interstate highway, causing vehicles to swerve around him, until one finally hit him; (e) a 74-year-old male who apparently survived the intentional wrecking of his pickup truck but then fatally stepped into the path of an oncoming tractor-trailer; and (f) a 38-year-old male with a history of depression who drove at speeds in excess of 100 mph on an interstate highway before crossing the median and hitting another vehicle head-on.

Motor vehicle crashes are almost always attributed to accidental causes. Even when evidence of suicidal intent is noted by investigators (e.g., suicidal threats, hopelessness, other recent attempts), there can be a strong disinclination to attribute a motor vehicle fatality to suicide. This may be partially an attempt to spare the family of the victim the "stigma" of suicide. This social pressure would be particularly powerful in cases of vehicular suicide that accidentally caused the deaths of other drivers. Misattribution could also be the consequence of base rate bias: if 95% of crashes are nonsuicidal, then the "best guess" in ambiguous cases is to avoid a determination of suicidal intent. Finally, the results of even the most thorough "psychological autopsy" are often equivocal. All of these factors can contribute to a significant undercount of vehicular suicides.

The most thorough study of its kind (Ohberg, Penttila, & Lonnqvist, 1999) found that initial determinations by accident investigators tended to seriously undercount the number of fatal motor vehicle crashes that actually involved suicide. The authors reviewed all motor vehicle fatalities in Finland from 1987 to 1991 and found, after careful investigation that included elements of a psychological autopsy, that nearly 6% of these were suicides. This is in contrast to the official driver suicide rate of only 2.6% (although many of these crashes had been considered by the initial investigators to be "probable" suicides or "undetermined"). Thus, in Finland during this time period, 1.2% of all suicides were driving related. If a similar rate held true for the United States, then there would have been about 500 driver suicides in 2010 (which is comparable to the number of people who died that year by intentionally putting themselves into the path of an oncoming train). Most important, the deaths of more than half of the vehicular suicide decedents had been initially misclassified as accidental.

Most of these suicide crashes involved men younger than 35, with the highest rate being among those age 15 to 24 (Ohberg, Penttila, & Lonnqvist, 1997). The typical driver suicide was not a single car accident but rather a "head-on collision between two vehicles with a large weight disparity" (p. 468). Although murder-suicide was not apparently intended, the death of some other person occurred in 4% of the cases (e.g., the driver of the targeted truck or the driver of a third vehicle that the targeted truck hit after it lost control). Of the drivers who died by suicide, 76% had recently experienced a significant life stressor, 43% had been treated for a mental disorder, and 13% had a prior suicide attempt (versus 25%, 1.2%, and 0% for a control group of nonsuicidal MVA fatalities).

Suicide by means of a motor vehicle is more common than is generally realized. Vehicular suicide could be conservatively estimated to account for approximately 1% of all completed suicides in the United States. As many as 6% of all motor vehicle fatalities could potentially be attributable to suicide. Significant suggestions of vehicular suicide are (a) younger male victim, (b) recently suffered a loss and/or has a mental health history, and (c) a head-on collision between victim's vehicle and much larger truck.

EQUIVOCAL AND SUBINTENTIONED DEATHS

Much of the difficulty in accurately determining the cause of death in a motor vehicle fatality (suicide or accident) is that many deaths are "equivocal." Shneidman (1996) observed that all suicides are marked by ambivalence and that no one is 100% committed to his or her death. Others have observed that many "accidental" deaths and injuries appear to mask an unconscious self-destructive intent. Durkheim (1897/1951) observed that suicides "are merely the exaggerated forms of common practices" (p. 45). In his view, the driver who crosses the double yellow line to pass a slower vehicle, the man with hypertension who is inconsistent in taking his medication, and even the creative producer who works long into the night all exhibit suicidal behavior. Menninger (1938) declared that "In the end each man kills himself in his own selected way, fast or slow, soon or late" (p. vii). Firestone and Seiden (1987) powerfully suggested that death anxiety is the source of what they call "microsuicides" (e.g., disengagement from life, withdrawal from others, accident proneness, compulsive working, overeating, drug/alcohol abuse, emotional withholding). Through these microsuicides, we "achieve an illusion of mastery over life and death by committing small suicides on a daily basis" (p. 33). Vehicular suicide, especially as a consequence of reckless or careless driving, seems to be an exemplary form of equivocal death, in that the intent of the crash is concealed not only from others but also from oneself.

Shneidman (1977) estimated that 10% to 15% of deaths are "equivocal"; that is, they do not fall neatly into one of the categories of natural, accidental, suicide, or homicide (NASH) used by many coroners. Further, he found that 36% of all deaths involve some psychological component. Shneidman asked coroners to rate 974 deaths on a scale of "imputed lethality," ranging from high (the decedent definitely wanted to die), to medium (the decedent played an important role in effecting or hastening his own death), to low (the decedent played some small but not insignificant role in effecting or hastening his own demise), to absent (the decedent played no role in effecting his own death). Unsurprisingly, all of the suicide deaths in this sample were rated as high lethality, but, far more interestingly, of the accidental deaths, 1% were rated as high lethality, 44% as medium, and 22% as low. In a substantial percentage of accidental deaths, the decedent was deemed to have facilitated his or her own death by self-neglectful or otherwise imprudent actions that may have reflected a partial, covert, or unconscious desire for death.

Several researchers have suggested that some MVAs are covert or unconscious (i.e., subintentioned) forms of suicidal behavior. Selzer and Payne (1962) found that suicidal veterans engaged in psychotherapy reported more than twice as many past vehicular accidents as nonsuicidal patients (2.7 vs. 1.3; $N = 60$). Only one of these suicidal patients believed that their past MVAs were suicide-related. Interestingly, of the 12 patients who had previously attempted suicide, only one of their 21 total suicide attempts involved the deliberate crashing of a motor vehicle. As the authors noted, "The automobile lends itself admirably to attempts at self-destruction because of the frequency of its use, the generally accepted inherent hazards of driving, and the fact that it offers the individual an opportunity to imperil or end his life without consciously confronting himself with his suicidal intents" (p. 239). Furthermore, "[m]ore conventional modes of suicide do not offer as dramatic an opportunity for the gratification of destructive and aggression impulses" (p. 239). If, as in Freud's view, suicide represents aggression turned against the self, then suicide by motor vehicle affords potential victims a gratifyingly fiery, loud, and calamitous end.

"Accidental" death in a motor vehicle crash might be preferred by ambivalent persons who dread the "stigma" of suicide or who "could never do that to my family." Pompili, Girardi, Tatarelli, and Tatarelli (2006) compared 30 survivors of single-car accidents to healthy controls who had never had an accident. The single-car accident survivors had experienced more life stressors during the previous year, and they were more likely to endorse ever having felt "tired of living" or that "life was not worth living" (53.3% vs. 23.3%). The single-car accident survivors had poorer coping skills than controls, but they indicated a stronger sense of family responsibility. In fact, the primary reason given by the single-car accident survivors as to why they would not kill themselves was because they felt a responsibility for their families and a desire not to cause them distress. The control group, however, was opposed to suicide on the grounds that they were attached to life and that they had the resources necessary to handle their problems.

ASSESSMENT AND TREATMENT

Risky driving behaviors are more prevalent than is commonly thought. Taken together, the research findings presented here underscore the importance of carefully assessing risk-taking behaviors among veterans. Routine screening of veterans could be useful but only if careful and empathetic follow-up assessment and intervention is available. Given the high base rates for postdeployment risk taking, it may be prudent to simply assume that a veteran is engaging in such behaviors until contrary evidence emerges. In 2004, the first author was a cotherapist of an early Iraq War veterans' psychotherapy group. On the first night, one of the members was late because he had very nearly been in an accident on his motorcycle. This led another group member to relate how he had had an accident recently in his civilian job as a truck driver. Very quickly, each and every combat veteran present shared his or her story of a MVA or near-miss that he had experienced in the few months since returning from Iraq.

This spontaneous revelation afforded the therapists an opportunity to explore with the veterans the potential reasons behind these accidents and near-accidents. Some veterans felt that it was because they had not driven much during their year-long deployments. Some shared that they felt tense and angry while driving. Some, particularly those whose military duties that involved driving, noted that when they got behind the wheel they sometimes felt like they were "back in the desert." One veteran observed candidly that he had been going for long, high-speed drives late at night just to "get away from my family for a while."

This example shows that "[n]uanced assessment of these behaviors yields clinically relevant material such as trauma-related cognitions and dysfunctional coping strategies, which may inform clinical interventions and treatment strategies" (Strom et al., 2012, p. 394). In other words, it is critical not only to screen for these behaviors but also to probe for the underlying motives that drive them. Merely instructing a veteran to desist from an unwanted behavior is futile. Helping a veteran find substitutive means of gratifying critical needs is therapeutic. In his clinical practice, the first author has found that simply asking a recent veteran, "What's it been like driving since you've been back?" is almost certain to elicit a host of clinically relevant material. In the event of a neutral reply (e.g., "fine"), following up with specific questions regarding accidents, near-misses, close calls, being pulled over for speeding (not necessarily getting a speeding ticket, as some police officers are disinclined to ticket recently returned veterans), driving in order to get away from other people or to blow off steam, feeling tense or angry while driving, and so on is in order. One patient of the first author denied any problems with driving since returning from a tour of duty as a helicopter pilot in Iraq. One day, after the patient showed up soaking wet to a therapy session, it became clear that he had been walking the more than three miles from his apartment to the medical center. The reason he walked was because getting into his car reminded him too much of climbing into the cockpit of his helicopter.

Battlemind training has been shown to be an effective form of postdeployment early intervention (Adler et al., 2009). The intervention, which is usually delivered by service personnel in a group setting to members of the same unit, "emphasizes the transition from combat to home, and recognizes that this transition is a critical social-psychological task" (p. 929). The Battlemind approach minimizes the recounting of past traumatic experiences and thus sidesteps the problem of traumatic re-exposure and secondary trauma exposure in group settings. Battlemind training "positively reframes traditional postdeployment transition difficulties such as PTSD, depression, and sleep problems ... as being a natural consequence of having developed effective occupational coping skills related to combat" (p. 930). For example, "the essential combat skill of maintaining tactical awareness may lead to hypervigilance back home, contributing to sleep problems as well as PTSD symptoms of hyperreactivity, hyperalertness, and startling" (p. 930).

Battlemind training contrasts the effective skill of combat driving, which is necessary to avoid danger in the war zone, to overly aggressive driving back home. The training encourages veterans to notice if they or their buddies have been too driving fast, have been involved in accidents, or are more easily angered while driving. Veterans are encouraged to seek help if needed. While the training serves an important function in alerting veterans that increased risky driving is a sign that their postdeployment transition is not yet entirely successful, it presents no specific skills that might reduce those behaviors.

Zimbardo, Keough, and Boyd (1997) presented a portrait of the risky driver: "high risk takers in other areas, unconventional or somewhat deviant, mildly antisocial, sensation seeking, overconfident in their driving ability, engrossed in the thrill of the present moment, and male" (pp. 1020–1021). This portrait is generally consistent with the prototypical veteran of Iraq or Afghanistan. The authors present research that supports "present time perspective" (versus past or future time perspective) as a risk factor for dangerous driving. It is certainly useful to consider whether some veterans lose themselves in driving in order to negate or block out their past experiences. Others may engage in risky driving because the future seems overwhelming or hopeless. From this perspective, it follows that helping veterans to process their past experiences and foster hope for the future could reduce their driving risk.

Strom and colleagues (2013) reported on a pilot study of an eight-session treatment for driving-related anger and risky driving in combat veterans. Of the nine veterans who completed the treatment, eight (89%) experienced reliable change in driving-related aggression and six (67%) experienced reliable change in driving-related anger. The first two sessions presented psychoeducation on the possible causes of risky driving behavior. The third session focused on identifying more adaptive coping behaviors and on breathing relaxation training. The fourth and fifth sessions introduced the ABCD model of cognitive restructuring. The sixth and seventh sessions provided the rationale for in vivo exposure activities and encouraged their use. The eighth session focused on termination and relapse prevention.

Psychiatric medication could be helpful in this population. Drivers with a history of suicidal ideation but who are not currently taking antidepressant medication are more than four times more likely to be injured in a motor vehicle crash than drivers with no history of suicidal ideation (Lam, Norton, Connor, & Ameratunga, 2005). Grigsby (1991) pointed out that one reason combat veterans may be disinclined to take psychiatric medication is because the drugs often work as intended; that is, they reduce the intensity and frequency of combat memories. Postdeployment existence can seem bleak and boring in contrast to the life-or-death vividness of combat. Memories of combat are both distressing and exciting, according to Grigsby, and serve to relieve the anhedonia and meaninglessness that mark some veterans' lives. Helping veterans to find other means of relieving anhedonia (e.g., antidepressant medication, life-enhancing activities) is a critical treatment goal. Equally important is helping them to discover meaning in their combat experiences and purpose in their present existence (Frankl, 1946/2006).

The clinician should be aware that reckless driving could have anxiolytic properties for the patient, as suicidal ideation often does. As Nietzsche (1886/1966) observed, "The thought of suicide is a powerful comfort; it helps one through many a dreadful night" (p. 91). The knowledge that relief from tension or boredom is only as far away as your vehicle can be a tremendous source of comfort. The mostly submerged thought that driving recklessly could result in a fatal wreck, and thereby put a permanent end to one's sea of troubles, can have a similar effect. The suicidal ideation and the reckless driving are both reinforced by a relative, temporary reduction in anxiety and by this mechanism become more deeply ingrained and frequent. Again, it is imperative that the clinician help the patient to develop alternative means of stress reduction and to enhance his or her distress tolerance.

CONCLUSION

For the suicidal patient, the means of one's self-destruction are nearly always at one's disposal. Vehicular suicide, like suicide by other means, is marked by ambivalence—a struggle between the desire to live and the desire to end one's intolerable psychological pain by means of self-annihilation (Shneidman, 1996). There are many factors that contribute to the excess mortality from MVAs consistently observed among recently returned veterans. Among these are a recalibration of "what is dangerous," a habituation to danger and fear (enhanced acquired capability for suicide), increased aggression, reduced tolerance for boredom and frustration, and interpersonal difficulties.

Each veteran's response to combat exposure is different, as these individuals have unique personal histories, family structures, educational experiences, temperaments, personality characteristics, coping strategies, capacities for resilience, and so on. Indeed, even members of the same unit are exposed to different combat stressors: a near-miss for one could have gone completely unnoticed by another; seeing a burned doll lying in a bombed-out house could be traumatic for a soldier who is also a father but a neutral event for his combat buddy. Postdeployment experiences also differ widely, with some veterans more likely than others to avoid alcohol or other drugs, to find support in family and civilian friends, and to have the opportunity for meaningful work. Because each veteran is unique, every treatment must be unique. The challenge to clinicians is to formulate a case conceptualization and treatment plan *for this individual veteran*, one that accounts for his or her history, combat experiences, present circumstances, strengths and capabilities, and current symptoms.

FUTURE DIRECTIONS

1. Due to their very nature, subintentioned deaths by MVA are difficult to detect. It is possible that the scope of this problem among veterans is greater than we expect. Approaching the MVA fatalities of active duty service members and recent veterans as equivocal deaths, and investigating the circumstances surrounding these deaths as fully as possible, is a necessary first step.
2. More research is needed on psychoeducational group interventions for postdeployment risky driving. These interventions should be evaluated not only for their ability to effect behavioral change but also for their ability to accurately identify high-risk participants and refer them for more intensive treatment.
3. Additional clinician-training on "microsuicides," subintentioned suicide, and risky postdeployment behaviors is needed across the medical and allied health professions. In general, suicide assessment is poorly done, and assessing the types of behaviors described here requires clinical sensitivity far greater than the typical clinician's inquiry: "You haven't been thinking of hurting yourself, have you?"

REFERENCES

Adler, A. B., Bliese, P. D., McGurk, D., Hoge, C. W., & Castro, C. A. (2009). Battlemind debriefing and Battlemind training as early interventions with soldiers returning home from Iraq: Randomization by platoon. *Journal of Consulting and Clinical Psychology*, 77(5), 928–940.

Beck, A. T., & Steer, R. A. (1988). *Beck Hopelessness Scale: Manual*. San Antonio, TX: Psychological Corporation.

Bongar, B., & Sullivan, G. R. (2013). *The suicidal patient: Clinical and legal standards of care* (3rd ed.). Washington, DC: American Psychological Association.

Boyle, C. A., Decoufle, P., Delaney, R. J., DeStefano, F., Flock, M. L., Hunter, M. I., . . . Worth, R. M. (1987). *Postservice mortality among Vietnam veterans*. Atlanta, GA: Centers for Disease Control and Prevention.

Bryan, C. J., Cukrowicz, K. C., West, C. L., & Morrow, C. E. (2010). Combat experience and the acquired capability for suicide. *Journal of Clinical Psychology*, 66, 1044–1056.

Bryan, C. J., Stone, S. L., & Rudd, M. D. (2011). A practical, evidence-based approach for means-restriction counseling with suicidal patients. *Professional Psychology: Research and Practice*, 42, 339–346.

Bullman, T. A., & Kang, H. K. (1994). Posttraumatic stress disorder and the risk of traumatic deaths among Vietnam veterans. *The Journal of Nervous and Mental Disease*, 182, 604–610.

Caldwell, A. B. (2001). What do the MMPI scales fundamentally measure? Some hypotheses. *Journal of Personality Assessment*, 76(1), 1–17.

Cypel, Y., & Kang, H. (2008). Mortality patterns among women Vietnam-era veterans: Results of a retrospective cohort study. *Annals of Epidemiology*, 18(3), 244–252.

Durkheim, E. (1897/1951). *Suicide: A study in sociology* (J. A. Spaulding & G. Simpson, Trans.) New York: The Free Press.

Fear, N. T., Iversen, A. C., Chatterjee, A., Jones, M., Greenberg, N., Hull, L., . . . Wessley, S. (2008). Risky driving among regular armed forces personnel from the United Kingdom. *American Journal of Preventative Medicine*, 30, 529–544.

Firestone, R. W., & Seiden, R. H. (1987). Microsuicide and suicidal threats of everyday life. *Psychotherapy*, 24, 31–39.

Frankl, V.E. (2006). *Man's search for meaning*. Boston: Beacon Press. (Original work published 1946)

Grigsby, J. (1991). Combat rush: Phenomenology of central and autonomic arousal among war veterans with PTSD. *Psychotherapy*, 28, 354–363.

Hamburger, E. (1969). Vehicular suicidal ideation. *Military Medicine*, 134(6), 441–444.

Joiner, T. E. (2005). *Why people die by suicide*. Cambridge, MA: Harvard University Press.

Kang, H. K., & Bullman, T. A. (1996). Mortality among U.S. veterans of the Persian Gulf War. *The New England Journal of Medicine*, 335(20), 1498–1504.

Killgore, W. D. S., Cotting, D. I., Thomas, J. L., Cox, A. L., McGurk, D., Vo, A. H, . . . Hoge, C. W. (2008). Post-combat invincibility: Violent combat experiences are associated with increased risk-taking following deployment. *Journal of Psychiatric Research*, 42, 1112–1121.

Knapik, J. J., Marin, R. E., Grier, T. L., & Jones, B. H. (2009). A systematic review of post-deployment injury-related mortality among military personnel deployed to conflict zones. *BMC Public Health*, 9, 231.

Kroke, P. E., & Sullivan, G. R. (2013, May). *Driving-related suicidal ideation in a sample of ROTC cadets*. Paper presented at the annual meeting of the Association for Psychological Science, Washington, DC.

Kuhn, E., Drescher, K., Ruzek, J., & Rosen, C. (2010). Aggressive and unsafe driving in male veterans receiving residential treatment for PTSD. *Journal of Traumatic Stress*, 23(3), 399–402.

Lam, L. T., Norton, R., Connor, J., & Ameratunga, S. (2005). Suicidal ideation, antidepressive medication and car crash injury. *Accident Analysis and Prevention*, 37(2), 335–339.

Linehan, M. M., Goodstein, J. L., Nielsen, S. L., & Chiles, J. A. (1983). Reasons for staying alive when you are thinking about killing yourself: The Reasons for Living Inventory. *Journal of Consulting and Clinical Psychology*, 51, 276–286.

Menninger, K. (1938). *Man against himself*. New York: Harcourt, Brace & World.

Murray, D., & de Leo, D. (2007). Suicidal behavior by motor vehicle collision. *Traffic Injury Prevention*, 8(3), 244–247.

Nietzsche, F. (1966). *Beyond good and evil* (W. Kaufmann, Trans.). New York: Vintage Books. (Original work published 1866)

Ohberg, A., Penttila, A., & Lonnqvist, J. (1999). Driver suicides. *The British Journal of Psychiatry*, 171, 468–472.

Okpala, N. C., Ward, N. J., & Bhullar, A. (2007). Seatbelt use among military personnel during operational deployment. *Military Medicine*, 172, 1231–1233.

Peck, D. L., & Warner, K. (1995). Accident or suicide? Single-vehicle accidents and the intent hypothesis. *Adolescence*, 30(118), 463–472.

Pompili, M., Girardi P., Tatarelli, G., & Tatarelli, R. (2006) Suicidal intent in single-car accident drivers: Review and new preliminary findings. *Crises*, 27, 92–99.

Selzer, M. L., & Payne, C. E. (1962). Automobile accidents, suicide, and unconscious motivation. *The American Journal of Psychiatry*, 119, 237–240.

Schmidt, C. W., Shaffer, J. W., Zlotowitz, H. I., & Fisher, R. S. (1977). Suicide by vehicular crash. *The American Journal of Psychiatry*, 134, 175–178.

Sheppard, S. C., & Earleywine, M. (2013). Using the unmatched count technique to improve base rate estimates of risky driving behaviors among veterans of the wars in Iraq and Afghanistan. *Injury Prevention*, 19, 382–386.

Shneidman, E. S. (1977). The psychological autopsy. In L. I. Gottschalk et al. (Eds.), *Guide to the investigation and reporting of drug abuse deaths* (pp. 179–210). Washington, DC: US Government Printing Office.

Shneidman, E. S. (1996). *The suicidal mind*. New York: Oxford University Press.

Stevenson, C. (2009). Evolving mechanisms and patterns of blast injury and the challenges for military first responders. *Journal of Military and Veterans Health*, 17, 20–24.

Strom, T. Q., Leskela, J., James, L. M., Thuras, P. D., Voller, E., Weigel, R., . . . Holz, K.B. (2012). An exploratory examination of risk-taking behavior and PTSD symptom severity in a veteran sample. *Military Medicine*, 177(4), 390–396.

Strom, T., Leskela, J., Possis, E., Thuras, P., Leuty, M. E., Doane, B. M., . . . Rosenzweig, L. (2013). Cognitive-behavioral group treatment for driving-related anger, aggression, and risky driving in combat veterans: A pilot study. *Journal of Traumatic Stress*, 26, 405–408.

Sullivan, G. R., & Kimsey, B. F. (2008, August). *Dangerous driving after war zone deployment*. Paper presented at the 116th Annual Convention of the American Psychological Association, Boston.

Thomas, T. L., Kang, H. K., & Dalager, N. A. (1991). Mortality among women Vietnam veterans, 1973–1987. *American Journal of Epidemiology*, 134(9), 973–980.

Thomsen, C. J., Stander, V. A., McWhorter, S. K., Rabenhorst, M. M., & Milner, J. S. (2010). Effects of combat deployment on risky and self-destructive behavior among active duty military personnel. *Journal of Psychiatric Research*, 45, 1321–1331.

Zimbardo, P. G., Keough, K. A., & Boyd, J. N. (1997). Present time perspective as a predictor of risky driving. *Personality and Individual Differences*, 23(6), 1007–1023.

10

Identifying MMPI-2 Risk Factors for Suicide

John J. Barreto
Roger L. Greene

INTRODUCTION

There has been a long history of research with the Minnesota Multiphasic Personality Inventory (MMPI) and the Minnesota Multiphasic Personality Inventory-2 (MMPI-2) in predicting suicide. However, this research has found that MMPI/MMPI-2 clinical scale elevations (Clopton, 1979; Clopton, Post, & Larde, 1983; Spirito, Faust, Myers, & Bechtel, 1988; Watson, Klett, Walters, & Vassar, 1984), MMPI-2 codetypes (Craig & Olson, 1990; Daigle, 2004), scales specifically created to assess suicide (Farberow & Devries, 1967), simultaneous item analysis (cf. Clopton, Pallis, & Birtchnell, 1979; Leonard, 1977), and MMPI/MMPI-2 expertise (Clopton & Baucom, 1979) have provided inconsistent results, at best, in the prediction of suicide. In fact, several reviews of MMPI/MMPI-2's utilization in predicting suicide have come to the same conclusion that the MMPI/MMPI-2 does not have any scale, scale combination, item, or group of items to date that has been proven to differentiate reliably suicide attempters from nonattempters (Eyman & Eyman, 1991; Friedman, Archer, & Handel, 2005; Greene, 2011; Lester, 1971).

LOW BASE RATE ISSUE

Base rates (prevalence) or the frequencies by which a given behavior or symptom occurs in a given group of individuals or patients are the most vital guides in determining the potential accuracy of any clinical decision-making. The frequency of suicide, for example, is less than 5% to 10% in even the most high-risk groups for suicide. Since 90% to 95% of clients in most settings will not be suicidal, all psychological measures are likely to identify individuals who are not suicidal as suicidal (i.e., false positive predictions of suicide; cf. Finn, 2009). Consequently, the focus becomes one of identifying the risk factors for suicide rather than making a prediction of suicide from a single scale or group of MMPI-2 items.

RISK FACTORS IDENTIFIED BY MMPI-2

The MMPI-2, however, is useful in identifying potential risk factors for suicide rather than as a predictor of suicide (Friedman et al., 2005). As with other populations, the literature attempting to identify a single or the most significant suicide risk factors among military personnel is endless. One of the advantages of the MMPI-2 is that it measures a number of different facets of psychopathology simultaneously. While research has focused extensively on single causes for suicidal behavior, Greene (2011) suggests that future research should discriminate among the various causes and types of suicide to determine whether specific MMPI-2 scale patterns can assist in successfully identifying subtypes of individuals prone to suicide. The MMPI-2 may identify pertinent diagnostic and psychosocial suicidal risk factors that can be used to develop appropriate interventions and to identify individuals who may be vulnerable to the stressors of military life.

Suicide Items

Glassmire, Stolberg, Greene, and Bongar (2001) suggested that endorsement of suicide-related MMPI-2 items (150, 303, 506, 520, 524, and 530) should be major considerations in screening for suicidality. MMPI-2 items 506 and 520 with explicit suicidal content, known as the "I mean business" items (Sepaher, Bongar, & Greene, 1999), are the most direct inquiries of suicidal risk. Greene (2011) reported that when a client has a T score of 75 or higher on Scale 2 (D) or the content scale of Depression (DEP) that there is better than a 50% probability that the client will endorse at least one of these suicide-related MMPI-2 items. When a client has a T score of 75 or higher on Scale 2 (D) or the content scale of Depression (DEP) and does not endorse any of the suicide items, further assessment is warranted in identifying the reasons why the client is *not* experiencing or engaging in suicidal thoughts and/or behavior. That is, the clinician should be able to identify the protective factors that reduce suicide risk.

Although suicide risk may be more readily reported in a clinical interview, Glassmire et al. (2001) found that nearly 20 percent of clients who did not report suicidal ideation in a structured telephone intake interview endorsed at least one suicide item on the MMPI-2. Research also suggests that military personnel may not openly report suicidality or other psychiatric symptoms due to the potential concern that seeking help for mental health issues implies weakness (Hoge et al., 2004). Clearly, clinicians need to inquire directly about suicide risk in the clinical interview and to investigate the potential risk factors identified by the MMPI-2. Even though the client did not report any suicidal risk factors in the interview, the clinician still must review carefully the suicide-related items on the MMPI-2.

Scales

While Scale 2 [Depression (D)] has been considered to be the most likely MMPI scale to predict suicide, several studies have not found any difference in item endorsements of Scale 2 (D) between suicidal and nonsuicidal individuals (Clopton & Jones, 1975; Farberow, 1956; Simon & Gilberstadt, 1958). In fact, Sepaher et al. (1999) found that clients with a spike 2 profile, in which Scale 2 is the only clinical scale elevated above a T score of 65, are the least likely to endorse the suicide-related MMPI-2 items. However, Dahlstrom, Welsh, and Dahlstrom (1972) suggest that, when a client has a spike 2 profile but does not report thoughts and feelings of depression, the risk of suicide is increased.

Scale 5 (Mf) was developed to assess gender-related behaviors and attitudes. Scale 5 (Mf) elevations also have been suggested as an indicator in identifying serious suicide attempters (Sendbuehler, Kincel, Nemeth, & Oertel, 1979). For example, Meyer (1989) suggests that women who elevate Scale 5 and have a low K scale have conflicted attitudes toward femininity. Moreover, when Scale 3 (Hs) and Scale 0 (Si) are elevated in these women, they are reporting self-identity issues, alienation, impulsivity, and a lack of psychosocial resources. Future research should examine the utility of Scale 5 (Mf) as a measure of masculinity in military samples, because masculinity has been suggested as a potential variable in the development of suicidal behavior among both male and female military personnel (Braswell & Kushner, 2012).

Codetypes

Codetypes are the cornerstone of summarizing and identifying key characteristics of an MMPI/MMPI-2 profile. Originally, the individual MMPI clinical scales were intended to distinguish psychopathological behavior from normal functioning. However, as more MMPI/MMPI-2 research was conducted, codetypes or scale combinations were found to differentiate among clients more consistently and adequately than the individual clinical scales (Greene, 2011). MMPI-2 codetypes are defined by the highest elevated clinical scales at or above a T score of 65. They also may be referred to as a spike codetype if only one elevated scale meets this criterion, a 2-point codetype for two elevated scales, and a 3-point codetype for three elevated scales. A 2-7/7-2 codetype implies a 2-point codetype in which Scales 2 (D) and 7 (Pt) are the two highest clinical scales in an MMPI-2 profile.

Being aware of an endorsement of a suicide item in the context of an MMPI-2 codetype is particularly useful. Sepaher et al. (1999) suggested that clients with 1-3/3-1 codetypes, who are typically reporting multiple somatic symptoms, and also endorse the suicide-related MMPI-2 items (506 and 520) may be reporting serious suicidal intention. In contrast, individuals with 6-8/8-6 codetypes, who are typically reporting psychotic symptoms but do not endorse these suicidal items may actually be hiding suicidal ideation or intent (Sepaher et al., 1999). Sepaher et al. provided the frequency with which the suicide-related MMPI-2 items were endorsed in all 55 spike and 2-point codetypes. Greene (2011, Table 7-39, pp. 307–308) provided similar information for item 520 in the Caldwell (2007) clinical sample.

Being able to recognize the different behavior patterns of each codetype can guide the clinician in determining how closely suicidal ideation may need to be monitored and the level of support an individual may need. Moreover, MMPI-2 codetypes may provide several domains pertinent to assessing suicidal potential among military personnel, which include cognitions, moods, interpersonal functioning, and treatment considerations.

Cognitions. Familiarity with the maladaptive thinking often found in codetypes may help the clinician understand how clients perceive their circumstances. For example, individuals with 2-8/8-2 codetypes are likely to perceive themselves as failures, interpret disappointments as proof of their perceived worthlessness, and believe that their death would be of benefit to others. Caldwell (2006) suggests that individuals with this codetype are reporting limited problem-solving skills and are engrossed in anhedonia and hopelessness. Bryan, Cukrowicz, West, and Morrow (2010) suggested that suicidal ideation among combat veterans may actually result from their perception of combat-related events rather than the events themselves. Negative perception of events are what individuals with 2-8/8-2 codetypes are reporting.

Moods. Meyer (1989) suggests that the energy to commit suicide may be evident in an elevated Scale 9 (Ma). Moderate elevations in Scale 9 (Ma) codetypes suggest subthreshold manic symptoms to hypomania or, in severe cases, the presence of a severe mood disorder. For example, individuals with 6-9/9-6 codetypes are often diagnosed with bipolar disorder with extreme paranoia. Indeed, mania characterized by dysphoric symptoms (e.g., agitation, paranoia) has been found to increase suicidality (Cassidy, Murry, Forest, & Carroll, 1998; Goldberg et al., 1999).

Scale 2 codetypes, on the other hand, suggest the presence of depressive symptoms, such as self-deprecation, concentration difficulties, and apathy. The 2-7/7-2 codetype, in particular, has been the codetype most often associated with suicide due to reported symptoms of depressive rumination (cf. Carson, 1969; Dahlstrom et al., 1972; Graham, 1987; Meyer, 1989). However, identifying a third elevated scale or 3-point codetype also may provide a better indication of the severity or type of depressive symptomatology reported by a client. For example, individuals with 2-7-8 codetypes may be reporting peculiar, even psychotic, thinking, making them more susceptible to suicidal behavior. Individuals with 2-7-3 codetypes usually report vegetative symptoms of depression, while 2-7-4 codetypes tend to reflect dysthymia or chronic depression. Individuals with 2-7-8 and the 2-7-4 codetypes are usually indicating the most suicidal risk. This increase in suicidal risk may reflect that Scale 8 (Sc) is identifying an individual in a more mentally confused state and Scale 4 (Pd) may be identifying more social alienation and/or anger.

Interpersonal. Social isolation is a considerable risk factor in increasing suicidal behavior (Trout, 1980). Particularly during postdeployment, the lack of structure in civilian life may make military personnel more vulnerable to social isolation. Studies suggest that psychosocial difficulties appear to increase suicidal ideation, while ample accessibility to family and friends appears to reduce suicidal ideation in postdeployment (e.g., Lemaire & Graham, 2011; Pietrzak et al., 2010). Moreover, Cox et al. (2011) found that suicide notes of U.S. Air Force airmen contained more interpersonal difficulties than intrapersonal distress.

MMPI-2 codetypes can provide the clinician with insight into the client's interpersonal relations, alienation, social isolation, and withdrawal. For example, codetypes containing an elevation on Scale 4 (Pd) may be evidencing greater interpersonal conflict, such as with authority or family members. Individuals with 4-6/6-4 codetypes, for example, may be easily insulted and ill-tempered, causing them to isolate or to alienate others.

Treatment considerations. Since brief therapies used to deliver mental health services quickly are often the norm in the military (Ball & Peak, 2006), suicidal risk factors must be more readily identified and addressed in treatment. MMPI-2 codetypes may aid the clinician in tailoring treatment when working with potentially suicidal clients. For example, while individuals with 2-7/7-2 codetypes are reporting a high-level of depressive symptomatology, they also typically display more insight and take responsibility for their circumstances, making them better candidates for therapy. Butcher and Perry (2008) provide a thorough description of issues that may be encountered when treating individuals with specific codetypes.

Childhood adversity has been identified as a potential risk factor for suicide. For example, in a sample of active duty military personnel, Skopp, Luxton, Bush, and Sirotin (2011) found that childhood adversity predicted suicidal ideation and was higher among service members who reported prior suicide attempts. Johnson et al. (2002) suggested that maladaptive parenting and childhood maltreatment may create interpersonal difficulties that result in suicidal behavior in late adolescence and early adulthood. MMPI-2 codetypes often reflect personality characteristics originating from responses to childhood trauma (Caldwell, 2006). For example, individuals with 2-4/4-2 codetypes may have had an impaired bond to an emotionally ambivalent parent. They may believe that they will be hurt when they care for someone who may not really care for them. This interpersonal dynamic may suggest why these individuals will engage in suicidal behavior. Meyer (1989) suggests that an individual with a 2-4/4-2 codetype may engage in manipulative suicidal behavior. In other words, individuals with 2-4/4-2 codetypes may express or engage in suicidal behavior as a means of attaining care from others.

Client Feedback

Client feedback using personality assessment has been shown to be of benefit in improving the client's well-being (Fischer, 2000). Butcher and Perry (2008) also give a detailed description of providing MMPI-2 feedback to the client. Caldwell (2001) suggests that a clinician may be able to develop a more empathic approach by explaining to clients that their elevated scales on the MMPI-2 are possibly adaptive responses to trauma. For example, he suggests that a clinician may explain to clients with 1-3/3-1 codetypes that preoccupations with health concerns are often seen in individuals who have had life-threatening experiences. This feedback approach is more therapeutic and less threatening than telling a client, "You are converting your psychological distress into physical problems."

Comorbid Psychiatric Disorders

MMPI-2 codetypes also may aid in identifying potential comorbid psychiatric disorders. Comorbid mental disorders, especially involving depression, have been found to elevate suicidal risk (Suominen, Henriksson, Suokas, & Isometsä, 1996). Individuals with either 2-7-8, 2-7-4, or 2-4-8 codetypes are reporting depression and are typically indicating high suicidal risk, in addition to their frequent comorbid personality disorders. Individuals with 2-7-8 codetypes have been often identified as having great suicide potential (Clopton et al., 1983; Choquet, Facy, Davidson, & Philippe, 1983; Lester, 1971), if not the greatest (Caldwell, 2006). These individuals may be diagnosed with chronic major depression, as well as schizophrenia, schizoid, or schizotypal personality disorder (Gilberstadt & Duker, 1965; Lenzenweger, 1991; Marks, Seeman, & Haller, 1974). Indeed, depressive symptoms among individuals with schizophrenia have been found to further increase suicidal risk than either disorder alone (Addington & Addington, 1992; Dassori, Mezzich, & Keshavan, 1990; Prasad, 1986; Radomsky, Haas, Mann, & Sweeney, 1999). The 2-4-8 codetype may be indicative of borderline personality disorder with depressive symptomatology. Research also suggests that comorbid borderline personality disorder and depression increases suicidal risk (Bolton, Pagura, Enns, Grant, & Sareen, 2010; Soloff, Lynch, Kelly, Malone, & Mann, 2000). Individuals with 2-7-4 codetypes may be reporting symptoms of dependent personality disorder with depressive symptoms, which also has been found to increase suicidal risk (Bolton, Belik, Enns, Cox, & Sareen, 2008).

Further research is necessary in identifying codetypes among military personnel with high suicide potential. For example, comorbid posttraumatic stress disorder (PTSD) and depression have been found to increase suicidal risk among military personnel (Lemaire & Graham, 2011). While MMPI-2 codetypes related to PTSD have been investigated

(Glenn et al., 2002), research is needed in identifying PTSD-related codetypes with high suicide potential.

CONCLUSION AND FUTURE DIRECTIONS

In conclusion, while numerous questions remain in adequately assessing suicidal risk among military personnel, the MMPI-2 is able to identify risk factors that may not be evident in a clinical interview. Especially when suicide is of concern and time is of the essence, the MMPI-2 can identify specific issues that the clinician can address when developing an effective therapeutic alliance and a social support plan. Packman, Marlitt, Bongar, and Pennuto (2004) outline general areas that should be assessed in determining suicidal potential that include: personal characteristics, dispositional factors, situational factors, and current symptoms. The MMPI-2 can provide pertinent information in all these areas. Generally, the MMPI-2 enables the clinician to more efficiently identify cognitions, moods, interpersonal styles, and treatment issues among clients. Moreover, the clinician may be able to diagnose complex and/or potentially higher risk mental health issues, such as comorbid depression and personality disorders. Future MMPI-2 research also may help identify suicide risk factors specific to members of the military, such as PTSD subtypes with higher suicide risk and military masculinity as a potential suicidal risk factor.

REFERENCES

Addington, D. E., & Addington, J. M. (1992). Attempted suicide and depression in schizophrenia. *Acta Psychiatrica Scandinavica*, 85, 288–291.

Ball, J. D., & Peake, T. H. (2006). Brief psychotherapy in the U.S. military: Principles and applications. In. C. H. Kennedy & E. A. Zilmer (Eds.), *Military psychology: Clinical and operational applications* (pp. 61–73). New York: Guilford Press.

Bolton, J. M., Belik, S. L., Enns, M. W., Cox, B. J., & Sareen, J. (2008). Exploring the correlates of suicide attempts among individuals with major depressive disorder: Findings from the National Epidemiologic Survey on Alcohol and Related Conditions. *Journal of Clinical Psychiatry*, 69, 1139–1149.

Bolton, J. M., Pagura, J., Enns, M. W., Grant, B., & Sareen, J. (2010). A population-based longitudinal study of risk factors for suicide attempts in major depressive disorder. *Journal of Psychiatric Research*, 44, 817–826.

Braswell, H., & Kushner, H. I. (2012). Suicide, social integration, and masculinity in the U.S. military. *Social Science & Medicine*, 74, 530–536.

Bryan, C. J., Cukrowicz, K. C., West, C. L., & Morrow, C. E. (2010). Combat experience and the acquired capability for suicide. *Journal of Clinical Psychology*, 66, 1044–1056.

Butcher, J. N., & Perry, J. N. (2008). *Personality assessment in treatment planning: Use of the MMPI-2 and BTPI*. New York: Oxford University Press.

Caldwell, A. B. (2001). What do the MMPI scales fundamentally measure? Some hypotheses. *Journal of Personality Assessment*, 76, 1–17.

Caldwell, A. B. (2006). *Adaptation to what? An alternative diagnostic paradigm*. Unpublished manuscript. Retrieved from http://www.caldwellreport.com/)

Caldwell, A. B. (2007). [MMPI-2 data research file for clinical patients]. Unpublished raw data.

Carson, R. C. (1969). Interpretive manual to the MMPI. In J. N. Butcher (Ed.), *MMPI: Research development and clinical applications* (pp. 279–296). New York: McGraw.

Cassidy, F., Murry, E., Forest, K., & Carroll, B. J. (1998). Signs and symptoms of mania in pure and mixed episodes. *Journal of Affective Disorders*, 50, 187–201.

Choquet, M., Facy, F., Davidson, F., & Philippe, A. (1983). Signification de la premiere tentative de suicide. *Social Psychiatry*, 18, 89–94.

Clopton, J. R. (1979). The MMPI and suicide. In C. S. Newmark (Ed.), *MMPI: Clinical and research trends* (pp. 149–166). New York: Praeger.

Clopton, J. R., & Baucom, D. H. (1979). MMPI ratings of suicide risk. *Journal of Personality Assessment*, 43, 293–296.

Clopton, J. R., & Jones, W. C. (1975). Use of the MMPI in the prediction of suicide. *Journal of Clinical Psychology*, 31, 52–54.

Clopton, J. R., Pallis, D. J., & Birtchnell, J. (1979). Minnesota Multiphasic Personality Inventory profile patterns of suicide attempters. *Journal of Consulting and Clinical Psychology*, 47, 135–139.

Clopton, J. R., Post, R. D., & Larde, J. (1983). Identification of suicide attempters by means of MMPI profiles. *Journal of Clinical Psychology*, 39, 868–871.

Cox, D. W., Ghahramanlou-Holloway, M., Greene, F. N., Bakalar, J. L., Schendel, C. L., Nademin, M. E., . . . Kindt, M. (2011). Suicide in the United States Air Force: Risk factors communicated before and at death. *Journal of Affective Disorders*, 133, 398–405.

Craig, R. J., & Olson, R. E. (1990). MMPI characteristics of drug abusers with and without histories of suicide attempts. *Journal of Personality Assessment*, 55, 717–728.

Dahlstrom, W. G., Welsh, G. S., & Dahlstrom, L. E. (1972). *An MMPI handbook*: Vol. I. *Clinical interpretation* (rev. ed.). Minneapolis: University of Minnesota Press.

Daigle, M. (2004). MMPI inmate profiles: Suicide completers, suicide attempters, and non- suicidal controls. *Behavioral Sciences and the Law*, 22, 833–842.

Dassori, A. M., Mezzich, J. E., & Keshavan, M. M. (1990). Suicidal indicators in schizophrenia. *Acta Psychiatrica Scandinavica*, 81, 409–413.

Eyman, J. R., & Eyman, S. K. (1991). Personality assessment in suicide prediction. *Suicide and Life-Threatening Behavior*, 21, 37–55.

Farberow, N. L. (1956). Personality patterns of suicidal mental hospital patients. In G. S. Welsh & W. G. Dahlstrom (Eds.), *Basic readings on the MMPI in psychology and medicine* (pp. 427–432). Minneapolis: University of Minnesota Press.

Farberow, N. L., & Devries, A. G. (1967). An item differentiation analysis of MMPIs of suicidal neuropsychiatric hospital patients. *Psychological Reports*, 20, 607–617.

Finn, S. E. (2009). Incorporating base rate information in daily clinical decision making. In J. N. Butcher (Ed.), *Oxford handbook of personality assessment* (pp. 140–149). New York: Oxford University Press.

Fischer, C. T. (2000). Collaborative, individualized assessment. *Journal of Personality Assessment*, 74, 2–14.

Friedman, A. F., Archer, R. P., & Handel, R. W. (2005). Minnesota Multiphasic Personality Inventories (MMPI/MMPI-2, MMPI-A) and suicide. In R. I. Yufit & D. Lester (Eds.), *Assessment, treatment, and prevention of suicidal behavior* (pp. 63–91). Hoboken, NJ: Wiley.

Gilberstadt, H., & Duker, J. (1965). *A handbook for clinical and actuarial MMPI interpretation*. Oxford: Saunders.

Glassmire, D. M., Stolberg, R. A., Greene, R. L., & Bongar, B. (2001). The utility of MMPI-2 suicide items for assessing suicidal potential: Development of a suicidal potential scale. *Assessment*, 8, 281–290.

Glenn, D. M., Beckham, J. C., Sampson, W. S., Feldman, M. E., Hertzberg, M. A., & Moore, S. D. (2002). MMP-2 profiles of Gulf and Vietnam combat veterans with chronic posttraumatic stress disorder. *Journal of Clinical Psychology*, 58, 371–381.

Goldberg, J. F., Garno, J. L., Portera, L., Leon, A. C., Kocsis, J. H., & Whiteside, J. E. (1999). Correlates of suicidal ideation in dysphoric mania. *Journal of Affective Disorders*, 56, 75–81.

Graham, J. R. (1987). *The MMPI: A practical guide* (2nd ed.). New York: Oxford University Press.

Greene, R. L. (2011). *MMPI-2/MMPI-2-RF: An interpretive manual* (3rd ed.). Boston: Allyn & Bacon.

Hoge, C. W., Castro, C. A., Messer, S. C., McGurk, D., Cotting, D. I., & Koffman, R. L. (2004). Combat duty in Iraq and Afghanistan, mental health problems and barriers to care. *The New England Journal of Medicine*, 351, 13–22.

Johnson, J. G., Cohen, P., Gould, M. S., Kasen, S., Brown, J., & Brook, J. S. (2002). Childhood adversities, interpersonal difficulties, and risk for suicide attempts during late adolescence and early adulthood. *Archives of General Psychiatry*, 59, 741–749.

Lemaire, C. M., & Graham, D. P. (2011). Factors associated with suicidal ideation in OEF/OIF veterans. *Journal of Affective Disorders*, 130, 231–238.

Lenzenweger, M. F. (1991). Confirming schizotypic personality configurations in hypothetically psychosis-prone university students. *Psychiatry Research*, 37, 81–96.

Leonard, C. V. (1977). The MMPI as a suicide predictor. *Journal of Consulting and Clinical Psychology*, 45, 367–377.

Lester, D. (1971). MMPI scores of old and young completed suicides. *Psychological Reports*, 28, 146.

Marks, P. A., Seeman, W., & Haller, D. L. (1974). *The actuarial use of the MMPI with adolescents and adults*. Baltimore: Williams & Wilkins.

Meyer, R. G. (1989). *The clinician's handbook: The psychopathology of adulthood and adolescence* (2nd ed.). Needham Heights, MA: Allyn & Bacon.

Packman, W. L., Marlitt, R. E., Bongar, B., & Pennuto, T. O. (2004). A comprehensive and concise assessment of suicide risk. *Behavioral Sciences & the Law*, 22, 667–680.

Pietrzak, R. H., Goldstein, M. B., Malley, J. C., Rivers, A. J., Johnson, D. C., & Southwick, S. M. (2010). Risk and protective factors associated with suicidal ideation in veterans of Operations Enduring Freedom and Iraqi Freedom. *Journal of Affective Disorders*, 123, 102–107.

Prasad, A. J. (1986). Attempted suicide in hospitalised schizophrenics. *Acta Psychiatrica Scandinavica*, 74, 41–42.

Radomsky, E. D., Haas, G. L., Mann, J. J., & Sweeney, J. A. (1999). Suicidal behavior in patients with schizophrenia and other psychotic disorders. *The American Journal of Psychiatry*, 156, 1590–1595.

Sendbuehler, J. M., Kincel, R. L., Nemeth, G., & Oertel, J. (1979). Dimension of seriousness in attempted suicide: Significance of the Mf scale in suicidal MMPI profiles. *Psychological Reports, 44*, 343–361.

Sepaher, I., Bongar, B., & Greene, R. L. (1999). Codetype base rates for the "I mean business" suicide items on the MMPI-2. *Journal of Clinical Psychology, 55*, 1167–1173.

Simon, W., & Gilberstadt, H. (1958). Analysis of the personality structure of 26 actual suicides. *The Journal of Nervous and Mental Disease, 127*, 555–557.

Skopp, N. A., Luxton, D. D., Bush, N., & Sirotin, A. (2011). Childhood adversity and suicidal ideation in a clinical military sample: Military unit cohesion and intimate relationships as protective factors. *Journal of Social and Clinical Psychology, 30*, 361–377.

Soloff, P. H., Lynch, K. G., Kelly, T. M., Malone, K. M., & Mann, J. J. (2000). Characteristics of suicide attempts of patients with major depressive episode and borderline personality disorder: A comparative study. *The American Journal of Psychiatry, 157*, 601–608.

Spirito, A., Faust, D., Myers, B. & Bechtel, D. (1988). Clinical utility of the MMPI in the evaluation of adolescent suicide attempters. *Journal of Personality Assessment, 52*, 204–211.

Suominen, K., Henriksson, M., Suokas, J., & Isometsä, E. (1996). Mental disorders and comorbidity in attempted suicide. *Acta Psychiatrica Scandinavica, 94*, 234–240.

Trout, D. L. (1980). The role of social isolation in suicide. *Suicide and Life-Threatening Behavior, 10*, 10–23.

Watson, C. G., Klett, W. G., Walters, C. & Vassar, P. (1984). Suicide and the MMPI: A cross-validation of predictors. *Journal of Clinical Psychology, 40*, 115–119.

11

Ethical Issues in the Treatment of Suicidal Military Personnel and Veterans

W. Brad Johnson

Gerald P. Koocher

INTRODUCTION

Each year, suicide attempts are associated with nearly 500,000 emergency room visits in the United States, and over 1 million people die by suicide worldwide (Bernert & Roberts, 2012). Suicide consistently ranks among the top causes of death among military service members (Jones, Kennedy, & Hourani, 2006). Among active duty military members serving in Operation Iraqi Freedom (OIF) and Operation Enduring Freedom (OEF), suicide rates have doubled since the beginning of the wars in Iraq and Afghanistan (U.S. Army, 2010).

Few events elicit as many powerful emotions in mental health professionals (MHPs) as a client's suicide attempt or completion (Barnett & Johnson, 2008). Stressful challenges associated with managing suicidal clients include (a) the MHP's inability to accurately predict low base rate behaviors (e.g., suicide attempts and completions), (b) decisions about inpatient admission, (c) powerful countertransference feelings (e.g., anxiety, helplessness, anger), and (d) the potential life-or-death implications of treatment (Jobes, Rudd, Overholser, & Joiner, 2008). In military settings, there are often additional concerns such as impact of a service member's suicidal ideation on the larger military mission, implications for the client's career, access to lethal means of suicide (e.g., firearms, explosives), and stigma associated with mental health care.

In this chapter we consider the ethical implications of addressing suicidal behavior in military service members and veterans. We briefly summarize key evidence bearing on suicide among service members and veterans and then consider the most prevalent feelings generated in the MHP when managing suicidal clients in mental health practice. Next, we consider the key ethical issues bearing on clinical treatment with suicidal clients and then describe the ethical challenges unique to working with suicidal clients in military settings. We conclude by summarizing the elements of competent practice when working with suicidal service members and veterans.

SUICIDAL BEHAVIOR IN MILITARY SERVICE MEMBERS AND VETERANS

For decades, military suicide rates approximated those in the general population (between 10 and 13 deaths per 100,000; Jones et al., 2006). But during the past decade, suicide rates in the U.S. military have steadily increased (Cox et al., 2011; Reger et al., 2015). In 2011, 301 U.S. service members died by suicide and 915 service members made at least one suicide attempt (Department of Defense [DOD], 2012). The suicide toll increased to 349 in 2012 (Briggs, 2013): one suicide every 25 hours. Multiyear comparisons of military suicide data reveal consistent dispositional/personal factors among these suicidal service members (DOD, 2012). Most were Caucasian, male, enlisted, and under 25 years old. In 2011 (DOD, 2012); service members most frequently used firearms to end their lives or died by hanging. Firearms were present in the home or immediate environment for more than half of these decedents. Mood disorders were reported for 20% and known substance abuse for 24% of service members who died by suicide. In contrast to civilian populations, most service members (74%) did not communicate their suicidal intent prior to dying by suicide. The fact that service members often fail to disclose suicidal urges or plans is of particular concern and may relate to adherence to social and military injunctions for masculine behavior and stigma associated with seeking mental health care (Burns & Mahalik, 2011; Langhinrichsen-Rohling, Snarr, Slep, Heyman, & Foran, 2011).

Research on suicide risk factors in the military reveal several consistent predictors. These include individual factors (e.g., depressive symptoms, alcohol problems, physical injury or illness, isolation, number of lifetime traumatic events, diminished coping efficacy, poor job performance, processing for military discharge, relational factors [e.g., conflict with significant others, break-up, divorce], workplace factors [e.g., dissatisfaction with military, conflict with superiors], and community factors [e.g., mental health resources, unit cohesion, perceived support]; Jones et al., 2006; Langhinrichsen-Rohling et al., 2011; Nelson et al., 2011).

Recent research has confirmed additional military-centric risk factors for suicide. These unique risk factors include the perception that one is a burden on others (e.g., one's comrades or military unit), the perception that one does not belong, fearlessness about death, and high pain tolerance (Bryan, Clemans, & Hernandez, 2012). Hopelessness, perceived burdensomeness, and thwarted belongingness were the factors most often communicated in the suicide notes of military members (Cox et al., 2011). Finally, one recent study revealed that disturbed sleep, often occasioned by unrelenting military operations, may confer suicide risk independent of depressed mood (Ribeiro et al., 2012). In a sample of active duty military members, self-reported insomnia symptoms were associated with suicidal ideation, even after accounting for symptoms of depression, hopelessness, post-traumatic stress disorder (PTSD) symptoms, and drug and alcohol abuse.

Among veterans, combat experience and resulting PTSD and mild traumatic brain injury appear to independently increase risk for suicide, above and beyond depression alone (Barnes, Walter, & Chard, 2012; Kleespies et al., 2011). A recent study of college student combat veterans ($N = 628$), revealed that a full 46% reported thinking about suicide (Rudd, Goulding, & Bryan, 2011). Ten percent of these veterans thought of suicide often and 7.7% had made a suicide attempt. PTSD was a significant suicide risk factor; 82% of those making a suicide attempt reported significant symptoms of PTSD. Unemployment rates as high as 10% among recent veterans (Plumer, 2013) add to the problem by eroding the ability to support oneself and one's family, even after surviving all manner of adversity in a war zone. Finally, strong social relationships and being married appear to protect veterans from suicidal behavior (Jakupak et al., 2010), but this protective effect is significantly diminished by PTSD.

THE MENTAL HEALTH PRACTITIONER AND THE SUICIDAL CLIENT

Client suicide is a genuine occupational hazard for most clinicians. A substantial portion of those who die by suicide in the United States each year—possibly half—are in active treatment at the time of their death (Rudd et al., 2009). Suicide is the most frequent of all mental health emergencies, with MHPs having better than a one in five chance of losing a client to suicide (Bongar & Sullivan, 2005).

Suicidal client behavior often elicits powerful feelings of anxiety and even anger in MHPs. A study of therapists' feelings about clients revealed that "the most widespread feeling was fear that a client would commit suicide (97.2% of sample), followed by fear that a client would get worse (90.0%)" (Pope & Tabachnick, 1993, p. 149). In this study, over half of therapists indicated having felt so afraid about a client that it affected the therapist's eating, sleeping, or concentration. Such feelings about a client's potential suicidality, when unacknowledged or inadequately addressed, may have both negative treatment outcomes and negative implications for the MHP's emotional and physical well-being (Pope & Tabachnick, 1993). In spite of helplessness, anger, or anxiety about suicidal clients, MHPs must be exceptionally careful not to become paralyzed or punitive in their work with suicidal clients (Barnett & Johnson, 2008). Rather, they will ideally use best practices and a coherent decision-making strategy to assess and reduce suicide risk, intervening when necessary to prevent client harm (Bernert & Roberts, 2012).

When MHPs have failed to adequately respond to client suicidality—perhaps resulting in a malpractice claim—it is generally the case that at least one of the following charges is leveled at the MHP: (a) failure to diagnose and safeguard, (b) failure to recognize a client's elevated suicide risk and not taking precautionary measures to protect the client, or (c) failure to use proper care and treatment (Jobes & Berman, 1993). Jobes and Berman reflect that the concepts of *foreseeability* and *reasonable care* are central to assessments of an MHP's competence, and perhaps liability, following a client suicide: "Foreseeability involves the reasonable and comprehensive assessment of risk.... Reasonable care involves the reliable and appropriate implementation of interventions or precautions based on the preceding assessment of risk" (p. 92). It is easy to empathize with the strong negative feelings associated with treating a suicidal client. Overestimating suicide risk may deprive clients of their rights (e.g., involuntary hospitalization), while underestimation of risk may jeopardize the safety of clients and increase liability for the MHP (Bernert & Roberts, 2012).

KEY ETHICAL ISSUES AND RISK MANAGEMENT STRATEGIES

Several ethical principles serve to guide an MHP's work with suicidal clients. Aspirational in nature, the principles focus on broad and fundamental professional norms and ideals (American Psychological Association [APA], 2010; Bernert & Roberts, 2012). As an MHP considers how best to respond to a suicidal person, the following principles should provide an ethical framework from which to reason. The principles are presented in no particular order; we consider each of them salient to assessment and treatment with suicidal clients.

- *Respect*—deep regard for the worth and dignity of all people
- *Beneficence and nonmaleficence*—striving to benefit those with whom we work while taking care to do no harm
- *Autonomy*—promoting the self-governance and independence of clients
- *Fidelity*—faithfulness to the interests of clients
- *Veracity*—adhering to truth and honesty in all communications with clients
- *Privacy*—protecting clients' personal information to the extent possible

Working in concert with these fundamental principles, MHPs must additionally attend to several specific ethical standards and risk management strategies when providing clinical services to a suicidal person.

Provide Clear and Ongoing Informed Consent

Informed consent to treatment requires the MHP to inform a client of all the procedures that they plan to use in the evaluation and management of suicide risk, clinical decision-making, and emergency assessment and referrals (Bernert & Roberts, 2012). Informed consent should be obtained using language that is reasonably understandable to the client (APA, 2010), and the process should be highly transparent and collaborative. Effective informed consent is a continuous and evolving process throughout treatment. In the case of the suicidal client, it might best be described as *working together to keep the client safe* (Bernert & Roberts, 2012). Describing the unique salience of informed consent with a suicidal client, Rudd and colleagues (2009) emphasize that information regarding the risk of suicide and suicide attempts "should be shared in an effort to help the patient and family understand the true nature of risk during the treatment process, recognizing that shared responsibility during treatment is essential to reduce the likelihood of a suicide attempt or death" (p. 462). The following elements are crucial to the informed consent process with suicidal clients (Bernert & Roberts, 2012; Jones, Kennedy, & Hourani, 2006; Rudd et al., 2009):

- Risk for death or injury resulting from suicide or suicide attempts
- The nature of assessment and treatment procedures likely to be employed
- Circumstances prompting an emergency referral or hospitalization
- Circumstances prompting disclosure of confidential information
- The high probability of emotional distress while working on difficult issues in therapy

- Specific procedures to follow in a crisis situation
- The expectation for effective self-management on the part of the client

A collaborative and transparent informed consent process supports the ethical principles of respect, beneficence, autonomy, fidelity, and veracity, among others. A persistent and open process of informed consent can strengthen the therapeutic alliance with a suicidal client.

Ensure Competence to Assess and Treat Suicidal Clients

MHPs have ethical and legal obligations to ensure their competence when providing professional services. Competence does not equate to perfection—most MHPs are competent in some areas and not in others—and competence with specific client types, disorders, and context always falls along a continuum from less to more (Barnett & Johnson, 2008). Competence to provide services to suicidal clients requires appropriate knowledge, skills, and attitudes with particular sophistication in assessment of suicide risk and clinical management of suicidal behavior during the course of treatment. By way of summary, each of the following factors is an evidence-supported risk factor for suicide (Barnett & Johnson, 2008; Hawton, et al, 2015; Jones et al., 2006; May, Overholser, Ridley & Raymond, 2015; Pope & Vasquez, 2007):

- *Direct verbal warning*
- *Sex*: Men are approximately four times more likely than women to complete suicide.
- *Previous attempts*
- *Passive suicidal ideation or history of self-harm*
- *Mood Disorders*: Major depression and bipolar disorder are strong predictors.
- *Age*: Older clients and adolescents present the greatest risk.
- *Substance use*: Alcohol abuse or dependence is particularly salient.
- *A suicide plan and means*
- *Bereavement*
- *Relationship loss or turmoil*: Breakups and divorce increase risk.
- *Stressful life events*: Job loss and sexual assault are good examples.
- *Suicide in family members or important friends*
- *Impulsivity*
- *Hopelessness*
- *Poor social support*

In addition to facility with assessment of suicide risk, MHPs must also be competent to implement interventions or precautions based on the rigorous and continuing assessment of risk as well as an assessment of the client's competency to participate in treatment and management decisions. According to Jobes and Berman (1993), "reasonable care [of the suicidal client] includes the development of a treatment plan; the consideration of medication, hospitalization, or both; the need for referral or consultation; decisions about the patient's self-control, affect regulation, and need for close observation; and the implementation of a recommended treatment" (p. 92). MHPs who are competent to treat suicidal clients do not rely on inadequate interventions, such as the "no-harm contract," but must instead be knowledgeable about evidence-based treatments and risk management strategies (Bernert & Roberts, 2012). Competence to practice in this area might include several specific steps and considerations such as those summarized in the following (Bernert & Roberts, 2012; Bongar & Sullivan, 2005; Pope & Vasquez, 2007; Rudd et al., 2009):

- Actively treat the client's comorbid psychiatric and physical disorders. Because 90% of suicide victims had a mental illness at the time of death (Bongar & Sullivan, 2005), it is imperative to provide effective psychological treatments, especially for mood disorders.
- Collaborate with clients to construct a supportive environment. Enlist the commitment and engagement of family, friends, community resources, and mental health care providers.
- Collaborate with clients and consultants on the need for hospitalization. Always consider the potential benefits and drawbacks before hospitalizing a client.
- Ensure continuity of care for a suicidal client and take steps to avoid real or perceived abandonment.
- Work with clients and their significant others to restrict access to potentially lethal means for suicide. Means restriction "entails the actual process of limiting or removing access to potentially lethal methods for suicide and self-harm (e.g., locking up medications, removing a firearm

from the home)" (Bryan, Stone & Rudd, 2011, p. 340).
- Document all aspects of a client's care including any consultation received along the way.
- Communicate care, concern, and, above all, hope to the client.

These are some of the elements of competent care for the suicidal client. Ethical guidelines require MHPs themselves to continually assess their own proficiency and emotional reactions when providing professional services to the suicidal client (e.g., APA, 2010). In addition to appropriate education, supervised experience, and consultation, MHPs should maintain competence in this area through some form of continuing education in suicidology (Bernert & Roberts, 2012).

Standard 2 of the APA's Code of Conduct address the issues of competence, but also include a provision for providing services in emergencies. Standard 2.02 reads in part: "when psychologists provide services to individuals for whom other mental health services are not available and for which psychologists have not obtained the necessary training, psychologists may provide such services in order to ensure that services are not denied. The services are discontinued as soon as the emergency has ended or appropriate services are available."

One would hope that all military MHPs will have training in suicide risk assessment and prevention, but some will certainly be more experienced than others, and special population differences may come up (e.g., the necessity to address the needs of a suicidal child or other civilian family member seeking services as a military dependent). In such instances, ethical standards allow the MHP to act in a professional capacity on an emergency basis. Ideally, the MHP will seek necessary consultation from more qualified peers at his or her first opportunity.

Seek Consultation

Ethical, legal, and clinical standards of excellence in work with suicidal clients will more likely be achieved when an MHP routinely seeks professional consultation from colleagues, particularly those with considerable experience and forensic expertise (Bongar, 2002; Bongar & Sullivan, 2005). Elements for consideration and discussion during consultation regarding a suicidal client might include the overall management of the case, specific concerns regarding assessment or treatment, managing chronic or acute suicidality, the MHP's own feelings about the suicidal client, and the advisability of utilizing medications or hospitalization (Bongar & Sullivan, 2005).

Seeking consultation is a hallmark of ethical and professional excellence (APA, 2010; Koocher, & Keith-Spiegel, 2016). In one Veterans Affairs (VA) facility, psychologists established a consultation service specifically for MHPs working with suicidal veterans (Gutierrez et al., 2009). Designed to improve care to suicidal clients while decreasing the risk of negative and potentially fatal outcomes, the consultation service not only supports VA MHPs but also extends invitations to suicidal clients themselves, encouraging them to collaborate with their MHP in the consultation process.

SUICIDAL CLIENTS IN THE MILITARY: SOME UNIQUE ETHICAL CHALLENGES

Contextual understanding of the patient and the degree of control available to MHPs go to the heart of suicide prevention. Particularly when providing services to active duty military personnel, MHPs may occasionally find that their simultaneous commitment to the suicidal service member and another entity, such as unit's commanding officer or even the larger military, may present conflicts or incongruities Termed *mixed-agency* ethical dilemmas, such conflicts normally involve a sense of divided loyalty between a client and an organization (Howe, 2003; Johnson, 2013; Kennedy & Johnson, 2009). Mixed-agency dilemmas are often part of the day-to-day experience of military MHPs. Yet these challenges are exacerbated when the country is at war because of the routine need to determine service members' fitness for return to duty in combat zones (Kennedy & Johnson, 2009).

Although each of the primary mental health organizations promulgates a code of ethics with clear relevance to mixed-agency challenges, we focus on the APA's guidance on the topic. The APA's Ethical Principals of Psychologists and Code of Ethics (APA, 2010) and standard 1.03 in particular, is instructive for MHPs struggling with mixed-agency questions in military settings:

> 1.03 Conflicts Between Ethics and Organizational Demands
>
> If the demands of an organization with which psychologists are affiliated or for whom they are working are in conflict with this Ethics Code, psychologists clarify the nature of the conflict, make known their commitment to the Ethics Code, and take reasonable steps to resolve the conflict consistent with the General Principles and Ethical Standards of the Ethics Code. Under no circumstances may this standard be used to justify or defend violating human rights. (APA, 2010, p. 4)

This standard places clear responsibility for addressing mixed-agency dilemmas squarely with the military MHP (Johnson, 2013; Kennedy & Johnson, 2009). Thus when a military psychologist concerned about a client's potential for self-harm feels pressure from the client's command to clear the client for an important mission, the MHP has a duty to carefully assess the client's functioning, current risk for suicidal behavior, and the potential risk to civilians and other members of the unit before rendering a formal recommendation. At times, the psychologist may have an obligation to raise formal objections or take other action should a recommendation be ignored or if pressured to engage in behavior that might violate other aspects of the Code of Ethics. As this discussion indicates, MHPs employed by the military may experience considerable role stress related to the range of professional obligations owed to the individual service member or veteran, members of the client's unit or military team, the DOD, and even society as a whole (e.g., taxpayers, citizens, and those likely to be affected by the preparedness and efficacy of the military; Koocher & Keith-Spiegel, 2016).

When an MHP is employed by the DOD—particularly when the MHP is also a service member—elements of military service may exacerbate mixed-agency dilemmas (Johnson, 2008, 2013). Some of these elements are a function of the MHP's dual identity as both military officer and health care provider and others merely reflect the realities of practice within the broader military culture. MHPs should consider how each might intensify ethical challenges in work with suicidal clients. The particular ethical standards that come into play include avoiding harm (Standard 3.04), confidentiality and privacy (Standard 4), informed consent (Standard 3.10), multiple role relationships (Standard 3.05), third-party requests for services (Standard 3.07), and use of assessments (Standard 9).

- *Dual identities*: Military MHPs may struggle with the simultaneous and sometimes competing identities of licensed provider and commissioned military officer. After taking the oath of office, military officers are obligated to promote the fighting power and combat readiness of military personnel; this includes holding subordinates accountable for military standards for order and discipline. For instance, a uniformed MHP may struggle to balance holding a client accountable for acting out behavior (e.g., suicide gestures aimed at gaining release from active duty) with empathy and concern about the possibility of genuine depression.
- A *superordinate mission*: Military officers—including health care providers—are indoctrinated early into a military culture that places the immediate mission (e.g., deploying to sea, winning a battle) first and above all else. At times, an MHP may struggle to convey the gravity of concerns about a depressed or impulsive service member in a milieu in which the interests of individual service members must be subordinate to the mission and personal problems may be considered superfluous. When an MHP is sufficiently concerned about the potential for self-harm in a service member, he or she may have to convey this concern in a mission-oriented framework likely to command the attention of superior officers.
- *Identifying the "client" may prove difficult*: Professional psychologists and other MHPs are taught to clearly identify their client at the outset so that they can establish their obligations to clients and ascertain the nature and limits of the professional relationship through a process of informed consent (APA, 2010; Johnson, 2013). Yet, in military settings, the MHP may not enjoy the luxury of serving primarily as an agent for the organization or for the individual client. Individual clients are often referred by a senior officer in their chain of command, often an officer senior in rank to the MHP. Thus the

military MHP may struggle with perceived—even conflicting—ethical obligations to multiple parties in many cases. In the case of a suicidal sailor or soldier, an MHP may struggle to balance the client's privacy and confidentiality with the clear obligation to inform the commanding officer regarding the client's level of impairment or ability to deploy. In the case of a recently suicidal Marine, an MHP may feel caught between the client's wish to remain in the military and engaged with his unit and an obligation to safeguard the health and well-being of the client's comrades who might be harmed or placed at risk if the client were to decompensate or act-out while deployed.

- *Unanticipated role shifts*: Sooner or later, an MHP will find him- or herself in a situation involving an unpredicted and sudden shift in roles with a military client. Most often, an MHP will be ordered to conduct a formal evaluation for fitness for duty, a security clearance, ability to deploy, selection for special assignment, or even capacity to stand trial, with a current or former client (Johnson, 2008, 2013). Such role shifts have the potential to harm an ongoing clinical relationship; they are most risky when sudden, entirely unanticipated, and beyond the control of the MHP. For instance, an MHP may establish a positive therapeutic alliance with a suicidal client and note that both depression and thoughts about suicide are declining. When suddenly ordered to complete a fitness to deploy and carry a firearm assessment with this client, the MHP may be ethically bound to recommend against the client's readiness for duty resulting in injury to the therapy alliance and a possible recurrence of symptoms. In many circumstances, a military MHP will be the only provider available to conduct the assessment.
- *"Avoiding harm" is not so easy*: Although MHPs hold an ethical obligation to avoid harming their clients and others with whom they work, and to minimize harm where foreseeable and unavoidable (APA, 2010), much has been written about the difficulty inherent in eliminating harm to military clients in combat environments (Camp, 1993; Howe, 2003; Koocher & Keith-Spiegel, 2016). For instance, how will the MHP approach the decision to return a suicidal client to combat when that client's skills are essential for achieving a high-stakes mission? And might a career-aspiring service member be harmed vocationally if declared nondeployable resulting in loss of flight status and possibly an inability to earn further promotions?
- *MHPs may be embedded with clients*: Embedded practice occurs when an MHP is deployed as part of a unit or force when the MHP is simultaneously a member of the unit and legally or otherwise bound to place the mission foremost (Johnson, 2008). It is quite likely that a contemporary military MHP will be deployed as a member of an aircraft carrier crew or as an officer in an Army brigade. This embedded status may facilitate more immediate prevention, assessment, and treatment of suicidality among unit members. On the downside, it may create considerable distress for the military MHP who can never "get away" from a suicidal client and the associated anxiety about the client's potential for acting out. For instance, psychologist stationed aboard an aircraft carrier will rarely go through a routine "man overboard" drill without experiencing genuine panic about which client may have just attempted suicide (Johnson, Ralph & Johnson, 2005).
- *Career implications for clients*: Military MHPs must remain sensitive to the profound power they wield over all aspects of a military client's life and career. When managing a suicidal service member, the military MHP may feel the weight of career implications if the client is placed on limited duty, deemed nondeployable, or restricted from access to weapons. Though each of these actions may be quite appropriate in the context of minimizing risk of harm and allowing the client access to treatment, each action may also place a service member's subsequent promotions in jeopardy.

An important component of all these situations is the informed consent process and contextual understanding. The active duty patient has a right to reasonable clarity at the start of the professional relationship. Just as the medical corpsman does not stop to discuss the risk of infection while field dressing an open wound, the MHP working with an acutely suicidal comrade—while embedded in a war zone—would not be held to the same standard of obtaining consent that we would expect of a practitioner in a private office. The more important focus is the prevention of harm in the first instance and attending to detailed niceties of the ethical code secondarily.

DEGREE OF CONTROL AND RESPONSIBILITY OBLIGATIONS

A final but very important ethical obligation of MHPs serving active duty and veteran military personnel involves responsibility flowing from control. Patients with chronic physical illness are at increased risk for suicide (Ballard et al., 2008), and one national study in the United Kingdom reported that 16% of suicide cases involved psychiatric inpatients, 9% of whom were on locked wards (Gordon, 2002). The military command structure has substantial control over the lives and activities of active duty uniformed personnel. The institution has even tighter control over their patients hospitalized at a military or VA facility. In addition, this population has substantial experience in weapon use, as well as access, as active duty personnel or in civilian life. Given the level of control and number of risk elements typically present, MHPs working in such settings have an obligation to consider suicide risk and routinely inquire of all patients, particularly those evidencing depression or agitation.

One particular challenge that has not been well studied or reported involves the need for MHPs to become assertive on behalf of their clients, when the "system" proves unresponsive or fails to adequately address client's needs. When a psychotic Marine is returned to service and sent to a war zone without his medication, nothing good will come of it (Kraft, 2011). When thousands of veterans must wait two weeks or more before receiving help for mental health problems (Zoroya & Hoyer, 2013) their feelings of depression and hopelessness will increase. When front-line psychologists observe such conditions and do nothing, they violate the core spirit of the Ethical Principles of Psychologists and behave instead as oblivious bystanders. Unfortunately, it is unlikely that those bystanders will be held accountable for their inaction.

CONCLUSION AND FUTURE DIRECTIONS

Effective treatment of suicidal military and veteran clients may pose unique clinical and contextual challenges for MHPs. In this concluding section, we offer several recommendations for MHPs—military or civilian—who might work with suicidal service members.

First, MHPs must understand their ethical and legal obligations. Practitioners must be familiar with both their professional ethics code and the legal statutes relevant to suicidal behavior, involuntary commitment, exceptions to confidentiality, and appropriate risk management (Jobes & Berman, 1993). Within military or VA facilities, MHP must be conversant with DOD or VA policies governing management and care of suicidal clients and how these statutes and policies dovetail or conflict with ethical obligations.

Second, the MHP must provide rigorous informed consent. With suicidal clients, solid and ongoing informed consent discussions and documentation can be used to structure treatment, openly discuss risk factors, and create shared understanding about treatment ground rules, limits, and boundaries (Jobes et al., 2008). In light of the numerous threats to confidentiality when working with the suicidal military client, informed consent can also be used to prepare the client for the fact that senior members of his or her chain of command will likely be apprised of both the client's treatment involvement and current suicide risk (Johnson, 2008).

Third, the MPH should conduct adequate assessment of risk and assure his or her own competence to provide effective care for the suicidal service member. Suicidology is an increasingly complex and nuanced area of clinical expertise (Jobes & Berman, 1993). An adequate assessment of suicide risk includes client history, relational aspects of suicide risk, cognitive risk factors (e.g., hopelessness), environmental risk factors, and more proximate warning signs (e.g., reckless behavior, dramatic mood changes, storing ammunition); these and other factors must be considered initially and periodically thereafter when caring for the suicidal client (Jobes et al., 2008). Remember that active duty military personnel and veterans will be primarily male and have had weapons training. Male gender and access to firearms are associated with greater risk for completed suicide.

Finally, the MHP should develop a consortium of collegial consultants with expertise in the assessment and treatment of suicidality in military populations. A consultant who remains up-to-date with empirically supported treatments for suicidality, as well as with a consultant with expertise in the area of military clinical psychology, could prove invaluable in making difficult triage and treatment decisions.

REFERENCES

American Psychological Association. (2010). *Ethical principles of psychologists and code of conduct.* Retrieved from http://www.apa.org/ethics/code/index.aspx

Ballard, E., Pao, M., Henderson, D., Lee, L., Bostwick, J.M., & Rosenstein, D. L. (2008). Suicide in the medical setting. *Joint Commission Journal on Quality and Patient Safety, 34*(8), 474–481.

Barnes, S. M., Walter, K. H., & Chard, K. M. (2012). Does a history of mild traumatic brain injury increase suicide risk in veterans with PTSD? *Rehabilitation Psychology, 57,* 18–26.

Barnett, J. E., & Johnson, W. B. (2008). *The ethics desk reference for psychologists.* Washington, DC: American Psychological Association.

Bernert, R. A., & Roberts, L. W. (2012). Ethical considerations in the assessment and management of suicide risk. *Focus: Journal of Lifelong Learning in Psychiatry, 10,* 467–472.

Bongar, B. (2002). *The suicidal patient: Clinical and legal standards of care* (2nd ed.). Washington, DC: American Psychological Association.

Bongar, B., & Sullivan, G. R. (2005). Treatment and management of the suicidal patient. In G. P. Koocher, J. C. Norcross, & S. S. Hill (Eds.), *Psychologist's desk reference* (2nd ed., pp. 240–245). New York: Oxford University Press.

Briggs, B. (2013, 14, January). Military suicide rate hit record high in 2012. *NBC News.* Retrieved from http://usnews.nbcnews.com/_news/2013/01/14/16510852-military-suicide-rate-hit-record-high-in-2012?lite

Bryan, C. J., Clemans, T. A., & Hernandez, A. M. (2012). Perceived burdensomeness, fearlessness of death, and suicidality among deployed military personnel. *Personality and Individual Differences, 52,* 374–379.

Bryan, C. J., Stone, S. L., & Rudd, M. D. (2011). A practical, evidence-based approach for means-restriction counseling with suicidal patients. *Professional Psychology: Research and Practice, 42,* 339–346.

Burns, S. M., & Mahalik, J. R., (2011). Suicide and dominant masculinity norms among current and former United States military servicemen. *Professional Psychology: Research and Practice, 42,* 347–353.

Camp, N. M. (1993). The Vietnam War and the ethics of combat psychiatry. *The American Journal of Psychiatry, 150,* 1000–1010.

Cox, D. W., Ghahramanlou-Holloway, M., Greene, F. N., Bakalar, J. L., Schendel, C. L., Nademin, M. E., . . . Kindt, M. (2011). Suicide in the United States Air Force: Risk factors communicated before and at death. *Journal of Affective Disorders, 133,* 398–405.

Department of Defense. (2012). *2011 Department of Defense suicide event report (DoDSER).* Retrieved from https://t2health.org/programs/dodser

Gordon, H. (2002). Suicide in secure psychiatric facilities. *Advances in Psychiatric Treatment, 8,* 408–417. doi:10.1192/apt.8.6.408.

Gutierrez, P. M., Brenner, L. A., Olson-Madden, J. H., Breshears, R. E., Homaifar, B. Y., Betthauser, L. M., . . . Adler, L. E. (2009). Consultation as a means of veteran suicide prevention. *Professional Psychology: Research and Practice, 40,* 586–592.

Hawton, K., Bergen, H., Cooper, J., Turnbull, P., Waters, K., Ness, J., & Kapur, N. (2015). Suicide following self-harm: Findings from the Multicentre Study of Self-Harm in England, 2000–2012. *Journal of Affective Disorders, 175,* 147–151. doi:http://dx.doi.org/10.1016/j.jad.2014.12.062

Howe, E. G. (2003). Mixed agency in military medicine: Ethical roles in conflict. In D. E. Lounsbury & R. F. Bellamy (Eds.), *Military medical ethics* (Vol. I, pp. 331–365). Falls Church, VA: Office of the Surgeon General, US Department of the Army.

Jakupak, M., Vannoy, S., Imel, Z, Cook, J. W., Fontana, A., & Rosenheck, R. (2010). Does PTSD moderate the relationship between social support and suicide risk in Iraq and Afghanistan War veterans seeking mental health treatment? *Depression and Anxiety, 27,* 1001–1005.

Jobes, D. A., & Berman, A. L. (1993). Suicide and malpractice liability: Assessing and revising policies, procedures, and practice in outpatient settings. *Professional Psychology: Research and Practice, 24,* 91–99.

Jobes, D. A., Rudd, M. D., Overholser, J. C., & Joiner Jr., T. E. (2008). Ethical and competent care of suicidal patients: Contemporary challenges, new developments, and considerations for clinical practice. *Professional Psychology: Research and Practice, 39,* 405–413.

Johnson, W. B. (2008). Top ethical challenges for military clinical psychologists. *Military Psychology, 20,* 49–62.

Johnson, W. B. (2013). Mixed-agency dilemmas in military psychology. In B. Moore & J. Barnett (Eds.). *The military psychologist's desk reference* (pp. 112–116). New York: Oxford University Press.

Johnson, W. B., Ralph, J., & Johnson, S. J. (2005). Managing multiple roles in embedded environments: The case of aircraft carrier psychology.

Professional Psychology: Research and Practice, 36, 73–81.

Jones, D. E., Kennedy, K. R., & Hourani, L. L. (2006). Suicide prevention in the military. In C. H. Kennedy & E. A. Zillmer (Eds.), *Military psychology: Clinical and operational applications* (pp. 130–162). New York: Guilford Press.

Kraft, H. K. (2011). Psychotic, homicidal and armed: The delicate balance between personal safety and effectiveness in a combat environment. In W. B. Johnson & G. P. Koocher (Eds.), *Ethical conundrums, quandaries and predicaments in mental health practice: A casebook from the files of experts* (pp. 189–196). New York: Oxford University Press.

Kennedy, C. H., & Johnson, W. B. (2009). Mixed agency in military psychology: Applying the American Psychological Association Ethics Code. *Professional Psychology: Research and Practice, 6*, 22–31.

Kleespies, P. M., AhnAllen, C. G., Knight, J. A., Presskreischer, B., Barrs, K. L., Boyd, B. L., & Dennis, J. P. (2011). A study of self-injurious and suicidal behavior in a veteran population. *Psychological Services, 8*, 236–250.

Koocher, G. P., & Keith-Spiegel, P. (2016). *Ethics in psychology: Professional standards and cases* (4th ed.). New York: Oxford University Press.

Langhinrichsen-Rohling, J., Snarr, J. D., Slep, A. M. S., Heyman, R. E., & Foran, H. M. (2011). Risk for suicidal ideation in the U. S. Air Force: An ecological perspective. *Journal of Consulting and Clinical Psychology, 79*, 600–612.

May, C. N., Overholser, J. C., Ridley, J. & Raymond, D. (2015) Passive suicidal ideation: A clinically relevant risk factor for suicide in treatment-seeking veterans. *Illness, Crisis, & Loss, 23*(3), 261–277. doi:http://dx.doi.org/10.1177/1054137315585422

Nelson, C., Cyr, K. S., Corbett, B., Hurley, E., Gifford, S., Elhai, J. D., & Richardson, J. D. (2011). Predictors of posttraumatic stress disorder, depression, and suicidal ideation among Canadian Forces personnel in a national Canadian military health survey. *Journal of Psychiatric Research, 45*, 1483–1488.

Plumer, B. (2013). The unemployment rate for recent veterans is incredibly high. *Wonkblog: The Washington Post*. Retrieved from http://www.washingtonpost.com/blogs/wonkblog/wp/2013/11/11/recent-veterans-are-still-experiencing-double-digit-unemployment/

Pope, K. S., & Tabachnick, B. G. (1993). Therapists' anger, hate, fear, and sexual feelings: National survey of therapist responses, client characteristics, critical events, formal complaints, and training. *Professional Psychology: Research and Practice, 24*, 142–152.

Pope, K. S., & Vasquez, M. J. T. (2007). *Ethics in psychotherapy and counseling* (3rd ed.). San Francisco, CA: Jossey-Bass.

Reger, M. A., Smolenski, D. J., Skopp, N. A., Metzger-Abamukang, M. J., Kang, H. K., Bullman, T. A., . . . Gahm, G. A. (2015). Risk of suicide among US military service members following Operation Enduring Freedom or Operation Iraqi Freedom deployment and separation from the US military. *JAMA Psychiatry, 72*(6), 561–569. doi:http://dx.doi.org/10.1001/jamapsychiatry.2014.3195

Ribeiro, J. D., Pease, J. L., Gutierrez, P. M., Silva, C., Bernert, R. A., Rudd, M. D., & Joiner, T. E. Jr. (2012). Sleep problems outperform depression and hopelessness as cross-sectional longitudinal predictors of suicidal ideation and behavior in young adults in the military. *Journal of Affective Disorders, 136*, 743–750.

Rudd, M. D., Joiner, T., Brown, G. K., Cukrowwicz, K., Jobes, D. A., Silverman, M., & Cordero, L., (2009). Informed consent with suicidal patients: Rethinking risks in (and out) of treatment. *Psychotherapy: Theory, Research, Practice, and Training, 46*, 459–468.

Rudd, M. D., Goulding, J., & Bryan, C. J. (2011). Student veterans: A national survey exploring psychological symptoms and suicide risk. *Professional Psychology: Research and Practice, 42*, 354–360.

US Army. (2010). *Health promotion, risk reduction, suicide prevention report*. Washington, DC: Author.

Zoroya, G., & Hoyer, M. (2013, November 4). Many veterans still wait weeks for mental health care. *USA Today*. Retrieved from http://www.usatoday.com/story/news/nation/2013/11/04/veterans-mental-health-treatment/3169763/

12

Evidence-Based Treatments for PTSD

Clinical Considerations for PTSD and Comorbid Suicidality

Afsoon Eftekhari
Sara J. Landes
Katherine C. Bailey
Hana J. Shin
Josef I. Ruzek

Posttraumatic stress disorder (PTSD) is a complex condition that can cause significant impairment to functioning. Prevalence rates have been estimated at 7% in the general population (Kessler, Chiu, Demler, & Walters, 2005), and 10% to 30% in military personnel (Kulka et al., 1990; Litz & Schlenger, 2009). Suicidality is a broad term that can refer to suicidal ideation (thoughts about suicide), having a plan to commit suicide, suicide attempts, and completed suicide. Suicidal ideation can be distinguished between active and passive suicidal ideation (for the purposes of this chapter, we focus on active suicide ideation). A nationally representative survey of individuals ages 15 to 54 years found that 13.5% reported lifetime ideation, 3.9% reported a plan, and 4.6% reported a suicide attempt (Kessler, Borges, & Walters, 1999). Among veterans and service members, Bossarte and colleagues (2012) used data from a national health survey and found that the prevalence of suicidal behaviors among veterans was similar to that of adults in the general population, although other reports have indicated increases in completed suicides among service members and veterans. One report of Army suicides found an increase in suicides from 9 per 100,000 in 2001 to 22 per 100,000 in 2009, a rate even higher than that among civilians after 2007 (Black, Gallaway, Bell, & Ritchie, 2011).

Suicide risk appears to be heightened with individuals who have posttraumatic stress disorder (PTSD). In the U.S. National Comorbidity Survey, Kessler and colleagues (1999) found that those with PTSD were six times more likely than matched controls to attempt suicide. In another epidemiological study, Davidson and colleagues (1991) found that those with PTSD were 8.2 times more likely to have attempted suicide than those without PTSD when controlling for depression. In a sample of civilians with chronic PTSD, Tarrier and Gregg (2004) found that over half of the sample (56.4%) reported some component of suicidality (e.g., ideation, having a plan, making a suicide attempt), and, compared to an epidemiological study, this proportion was significantly greater than that found in the general population. Clearly with the high prevalence rates of suicide, particularly among those with PTSD, it is important to better understand the relationship between PTSD and suicide, as well as to examine treatment of PTSD with comorbid suicidal ideation.

The *Diagnostic and Statistical Manual of Mental Disorders* (fifth edition [DSM-5]; American

Psychiatric Association, 2013) defines PTSD as a psychiatric disorder that develops after exposure to actual or threatened death, serious injury, or sexual violence. The person may have directly experienced the traumatic event, witnessed the event in person as it occurred to others, or learned about a violent or accidental event(s) that occurred to close family member(s) or friend(s). The event can also include repeated exposure to aversive details of trauma in the work context (e.g., first responders, police officers). This initial criterion is necessary in order to proceed with examination of the subsequent criteria when determining whether someone meets PTSD diagnosis. Once this initial criterion is met, an individual is considered to meet the PTSD criteria if he or she has symptoms in the following areas: (a) *intrusions*: this can occur through intrusive memories, nightmares, flashbacks, or psychological and/or physical reactivity to triggers; (b) *persistent avoidance*: this involves avoidance of trauma related thoughts, feelings, triggers, and situations; (c) *negative alterations*: this involves dissociative amnesia; persistent and exaggerated negative beliefs or expectations about oneself, others, or the world; persistent cognitive distortions about the cause or consequence of the event leading to feelings of blame toward oneself or others; persistent negative emotional state; marked loss of interest; feelings of detachment; or persistent inability to experience positive emotions; and (d) *marked alterations* in arousal and reactivity associated with the event, including sleep disturbance, irritability, difficulty concentrating, reckless or self-destructive behavior, exaggerated startle, and hypervigilance. To meet PTSD diagnosis, symptoms must cause impairment to functioning and be present for at least one month after exposure to the traumatic stressor.

PTSD commonly co-occurs with various other diagnoses, including depression, substance use, generalized anxiety disorder, panic, and obsessive-compulsive disorder. For the purposes of this chapter, we focus on the relationship between PTSD and depression, as this diagnosis is most associated with increased suicide risk. The National Comorbidity Survey–Revised reported a strong association between PTSD and major depressive disorder (MDD; $r = .5$; Kessler et al., 2005). About two-thirds of trauma survivors with PTSD also meet criteria for MDD (Yehuda, Halligan, Golier, Grossman, & Bierer, 2004). Although individually PTSD and MDD are unique risk factors for suicidal behavior, together they have been found to be associated with higher rates of suicidal behavior (Cougle, Resnick, & Kilpatrick, 2009; Panagioti, Gooding, Dunn, & Tarrier, 2011; Krysinska & Lester, 2010). In a national household probability sample of women, cross-sectional analyses indicated that lifetime comorbidity of MDD and PTSD was associated with higher prevalence of suicidal ideation than either alone (Cougle et al., 2009). This is consistent with a study of Operation Enduring Freedom/Operation Iraqi Freedom vetrerans (Jakupcak et al., 2009). However, rates of suicide attempts among those with comorbid MDD and PTSD were equivalent to those with PTSD alone (Cougle et al., 2009).

EVIDENCE-BASED TREATMENTS FOR PTSD

Fortunately there are many well established evidence-based psychosocial and pharmacological treatments for PTSD. While a brief review of psychotropic medications is presented, the focus of the chapter is on evidence-based psychosocial treatments for PTSD and management of co-occurring PTSD and suicidal ideation within the context of these treatments. Psychosocial treatments endorsed by several practice guidelines include prolonged exposure therapy (PE), cognitive-processing therapy (CPT), eye-movement desensitization and reprocessing (EMDR), and stress inoculation training (SIT; Benedek, Friedman, Zatzick, & Ursano, 2009; Management of Post-Traumatic Stress Working Group, 2010). Current U.S. Food and Drug Administration approved psychotropic medication with the strongest support include paroxetine, fluoxetine, and venlafaxine (Benedek et al., 2009; Hoskins et al., 2015; Management of Post-Traumatic Stress Working Group, 2010). Medications with evidence for some but less benefit than those listed previously are mirtazapine, GR205171 (a neurokinin-1 antagonist), amitriptyline, and phenelzine (Benedek, et al., 2009; Hoskins et al., 2015; Management of Post-Traumatic Stress Working Group, 2010).

Prolonged Exposure Therapy

Exposure-based therapies are among the most empirically validated treatments for PTSD (e.g., Foa, Rothbaum, Riggs, & Murdock, 1991; Foa et al., 1999; Foa et al.; 2005; Rothbaum, Astin, & Marstellar,

2005; Schnurr et al., 2007; Tarrier et al., 1999; Taylor et al., 2003). Exposure therapy is recommended as a first line treatment for PTSD in a number of U.S. and international practice guidelines including guidelines by the U.S. Department of Veterans Affairs (VA)/Department of Defense (DOD), American Psychiatric Association, UK National Institute for Health and Clinical Excellence, and the International Society for Traumatic Stress (Forbes et al., 2010). After conducting a large-scale review of randomized controlled trials for both the psychosocial and pharmacological treatment of PTSD, the Institute of Medicine (IOM) concluded that exposure therapy was the only treatment with sufficient data to support its efficacy for the treatment of PTSD (IOM, 2008). PE is one of the most used and best validated exposure therapy for PTSD. PE is a cognitive-behavioral exposure therapy designed to help the patient approach feared and avoided trauma-related stimuli. The primary goal is to help the patient process and work through the traumatic experience. The main components of PE include (a) imaginal exposure that involves systematic and repeated exposure to the trauma memory followed by processing of the exposure, (b) in vivo exposure that involves systematic and repeated engagement with nondangerous activities and situations that are avoided because of the trauma, (c) psychoeducation about treatment and common reactions to trauma, and (d) breathing retraining (Foa et al., 2007). The standard PE protocol involves ten 90-minute, individual treatment sessions. In most clinical settings, the number of sessions ranges on average from 8 to 15 to meet the clinical needs of individual patients.

Cognitive Processing Therapy

CPT (Resick & Schnicke, 1992) is a cognitive-behavioral treatment for PTSD that was originally developed to treat PTSD resulting from rape. CPT is based on an information processing theory of PTSD (Foa, Steketee, & Olasov-Rothbaum, 1989; Lang 1977). Information processing theory explains how information is encoded, stored, and recalled. CPT addresses symptoms of PTSD by dealing with the memories, thoughts, and beliefs involving the trauma. Treatment broadly includes cognitive restructuring, education about PTSD, and a written trauma account. Patients are educated about both PTSD symptoms and the theory and rationale behind the treatment. The cognitive work includes learning how to identify thoughts and feelings and how to challenge maladaptive core beliefs. The treatment manual includes modules on specific beliefs regarding safety, trust, power, esteem, and intimacy. In randomized clinical trials, CPT has been found to be effective and efficacious (e.g., Alvarez et al., 2011; Chard, 2005; Resick & Schnicke, 1992) and is recommended as a front-line treatment in various practice guidelines (IOM, 2007; Management of Post-Traumatic Stress Working Group, 2010). CPT-Cognitive only is an additional form of CPT in which the written trauma account is not included (Resick, Galovski, O'Brien, Scher, & Young-Xu, 2008). CPT and CPT-Cognitive only can be offered in multiple modalities, including individual, group, and concurrent group and individual delivery, all of which have been shown to have evidence of effectiveness in reducing the severity of PTSD symptoms.

Given the strong support for PE and CPT, both are being disseminated nationally within the Veterans Health Administration and in the DOD (Karlin et al., 2010).

Eye Movement Desensitization and Reprocessing

EMDR is also strongly recommended in several practice guidelines. As another form of trauma-focused intervention, the objective of EMDR is to assist patients to access and process traumatic memories while bringing them to an adaptive resolution (Shapiro, 2001). In EMDR, the processing of distressing memories is facilitated by adding eye movements or other methods of attention to other stimuli while accessing the traumatic memory. EMDR includes eight stages of treatment: patient history and treatment planning, preparation, assessment, desensitization and reprocessing, installation of positive cognition, body scan, closure, and reevaluation (Shapiro, 2001; Shapiro & Maxfield, 2002). Unique to EMDR are stages 4 to 6. During desensitization and reprocessing, the patient is asked to hold the distressing image in mind, along with associated negative cognitions and bodily sensations, and at the same time to visually track the therapist's fingers across the patient's complete field of vision in rhythmic sweeps of one full back and forth sweep per second. At the end of approximately 20 seconds, the patient is asked to let go of the memory, take a deep breath and describe any changes in image, sensations, thoughts, or emotions that might have occurred. In successive tracking activity, the patient concentrates on whatever changes or new associations have occurred. After the distressing images have been desensitized, the patient is instructed to hold a new positive/desired cognition in mind while tracking the therapist's fingers (i.e., to install a positive cognition). In the body scan, the patient identifies any continuing bodily tensions or discomfort and attends to these while tracking the therapist's fingers. The effectiveness of the EMDR intervention has been demonstrated in a number of outcome trials (e.g., Rothbaum et al., 2005; van der Kolk et al., 2007).

Stress Inoculation Training

SIT was originally developed for the treatment of anxiety disorders generally and then later modified for PTSD. Compared to the interventions described previously, it is less focused on detailed review of the traumatic experience. Instead, it emphasizes anxiety management and coping skills training. Patients are taught a range of skills, including relaxation training, breathing retraining, positive thinking and self-talk, assertiveness training (teaching the person how to express wishes, opinions, and emotions appropriately and without alienating others), and thought stopping (distraction techniques to overcome distressing thoughts by inwardly shouting "stop"; Foa et al., 1999). Two randomized controlled outcome trials with PTSD patients have shown SIT to be effective with women who have survived sexual assault (Foa et al., 1991; Foa et al., 1999).

Pharmacotherapy

The VA/DOD Clinical Practice Guidelines for management of PTSD offer recommendations for pharmacological treatment of PTSD symptoms (Management of Post-Traumatic Stress Working Group, 2010). The most strongly recommended class of medications for the treatment of PTSD are selective serotonin reuptake inhibitors (SSRIs), though that is due in part to lack of sufficient data regarding meta-analysis of other drug classes (Hoskins et al., 2015). One specific norepinephrine reuptake inhibitor (SNRIs), venlafaxine, has also shown superiority over placebo in reduction of PTSD severity symptoms (Benedek, et al., 2009; Hoskins et al., 2015; Management of Post-Traumatic Stress Working Group, 2010). Among SSRIs, fluoxetine and paroxetine have the strongest support (Benedek, et al., 2009; Hoskins et al., 2015; Management of Post-Traumatic Stress Working Group, 2010). Though older guidelines initially suggested that sertraline provided significant benefit, the most recent meta-analysis on pharmacotherapy for PTSD including 21 studies comparing SSRIs and SNRIs versus placebo does not demonstrate superiority of sertraline compared to placebo (Hoskins et al., 2015). Second-line monotherapy recommendations are mirtazapine, GR205171 (a neurokinin-1 antagonist), amitriptyline, and phenelzine (Benedek, et al., 2009; Hoskins et al., 2015; Management of Post-Traumatic Stress Working Group, 2010). Several other drugs have either shown no benefit or are contraindicated in the treatment of PTSD (i.e., anticonvulsants; atypical antipsychotics, especially risperidone; and benzodiazepines). Evidence does not support the use of anticonvulsants (specifically, tiagabine, valproate, topiramate) for the management of PTSD core symptoms. Atypical antipsychotics are also not effective as monotherapy. Risperidone is specifically not recommended because it did not reduce PTSD symptoms and was associated with negative adverse effects including weight gain, fatigue, somnolence, and hypersalivation (Krystal et al., 2011). Finally, benzodiazepines are contraindicated and even considered harmful, particularly with regard to increasing risk of aggressive behavior (Management of Post-Traumatic Stress Working Group, 2010). In regard to PTSD with suicidal ideation specifically, there are no known studies or empirical data regarding a specific pharmacological approach. There is a very small naturalistic study ($N = 14$) of depressed patients with suicidal ideation who received intravenous ketamine, which decreased suicidal ideation for up to 10 days (Larkin & Beautrais, 2011). There is also a study demonstrating that ketamine reduced suicidal ideation compared to placebo among 133 patients with treatment-resistant depression (Ballard et al., 2014). However, there are no known studies of medication to prevent suicide. Nevertheless, pharmacological interventions used to treat PTSD symptoms (i.e., sleep disturbance, depression, nightmares) may be indirectly beneficial in the treatment of suicide. Prescribing providers should consult the VA/DOD Clinical Practice Guidelines for management of PTSD for information on specific studies.

TREATMENT OF PTSD WITH COMORBID SUICIDALITY

Clearly, there are many excellent options for the treatment of PTSD. Co-occurring suicidality does not prohibit proceeding with these treatments, but it does involve additional considerations, including those related to assessment of PTSD, suicidal ideation, psychosocial history, and current psychosocial support. It is also important to consider the role of stressors such as active deployment, fear of stigma, and psychiatric history in treatment planning. The following is a more in-depth review of how to proceed with assessment.

Assessment of PTSD

Good PTSD assessment involves both diagnostic interview and ongoing assessment via self-report. Commonly used structured interviews include the Structured Interview for the DSM-IV Axis I Disorders (First, Spitzer, Gibbon, & Williams, 2002), Clinician Administered PTSD Scale (CAPS-5; Weathers, Blake, Schnurr, Marx, & Keane, 2013), and the PTSD Symptom Scale–Interview (Foa, Riggs, Dancu, & Rothbaum, 1993). Commonly used self-report measures for adults include the PTSD Checklist for DSM-5 (PCL-5; Weathers et al., 2013), the Posttraumatic Diagnostic Scale (Foa, Cashman, Jaycox, & Perry, 1997), the Trauma Symptom Checklist–40 (Eliot & Briere, 1992), the Mississippi Scale for Combat-related PTSD (Keane, Caddell, & Taylor, 1988), and the Los Angeles Symptom Checklist (King, King, Leskin, & Foy, 1995). A more comprehensive list of self-report measures for adult PTSD assessment can be found at the website for the National Center for PTSD.

In addition to the assessment of the diagnosis of PTSD, it is important to clinically assess the severity of particular symptoms of PTSD, such as reexperiencing and hyperarousal symptoms found in sleep disturbances and nightmares, that may affect level of suicidal ideation, impulsivity, or judgment. For example, prior research has found that greater sleep disturbances and nightmares were associated with greater suicidal ideation among trauma survivors with PTSD and depression (Krakow et al., 2000), and nightmares have been found to be independently associated with suicidal ideation after controlling for level of depression and gender (Bernert, Joiner, Cukrowicz, Schmidt, & Krakow, 2005). Ongoing assessment of the severity of hyperarousal symptoms, particularly regarding anger and irritability, also will provide important information regarding level of risk for aggressive behavior enacted on oneself, others, or property, which has been found to be significant among veterans with PTSD (Jakupcak et al., 2009; Jakupcak et al., 2007). It is yet unknown if particular PTSD-related avoidance symptoms are associated with greater risk for suicidality, although avoidant coping behaviors, such as substance abuse, should clearly be included as part of a thorough clinical assessment.

Assessment of Suicidality

At a minimum, a good suicide assessment involves a thorough clinical interview assessing history of suicide attempts (as past attempts are the best predictor of future suicide) and current suicidal ideation. Also, it is important to identify suicide risk factors that are present and those that are absent, as well as protective factors. The following is a more detailed discussion of these points.

Suicide screening typically begins at initial intake when the patient's history is assessed. Questions regarding suicide history include asking whether the patient has ever attempted suicide or deliberately harmed him- or herself in any way. Some patients may not know what is considered deliberate self-harm; examples can include cutting, scratching, burning, head banging, jumping from high places, or other behaviors done to cause harm to self. The Lifetime–Suicide Attempt Self-Injury Count (Linehan & Comtois, 1996) interview can be used to obtain more detailed information about deliberate self-harm; another version, the Suicide Attempt Self-Injury Interview (Linehan, Comtois, Brown, Heard, & Wagner, 2006) targets specific time periods.

The Patient Health Questionnaire (PHQ-9; Kroenke, Spitzer, & Williams, 2001) can be used to screen for current suicidality. The PHQ-9 includes a suicide risk item: "Thoughts you would be better off dead, or of hurting yourself in some way." Patients can also be asked directly if they are having thoughts about killing themselves. Some providers are concerned that asking about suicidal ideation will make a patient suicidal or will increase risk of suicidal thinking; this is not the case. In terms of manner, it is important to convey openness to hearing about suicidal thoughts or urges so the patient knows it is appropriate to discuss them. For example, it is better to say, "If you are ever suicidal in the future, please tell me; this is something I would want to know" versus "You aren't suicidal are you?" Creating open discussion is essential for developing an open rapport that leads to more accurate assessment before and after treatment.

If there is evidence of suicidal thoughts or actions, it is important to assess for the presence and absence of risk factors. To help remember suicide risk factors, the SAD PERSONS acronym can be used: **S**ex (male); **A**ge less than 19 or greater than 45 years; **D**epression (patient admits to depression or decreased concentration, sleep, appetite, and/or libido); **P**revious suicide attempt or psychiatric care; **E**xcessive alcohol or drug use; **R**ational thinking loss: psychosis, organic brain syndrome; **S**eparated, divorced, or widowed; **O**rganized plan or serious attempt; **N**o social support; and **S**ickness, chronic disease. The Suicide Status Form (SSF), described later, is a useful tool to help remember these risk factors and assess them systematically.

Assessment of protective factors is also important, as doing so can help establish risk and suggest ways of increasing safety. Protective factors to assess include having hope for the future, attachment to life, responsibility to others (e.g., having family who depend on them, pets to feed), attachment to therapy or therapy provider, a protective social network or family (e.g., having others in their social network who care about them), fear of suicide, fear of social disapproval for suicide, spirituality or belief that suicide is immoral, a commitment to live, and willingness to follow a crisis plan. Discussion of protective factors can also function to highlight to the patient that he or she has positives in his or her life. For example, discussion of family or friends who may be impacted by the death of the patient may remind him or her that others do care, which may result in an increase in willingness to follow a crisis plan and not commit suicide. Discussion of social supports may also help the provider identify possible others who can serve as a support in a time of crisis or who should be included in a patient's crisis plan.

In addition to clinical interview, there are many standard tools for assessing suicide risk. It is important that the instrument include a standardized list of risk factors, those that are present as well as absent. One tool that can be used for assessing suicide risk is the SSF, developed for use in a treatment that directly addresses suicidality, the Collaborative Assessment and Management of Suicidality (CAMS; Jobes, 2006). The SSF helps the provider to assess the following: presence of a suicide plan (including access to means such as guns or pills), any preparation for committing suicide (e.g., saying goodbye to family), rehearsal of suicide (e.g., practicing shooting a gun), history of suicidality, current suicide intent, impulsiveness, substance abuse, significant loss, interpersonal isolation, relationship problems, health problems, and legal problems.

Conceptualization

After a thorough suicide risk assessment is completed and the person's level of risk (i.e., low, moderate, or high) is identified, treatment planning for suicidality and PTSD should be integrated. Treatment guidelines for PTSD recommend that imminent suicidality be treated before initiating treatment for PTSD (Foa, Keane, Friedman, & Cohen, 2009). Further, clinicians have the ethical and legal responsibility to hospitalize veterans at imminent risk for suicide. Though there are few effective suicide prevention interventions, some CBT-based interventions, including Dialectical Behavior Therapy (DBT), problem-solving therapy, and cognitive therapy, have been shown to reduce suicide-related behaviors (Tarrier et al., 2008). Motivational interviewing (MI) appears to increase treatment engagement; however, it is unclear whether MI improves treatment engagement for those considering suicide. Only theoretical work suggests that MI and self-determination theory may be helpful for engaging suicidal patients in treatment (Britton, Patrick, Wenzel, & Williams, 2010). Given the lack of clear evidence for motivating suicidal patients to engage in treatment for suicidality and PTSD, clinicians may try MI to resolve ambivalence about living, with the intention of reducing motivation to die.

A safety plan should also be developed to decrease impulsivity, if and when suicidal ideation increases. A written safety plan negotiated between a clinician and a patient is a collaborative document that the patient can have with him or her when suicidal ideation emerges and to which the clinician and patient can refer and update throughout the course of treatment. A comprehensive step-by-step plan should identify the following: (a) the individual's internal warning signs for suicide risk (e.g., specific thoughts, feelings, or sensations that emerge related to suicide), (b) immediate coping activities to self-soothe (e.g., specific activities or places that may distract or calm oneself), (c) multiple supportive people to contact for support or distraction (e.g., family, friends), (d) professional care providers to contact for support and/or intervention (e.g., clinicians, clergy), (e) crisis contact information (e.g., emergency services, crisis hotlines), and (f) specific plans to intervene in a veteran's living context to promote safety, such as removing means to carry out a suicide plan from a veteran's environment (i.e., disposing of medication that may be abused, removing a firearm from the house, temporarily giving vehicle keys to a safe person). A safety plan should also address any foreseeable barriers that may hinder his or her likelihood of carrying out each step of the safety plan. Clinicians and patients should have ongoing frank and open discussions about the conditions in which a welfare check by law enforcement or voluntary/involuntary hospitalization will be enacted. Further guidelines and examples can be found in the VA/DOD Clinical Practice Guideline for Assessment and Management of Patients at Risk for Suicide (Assessment and Management of Risk for Suicide Working Group, 2013).

Both during and after the development of a crisis plan, clinicians should be actively engaging veterans with suicidality and PTSD in treatment and minimizing barriers to treatment. Though factors that increase engagement and decrease barriers may be unique for individual veterans, some common issues may arise. First, increasing hope is an important principle for engaging suicidal veterans in evidence-based treatment for PTSD. Providers should express confidence in treatment, particularly in using CBT-based approaches that have demonstrated effectiveness in reducing suicidality (Jacupcak & Varra, 2011). Veterans with PTSD and suicidality may experience shame and a sense of stigma related to seeking treatment. Education should be provided that PTSD is a risk factor for suicidality and related thoughts and behaviors are not uncommon and can benefit from treatment. Emphasizing skills-based aspects of treatment may help these veterans to focus on benefits of treatment rather than their experience of vulnerability. Further, it may be useful to redefine ideas about "strength" to include willingness to seek help when necessary rather than avoiding treatment altogether.

Finally, establishing a collaborative plan with the veteran, and, if relevant, with family members, and other treatment providers may also promote engagement in treatment. Limits of confidentiality should be discussed directly, informing the veteran that acute risks for safety may result in breaches of confidentiality. The veteran's participation in treatment planning is essential. If possible, treatment planning may be enhanced by family involvement and by including other relevant treatment providers. (See other chapters in this volume for specific recommendations on treatment of active/imminent suicidality.)

Once veterans have some motivation to live and participate in treatment, evidence-based PTSD treatments can be used. Though about half of studies (46%) included in a meta-analysis of psychotherapy for PTSD excluded potential participants for suicide risk (Bradley, Greene, Russ, Dutra, & Westen, 2005), the other half did not exclude participants with suicidality. This suggests that those studies likely included veterans with some degree of suicidality, and PTSD treatment was still deemed effective. Though no evidence-based guideline exists for treatment planning for those with PTSD and suicidality (outside of imminent suicide), the following suggestions are extrapolated from known and emerging research previously described in this chapter. If suicide risk seems high (i.e., current moderate to severe suicidal ideation with a plan and intent to commit suicide), there is one small, preliminary evaluation of a protocol using DBT and PE for PTSD with promising results to reduce symptoms of PTSD without exacerbation of suicidal behavior (Harned, Korslund, Foa, & Linehan, 2012). For those with moderate or minimal suicidality, evidence-based psychotherapy for PTSD such as PE, CPT, or psychotropic medication may be used, but tailoring treatment to the individual's circumstances is likely to improve outcomes. The following suggestions are one possible algorithm for tailoring treatment.

If a veteran is deemed to have high risk for suicide, evidence-based treatments should be postponed and suicidality addressed and treated first. Once stabilization is reached, then further assessment is needed to determine whether to begin PTSD care. If a veteran has a low to moderate risk for suicide (i.e., suicidal ideation with a plan but no intent) plus additional risk factors (e.g., substance abuse, depressive disorder, recent housing or legal problems), treatment for PTSD will likely be enhanced if suicidality and risk factors are first addressed. For example, the patient may benefit from medication or therapy to reduce depressive symptoms or from resolving housing or legal problems prior to proceeding with PTSD treatment. Furthermore, addressing risk factors may decrease general distress and subsequently result in decreased suicidal ideation. If a veteran has moderate risk for suicide but few additional risk factors that can be changed, then evidence-based treatment for PTSD could be implemented with frequent monitoring of suicidality and risk factors and a plan to interrupt treatment for PTSD to focus solely on suicidality if necessary (i.e., using a crisis plan, tools from DBT, cognitive therapy, or problem-solving therapy). If a veteran has low risk for suicide (i.e., suicidal ideation but no plan or intent) with limited risk factors for suicide, then evidence-based treatment for PTSD may be implemented with monitoring for suicidality if his or her risk profile changes.

Treatment of PTSD with Comorbid Suicidality

Once a thorough suicide assessment has been conducted and suicidality addressed, providers can determine if it is appropriate to proceed with PTSD care. If suicide risk is deemed low and manageable, it is appropriate and important to proceed with addressing comorbid psychiatric problems such as PTSD as this may further reduce suicide risk. In general, once determined that it is acceptable to proceed, all treatments are implemented as usual with special care and monitoring of suicidal ideation. There is no set standard or protocol on how to adapt PE, CPT, EMDR, or SIT for comorbid suicidal ideation. Care is given on a case-by-case basis taking the individual patient into consideration. Typically, treatment would proceed as usual with weekly assessment of suicidal ideation. Furthermore, it may be important to provide additional support within and between sessions once trauma processing begins. If the patient does express increased distress or suicidal ideation during the course of treatment, procedures may be titrated to ensure treatment is proceeding in a manner the patient can tolerate. For example, in PE, there may be more frequent contact between sessions once imaginal exposure begins. Similar approaches are taken in CPT, EMDR, and SIT. All treatments require good clinical judgment. Providers should continue with treatment as usual while continually monitoring safety.

Another positive aspect of evidence-based psychosocial treatments for PTSD is that they promote re-engagement with one's life. Hopefully as patients' symptoms improve, they will be more likely to engage with significant individuals in their lives. Given the importance of positive psychosocial relations as a protective factor for suicide risk, treatments that help to improve relationships can be highly beneficial for patients who are ready to engage with treatment (Jakupcak & Varra, 2011). If necessary, one can also consider adjunctive treatments, such as couple's therapy, to help ensure a strong support system during the course of treatment. Furthermore, sessions with significant others can promote positive involvement of family and friends in the care and life of the veteran. Also, given the important role of re-experiencing symptoms in suicide risk (Bell & Nye, 2007), trauma-focused interventions may be particularly beneficial for comorbid PTSD and suicidal ideation.

In proceeding with care, it is also important to establish strong rapport. Strong therapeutic rapport is a critical part of all evidence based treatments for PTSD as well as for a good suicide assessment. This rapport may be particularly important given that a core feature of PTSD is avoidance. Clinicians should ensure the patient is acknowledging all emotions and possible thoughts of harming themselves. Trauma work may remind veterans of shame, guilt, or powerlessness they felt at the time of the trauma, which could increase suicidality in veterans at risk. In some cases, suicidal ideation may serve as a method of avoiding trauma-focused interventions, though it is unlikely to be a deliberate strategy for the veteran. In addition to ongoing assessment, it is also important to provide education about how treatment is likely to affect their symptoms and give the patient a sense of control as treatment progresses. It may also be useful to explicitly discuss how suicidality may wax and wane throughout treatment to help patients understand the relationship between their mood and suicidal thoughts. Further, for patient with some suicidal risk, it is important to have a crisis plan in place and the flexibility to adapt treatment to address acute suicidality (refer to the crisis plan previously discussed). To instill hope, patients should be educated that suicidality is likely to decrease during the course of successful treatment.

If during treatment the veteran reports increased suicidal ideation, then assessment of severity should be made. In high-risk situations, treatment can be stopped until further stabilization is established (e.g., stop PTSD treatment and switch to CAMS or skills building until stabilization is reached). In most cases, some titration and extra support is sufficient to continue. If it is necessary to discontinue evidence-based treatment for PTSD completely, then the patient can be re-evaluated at a later point in time. The critical component is ongoing assessment of suicidality, PTSD, and supportive factors.

The following case examples act as guides for the clinical process of treating PTSD with comorbid suicidality using the most empirically validated treatments for PTSD.

Case Vignette 1: Female MST Survivor

Janice is a 60-year-old, Caucasian Air Force Vietnam-era veteran. She served as a nurse while stationed on a number of different bases stateside. She is a divorced mother of two adult children. She reports that she generally enjoyed her military experience and the sense of community that came along with it. However, along with her positive experiences, she also experienced military sexual trauma (MST). She states that at the time of her service, such experiences were not talked about or reported. She therefore had nowhere to turn when she was raped by one of her peers. She reports that she believes she has been struggling with PTSD for years but was afraid to talk about the incident. Most recently, her symptoms have been exacerbated after entering into a romantic relationship. She reports that while she cares about the man she is dating, she finds the intimacy and connection with others triggering. She reports having intrusive thoughts about her experience and feeling depressed when she does so. She is also experiencing nightmares about the event, having a hard time leaving the house to go places like the fitness center, the VA, or other places where there may be groups of men. Public places in general are difficult for her. She has also been feeling emotionally cut off and having a hard time connecting with others. She reports sleeping more than usual to escape negative feelings and the memories of her trauma. Janice further reports that she has been feeling more jumpy and on edge. Her kids have also noted an increase in her irritability. She says that she cries quite a bit and is afraid of losing her current relationship but also cannot tolerate the physical intimacy her partner seeks. Janice feels guilty toward her kids and her partner as well as shame over her PTSD symptoms and what happened to her. She reports that recently she has been feeling like a failure and having thoughts about ending her life. She says she began making specific plans one night when she could not sleep.

In applying the recommendations described earlier in this chapter, thorough assessment and conceptualization should be done prior to discussing primary and ancillary treatment options for this veteran. Although she identified the MST as the primary traumatic stressor, it is unknown whether she experienced any other traumas prior to or since the MST that may affect her current level of PTSD symptomatology, particularly if she experienced other incidents of sexual assault in her lifetime. The CAPS-5 and Life Events Checklist would be appropriate measures to obtain a trauma history and assess both her current and past levels of symptomatology and functioning since the MST. Periodic self-report assessment via the PCL-5 and questions regarding her suicidal ideation would be appropriate throughout the course of treatment. Because she recently identified a specific plan for suicide during a night when she was unable to sleep, the SSF is another tool that can be used to assess specific questions about this plan, including the means she would use, if she has a timeline to potentially carry it out, and any preparatory behaviors to carry out a plan (e.g., saying goodbye to family, giving away meaningful personal belongings or financial resources, rehearsal of suicide). Other risk factors regarding history of depression (including any past perinatal or postpartum depression), history of suicidality, impulsivity, substance abuse, significant issues of loss, health changes, or legal/financial issues also should be assessed. A safety plan is warranted during the initial session(s), and a referral for a medication evaluation as soon as possible would be recommended if she is not already receiving pharmacotherapy.

It also is important to assess her current level of depressive symptomatology (i.e., hypersomnia, poor mood, feelings of guilt and shame) and influential risk factors (cf. SADPERSONS mnemonic, including any gender-specific and age-related issues that can exacerbate depression symptoms (e.g., peri- or postmenopausal effect of hormonal changes on mood), and whether a referral to her medical providers would be warranted to rule out medical problems. To build conceptualization of her clinical presentation, functional analysis of the factors that are increasing or maintaining her suicidality also may clarify if her suicidal ideation is consistent with escapist avoidance of her PTSD-related anxiety or more consistent with depressive hopelessness. Janice also has strong protective factors, including the support of her adult children and her romantic partner. Her case would be conceptualized as low to moderate risk for suicide, depending on the specificity, lethality, and level of preparation of her suicide plan. Her presentation for treatment is a good sign of hope for prognosis if she is able to engage in evidence-based trauma-focused treatment.

During the assessment and initial treatment phase with Janice, emphasis should be on establishing a safe and strong therapeutic rapport, enhancing motivation for treatment, and instilling hope for herself and her relational functioning, particularly as she provides a brief disclosure of her trauma history with the clinician via the CAPS-5. Open, candid discussion about treatment options and the importance of monitoring her suicidal ideation should be discussed as a part of informed consent both during the initial phase and throughout the course of treatment. This open discussion may include contingency plans between provider/treatment team and patient if she increases in specificity or lethality of her suicidal plan/intent and decreases in engagement in treatment (e.g., hospitalization may be required if the plan is imminent and specific and/or a welfare check by law enforcement may be enacted if these plans are coupled with no-shows to treatment appointments). Obtaining any necessary releases of information to collaborate and consult with family members or significant others may assist with safety planning.

After the initial safety planning, the clinician and Janice can collaboratively discuss treatment options and her thoughts and feelings about each of the options, such as individual trauma work (e.g., PE, CPT, SIT, EMDR), trauma-focused group psychotherapy (i.e., CPT group), or couples therapy that may also assist with her lack of social and emotional connectedness, (e.g., Cognitive Behavioral Conjoint Therapy for PTSD; Monson & Fredman, 2012).

Some patients with PTSD may have preferences for one type of treatment over another, while others may defer to their providers and ask for "whatever works." Though clinicians may have their own preferences, it is important to balance patient preferences (e.g., individual versus group, CPT versus PE), patient characteristics (e.g., low literacy, low cognitive abstraction), and provider clinical judgment (Raza & Holohan, 2015). A significant assessment component to consider in providing trauma-focused treatment to a patient with comorbid suicidal ideation includes possible influence of diverse cultural/religious worldviews and values related to sexual assault, mental health treatment, and suicide. It is important to query how she, her family, and possibly her faith community regard these sensitive issues and discuss possible personal and social responses before starting a particular trauma-focused treatment protocol. The degree to which these influences are supportive or punitive may affect her level of treatment engagement and adherence, as well as affect her suicide risk. Candidly addressing these factors and problem-solving around likely barriers to engagement in advance may minimize risk of dropout of trauma-focused treatment, which has been found to occur among a minority of those who initiate trauma-focused treatment (Kehle-Forbes, Meis, Spoont, & Polusny, 2015).

Case Vignette 2: Male Combat Veteran

Joe is a 30-year-old Marine Corps veteran who served as a machine gunner in Iraq in 2003. He is married with two young children and is currently unemployed because of injuries he sustained while serving in the military. He reports that his deployment was a difficult and scary experience as his training did not prepare him for what he experienced while he was deployed. He reports that he witnessed a lot of death and lost many friends. Currently, he reports experiencing a great deal of guilt and anger over some of the things he experienced. He reports that he frequently has thoughts of a traumatic event come up as though it is happening again and after these moments he feels he must retreat away from everyone. He reports that he frequently locks himself in the basement, but this further causes problems with his wife. Although he attempts to push the memories away, he cannot do so and becomes depressed and hopeless. He also reports that he avoids anything that reminds him of the experience, including connecting with others. Joe reports that he has felt disconnected from his family, affecting his relationship with his wife and kids. Additionally, he has felt very jumpy and constantly on guard. He reports that one day he was feeling so guilty about some things that happened that he took his gun and locked himself in the basement with the intent to shoot himself. This scared his wife who told him he must seek care for the sake of their children.

Many of the assessment considerations for the first vignette also would apply to this example. Similarly, the assessment and conceptualization phase for this case will include assessment of his suicidal ideation, which appears to have occurred impulsively when he became overwhelmed with re-experiencing symptoms from traumatic events and losses he experienced in combat. His hyperarousal, jumpiness, and anger may affect his emotionally reactive response when reminded of the traumatic event, leading to an attempt to avoid these internal experiences through cognitive and behavioral escape (i.e., suicidal thoughts and gestures). In addition to a thorough assessment of PTSD symptoms and prior trauma history using the aforementioned measures, it will be important to assess for current levels of substance use or other risk-taking behaviors, if any, as these behaviors may increase his risk for behaving impulsively in the presence of suicidal ideation in the future. A psychiatric medication evaluation and, given his military occupation specialty as a machine gunner, a screening for traumatic brain injury may be warranted, as mild traumatic brain injury, sleep disturbances, irritability, and PTSD may co-occur with suicidal ideation (Bramoweth & Germain, 2013). Risk for lethality by suicide may be considered high for this case because he actually had the intent and means to kill himself during a time of emotional crisis and impulsivity in the presence of his wife and children. Those behaviors, in conjunction with his male gender, disabled and unemployed status, social isolation, and depression symptoms (hopelessness, guilt, poor mood), may make him a high risk for suicide. However, the support of his wife and children and his commitment and responsibility to them are significant protective factors.

A written safety plan should be collaboratively constructed with him during the assessment and initial treatment phase, including an immediate plan for him or his adult family members to remove his firearms from the home. He should be encouraged to include his wife in the discussion of the safety plan and have candid discussions about contingency plans for safety for himself and his family (e.g., calling police, exit plans, where he or family should stay during crisis, identifying the nearest hospital emergency department) during the early stages of treatment and adjusted over the course of treatment as clinically indicated.

Treatment options in the context of establishing stability and ongoing monitoring of suicidal ideation also should be discussed with Joe during the initial treatment phase, including building his motivation for trauma-focused treatment either individually (e.g., PE, CPT, SIT, EMDR) or in a group (i.e., CPT group). If the veteran and his clinician select PE, in vivo exercises that focus on improving his engagement with his family members and relationships outside the home may be recommended, not only focusing on trauma-specific feared situations. Trauma-related avoided stimuli that are objectively dangerous in the context of suicidal or homicidal ideation (e.g., firearms) should not be included in an in vivo exposure hierarchy until suicidal ideation or homicidal ideation have safely remitted for an extended period of time. If the veteran and his clinician select CPT, inclusion of psychoeducation and discussion about traumatic bereavement, grief, and loss (session 2a) would be recommended for this veteran, particularly a discussion of how his traumatic bereavement response may influence and be influenced by his PTSD symptoms and recent suicidal ideation (Resick, Monson, & Chard, 2014). If guilt and suicidal ideation stem from possible moral injury (i.e., psychological consequences of killing or atrocities during combat), CBT interventions that allow psychological and emotional reprocessing of the moral injury and exposure to a corrective life experience with a benevolent authority figure have been suggested (Litz et al., 2009); however, there is not yet tested evidence supporting effective interventions for recovery from morally injurious acts in the context of combat.

In addition to his own individual treatment, if Joe's wife is his caregiver, additional family support via a caregiver support group may be recommended. Clinical indication and appropriateness for couples therapy may be assessed on an ongoing basis, depending on their relationship history (ruling out active or recent intimate partner violence) and stability and motivation of both the patient and family member. It will be important to collaboratively discuss the pros and cons of concurrent versus sequential family treatment with his own trauma-focused treatment. If his wife or children need additional support for themselves, a referral to a community clinic may be recommended. Additionally, as trauma-focused treatment ensues, and as he improves in social and occupational functioning outside of the home, increasing prosocial activities where he is contributing to his family and community may assist with community reintegration and quality of life (Plach & Sells, 2013; Schnurr, Lunney, Bovin, & Marx, 2009).

CONCLUSION

The experience of PTSD symptoms involves perceptions of uncontrollability and helplessness. Intrusive memories seem often to arrive "out of the blue," and they are difficult to dismiss. Trauma-related negative beliefs (e.g., self-blame, guilt, moral transgression) are perceived to be facts, rather than appraisals. Ordinary ways of coping fail to diminish the symptoms, and the chronicity of symptoms leads to perceptions that nothing is likely to change. This situation often eventually leads to consideration of suicide as an escape option. Given this process, and the empirical association between PTSD and suicide, it is crucial that PTSD treatment providers assess for suicidal ideation and that, when mental health clinicians are working with a suicidal patient, they assess for PTSD.

As noted, studies of evidence-based treatments for PTSD have often included patients reporting suicidal ideation, with positive results. When these treatments are successful in reducing PTSD symptoms, they often decrease suicide ideation and risk. Therefore, when it is deemed appropriate following a comprehensive assessment, it is important to deliver evidence-based treatments to these patients. If PTSD is not effectively treated, many of the circumstances that increase risk of suicide will remain present. With suicidal PTSD patients, as mentioned earlier, it is prudent to consider some modifications to the treatment protocols, including increased surveillance of suicidal ideation. While careful attention to preparation for initiation of these treatments is a part of most evidence-based approaches, this clinical awareness will be even more necessary for patients at risk for suicide, who will need to understand how treatment will work and what kinds of responses they can be expected to have during the process. Also important will be assessment and modification of the environment of the patient. For example, many PTSD patients, feeling that the world is an extremely dangerous place, will have weapons available (e.g., in their car or by the bed), so that addressing issues of weapon storage prior to initiation of treatment will be required. There is a strong association between PTSD and a range of factors that increase risk for suicide, including depression, sleep problems and nightmares, problem alcohol or drug use, marital and familial problems, lack of social support, and co-occurring physical illness. Some of these factors may change as a result of PTSD treatment, but others may require additional targeted interventions in addition to the evidence-based PTSD treatment. Suicidality may increase when a number of factors interact and combine to create distress, so the treatment plan should focus not only on remediation of PTSD symptoms but also on strengthening social support and connection, improving relationship functioning, increasing participation in positive activities, reducing alcohol and drug consumption, managing financial stressors, and the like. Selection of these supplemental targets of intervention must depend on individualized assessment to determine the primary drivers of hopeless and suicidal thoughts.

Treatment with evidence-based protocols can often reduce PTSD symptoms and improve functioning. Clinicians can select among several treatment protocols that differ in focus and target, to a greater or lesser degree, trauma memories and reminders, trauma-related beliefs, and self-management of stress reactions. Combined with other treatment options that address additional risk factors for suicide, these treatments can often reduce risk of suicide and increase hope, sense of purpose, positive emotions, and the re-embracing of life.

REFERENCES

Alvarez, J., McLean, C., Harris, A. H., Rosen, C. S., Ruzek, J. I., & Kimerling, R. (2011). The comparative effectiveness of cognitive processing therapy for male veterans treated in a VHA posttraumatic stress disorder residential rehabilitation program. *Journal* of Consulting and Clinical Psychology, 79, 590–599. doi:10.1037/a0024466

American Psychiatric Association. (2013). Diagnostic and statistical manual of mental disorders (5th ed.). Washington, DC: Author.

Assessment and Management of Risk for Suicide Working Group. (2013). VA/DoD Clinical practice guideline for assessment and management of patients at risk for suicide. Retrieved from http://www.healthquality.va.gov/guidelines/MH/srb/VADODCP_SuicideRisk_Full.pdf.

Ballard, E. D., Ionescu, D. F., Vande Voort, J. L., Niciu, M. J., Richards, E. M., Luckenbaugh, D. A., ... Zarate, C.A. (2014). Improvement in suicidal ideation after ketamine infusion: Relationship to reductions in depression and anxiety. Journal of Psychiatric Research, 58, 161–166.

Bell, J. B., & Nye., E. C. (2007). Specific symptoms predict suicidal ideation in Vietnam combat veterans with chronic post-traumatic stress disorder. Military Medicine, 172, 1144–1147.

Benedek, D. M., Friedman, M. J., Zatzick, D, & Ursano, R. J. (2009). Guideline watch (March 2009): Practice guideline for the treatment of patients with acute stress disorder and post-traumatic stress disorder. Retrieved from http://psychiatryonline.org/pb/assets/raw/sitewide/practice_guidelines/guidelines/acutestressdisorderptsd-watch.pdf

Bernert, R. A., Joiner, T. E., Cukrowicz, K. C., Schmidt, N. B., & Krakow B. (2005). Suicidality and sleep disturbances. Sleep, 28, 1135–1141.

Black, S. A., Gallaway, S., Bell, M. R., & Ritchie, E. C. (2011). Prevalence and risk factors associated with suicides of Army soldiers. Military Psychology, 23, 433–451. doi:10.1080/08995605.2011.590409

Bossarte, R. M., Knox, K. L., Piegari, R., Altieri, J., Kemp, J., & Katz, I. R. (2012). Prevalence and characteristics of suicide ideation and attempts among active military and veteran participants in a national health survey. American Journal of Public Health, 102, S38–S40. doi:10.2105/AJPH.2011.300487

Bradley, R., Greene, J., Russ, F., Dutra, L., & Westen, D. (2005). A multidimensional meta-analysis of psychotherapy for PTSD. The American Journal of Psychiatry, 162, 214–227. doi:10.1176/appi.ajp.162.2.214

Bramoweth, A. D., & Germain, A. (2013). Deployment-related insomnia in military personnel and veterans. Current Psychiatry Reports, 15, 401–409. doi:10.1007/s11920-013-0401-4

Britton, P. C., Patrick, H., Wenzel, A., & Williams, G. C. (2010). Integrating motivational interviewing and self-determination theory with cognitive behavioral therapy to prevent suicide. Cognitive and Behavioral Practice, 18, 16–27.

Brown, T. A., Dinardo, P. A., & Barlow, D. H. (2004). Anxiety Disorders Interview Schedule Adult Version (ADIS-IV): Client Interview Schedule. New York: Oxford University Press.

Chard, K. M. (2005). An evaluation of cognitive processing therapy for the treatment of posttraumatic stress disorder related to childhood sexual abuse. Journal of Consulting and Clinical Psychology, 73, 965–971. doi:10.1037/0022-006X.73.5.965

Cougle, J. R., Resnick, H., & Kilpatrick, D. G. (2009). PTSD, depression, and their comorbidity in relation to suicidality: Cross-sectional and prospective analyses of a national probability sample of women. Depression and Anxiety, 26, 1151–1157. doi:10.1002/da.20621

Davidson, J. R. T., Hughes, D. C., Blazer, D. G., & George, L. K. (1991). Posttraumatic stress disorder in the community: An epidemiological study. Psychological Medicine, 21, 713–721.

Eliot, D. M., & Briere, J. (1992). Sexual abuse trauma among professional women: Validating the Trauma Symptom Checklist–40 (TSC-40). Child Abuse & Neglect, 16, 391–398.

First, M. B., Gibbon, M., Spitzer, R. L., & Williams, J. B. W. (1996). User's guide for the Structured Clinical Interview for DSM-IV Axis I disorders: SCID-I — Research version. Washington, DC: American Psychiatric Press.

First, M. B., Spitzer, R. L., Gibbon M., and Williams, J. B. W. (2002, November). Structured Clinical Interview for DSM-IV-TR Axis I Disorders, Research Version, Patient Edition. (SCID-I/P). New York: Biometrics Research, New York State Psychiatric Institute.

Foa, E., Cashman, L., Jaycox, L., & Perry, K. (1997). The validation of a self-report measure of PTSD: The Posttraumatic Diagnostic Scale. Psychological Assessment, 9, 445–451. doi:10.1037//1040-3590.9.4.445

Foa, E. B., Dancu, C. V., Hembree, E. A., Jaycox, L. H., Meadows, E. A., & Street, G. P. (1999). A comparison of exposure therapy, stress inoculation training, and their combination for reducing posttraumatic stress disorder in female assault victims. Journal of Consulting and Clinical Psychology, 67, 194–200. doi:10.1037//0022-006X.67.2.194

Foa, E. B., Hembree, E. A., Cahill, S. P., Rauch, S. A., Riggs, D. S., Feeny, N. C., ... Yadin, E. (2005). Randomized trial of prolonged exposure for post-traumatic stress disorder with and without cognitive restructuring: Outcome at academic and community clinics. Journal of Consulting and Clinical Psychology, 73, 953–964. doi:10.1037/0022-006X.73.5.953

Foa, E. B., Hembree, E., & Rothbaum, B. O. (2007). *Prolonged exposure therapy for PTSD: Emotional processing of traumatic experiences therapist guide.* New York: Oxford University Press.

Foa, E. B., Keane, T. M., Friedman, M. J., & Cohen, J. A. (2009). *Effective treatments for PTSD: Practice guidelines from the International Society for Traumatic Stress Studies* (2nd ed.). New York: Guilford Press.

Foa, E. B., Riggs, D., Dancu, C., & Rothbaum, B. (1993). Reliability and validity of a brief instrument for assessing post-traumatic stress disorder. *Journal of Traumatic Stress, 6,* 459–474. doi:10.1007/BF00974317

Foa E. B., Rothbaum B. O., Riggs D. S., & Murdock T. B. (1991). Treatment of posttraumatic stress disorder in rape victims: A comparison between cognitive-behavioral procedures and counseling. *Journal of Consulting and Clinical Psychology, 59,* 715–723. doi:10.1037//0022-006X.59.5.715

Foa, E., Steketee, G., & Olasov-Rothbaum, B. (1989). Behavioral/cognitive conceptualisations of post-traumatic stress disorder. *Behaviour Therapy, 20,* 155–176. doi:10.1016/S0005-7894(89)80067-X

Forbes, D., Creamer, M., Bission, J. I., Cohen, J. A., Crow, B. E., Foa, E. B., ... Ursano, R. J. (2010). A guide to guidelines for the treatment of PTSD and related conditions. *Journal of Traumatic Stress, 23,* 537–552. doi:10.1002/jts.20565

Harned, M., Korslund, K., Foa, E., & Linehan, M. (2012). Treating PTSD in suicidal and self-injuring women with borderline personality disorder: Development and preliminary evaluation of a dialectical behavior therapy prolonged exposure protocol. *Behavior Research and Therapy, 50,* 381–386. doi:10.1016/j.brat.2012.02.011

Hoskins, M., Pearce, J., Bethell, A., Dankova, L., Barbui, C., Tol, W. A., ... Bissen, J. I. (2015). Pharmacotherapy for post-traumatic stress disorder: Systematic review and meta-analysis. *The British Journal of Psychiatry, 206,* 93–100.

Institute of Medicine. (2008). *Treatment of posttraumatic stress disorder: An assessment of the evidence.* Washington, DC: National Academies Press.

Jakupcak, M., Conybeare, D., Phelps, L., Hunt, S., Holmes, H. A., Felker, B., ... McFall, M. E. (2007). Anger, hostility, and aggression among Iraq and Afghanistan war veterans reporting PTSD and subthreshold PTSD. *Journal of Traumatic Stress, 20,* 945–954. doi:10.1002/jts

Jakupcak, M., Cook, J., Imel, Z., Fontana, A., Rosenheck, R., & McFall, M. (2009). Posttraumatic stress disorder as a risk factor for suicidal ideation in Iraq and Afghanistan combat veterans. *Journal of Traumatic Stress, 22,* 303–306. doi:10.1002/jts.20423

Jakupcak, M., & Varra, E. M. (2011). Treating Iraq and Afghanistan war veterans with PTSD who are at high risk for suicide. *Cognitive and Behavioral Practice, 18*(1), 85–97.

Jobes, D. A. (2006). *Managing suicidal risk: A collaborative approach.* New York: Guilford Press.

Karlin, B. E, Ruzek, J. I., Chard, K. M., Eftekhari, A., Monson, C. M., Hembree, E. A., ... Foa, E. B. (2010). Dissemination of evidence-based psychological treatments for posttraumatic stress disorder in the Veterans Health Administration. *Journal of Traumatic Stress, 23,* 663–673. doi:10.1002/jts.20588.

Keane, T. M., Caddell, J. M., & Taylor, K. L. (1988) Mississippi Scale for Combat-Related Posttraumatic Stress Disorder: Three studies in reliability and validity. *Journal of Consulting and Clinical Psychology, 56,* 85–90. doi:10.1037//0022-006X.56.1.85

Kehle-Forbes, S. M., Meis, L. A., Spoont, M. R., & Polusny, M. A. (2015). Treatment initiation and dropout from prolonged exposure and cognitive processing therapy in a VA outpatient clinic. *Psychological Trauma: Theory, Research, Practice, and Policy, 8*(1), 107–114.

Kessler, R. C., Borges, G., & Walters, E. E. (1999). Prevalence of and risk factors for lifetime suicide attempts in the National Comorbidity Survey. *Archives of General Psychiatry, 56,* 617–626. doi:10.1001/archpsyc.56.7.617

Kessler, R. C., Chiu, W. T., Demler, O., & Walters, E. E. (2005). Prevalence, severity, and comorbidity of 12-month DSM-IV disorders in the National Comorbidity Survey Replication. *Archives of General Psychiatry, 62,* 617–627. doi:10.1001/archpsyc.62.6.617

King, L. A., King, D. W., Leskin, G. A., & Foy, D. W. (1995). The Los Angeles Symptom Checklist: A self-report measure of posttraumatic stress disorder. *Assessment, 2,* 1–17. doi:10.1177/1073191195002001001

Krakow, B., Artar, A., Warner, T. D., Melendrez, D., Johnston, L., Hollifield, M., ... Koss, M. (2000). Sleep disorder, depression and suicidality in female sexual assault survivors. *Crisis, 21,* 163–170. doi:10.1027//0227-5910.21.4.163

Kroenke, K., Spitzer, R. L., & Williams, J. B. W. (2001). The PHQ-9: Validity of a brief depression severity measure. *Journal of General Internal Medicine, 16,* 606–613. doi:10.1046/j.1525-1497.2001.016009606.x

Krysinska, K., & Lester, D. (2010). Post-traumatic stress disorder and suicide risk: A systematic review. *Archives of Suicide Research, 14,* 1–23. doi:10.1080/13811110903478997

Krystal, J. H., Rosenheck, R. A., Cramer, J. A., Vessicchio, J. C., Jones, K. M., Vertrees, J. E., ... Veterans Affairs Cooperative Study No. 504 Group. (2011). Adjunctive risperidone treatment for antidepressant-resistant symptoms of chronic military service related PTSD: A randomized trial. *JAMA: Journal of the American Medical Association, 306*, 493–502. doi:10.1001/jama.2011.1080

Kulka, R. A., Schlenger, W. E., Fairbank, J. A., Hough, R. L., Jordan, B. K., Marmar, C. R., ... Weiss, D. (1990). *Trauma and the Vietnam War generation: Report of findings from the National Vietnam Veterans Readjustment Study.* New York: Brunner/Mazel.

Lang, P. J. (1977). Imagery in therapy: An information processing analysis of fear. *Behavior Therapy, 8*, 862–886. doi:10.1016/S0005-7894(77)80157-3

Larkin, G. L., & Beautrais, A. L. (2011). A preliminary naturalistic study of low-dose ketamine for depression and suicide ideation in the emergency department. *International Journal of Neuropsychopharmacology, 14*, 1127–1131. doi:10.1017/S1461145711000629

Linehan, M. M., & Comtois, K. (1996). *Lifetime parasuicide history.* Unpublished manuscript. University of Washington, Seattle.

Linehan, M. M., Comtois, K. A., Brown, M. Z., Heard, H. L., & Wagner, A. (2006). Suicide Attempt Self-Injury Interview (SASII): Development, reliability, and validity of a scale to assess suicide attempts and intentional self-injury. *Psychological Assessment, 18*(3), 303–312.

Litz, B. T., & Schlenger, W. E (2009). Posttraumatic stress disorder in service members and new veterans of the Iraq and Afghanistan wars: A bibliography and critique. *PTSD Research Quarterly, 20*, 1–7.

Litz, B. T., Stein, N., Delaney, E., Lebowitz, L., Nash, W. P., ... Maquen, S. (2009). Moral injury and moral repair in war veterans: A preliminary model and intervention strategy. *Clinical Psychology Review, 29*, 695–706. doi:10.1016/j.cpr.2009.07.003

Management of Post-Traumatic Stress Disorder Working Group. (2010). VA/DoD clinical practice guideline for management of post-traumatic stress. Retrieved from http://www.healthquality.va.gov/ptsd/PTSD-FULL-2010a.pdf

Monson, C. M., & Fredman, S. J. (2012). *Cognitive-behavioral conjoint therapy for PTSD: Harnessing the healing power of relationships.* New York: Guilford Press.

Panagioti, M., Gooding, P. A., Dunn, G., & Tarrier, N. (2011). Pathways to suicidal behavior in posttraumatic stress disorder. *Journal of Traumatic Stress, 24*, 137–145. doi:10.1002/jts.20627

Plach, H. L., & Sells, C. H. (2013). Occupational performance needs of young veterans. *American Journal of Occupational Therapy, 67*, 73–81. doi:10.5014/ajot.2013.003871

Raza, G. T., & Holohan, D. R. (2015). Clinical treatment selection for posttraumatic stress disorder: Suggestions for researchers and clinical trainers. *Psychological Trauma: Theory, Research, Practice, and Policy, 7*(6), 547–554.

Resick, P. A., Monson, C. M., & Chard, K. M. (2014). *Cognitive processing therapy: Veteran/military version: Therapist's manual.* Washington, DC: Department of Veterans Affairs.

Resick, P. A, & Schnicke, M. K. (1992). Cognitive processing therapy for sexual assault victims. *Journal of Consulting and Clinical Psychology, 60*, 748–756. doi:10.1037//0022-006X.60.5.748

Resick, P. A., Galovski, T. E., O'Brien, U. M., Scher, C. D., & Young-Xu, Y (2008). A randomized clinical trial to dismantle components of cognitive processing therapy for posttraumatic stress disorder in female victims of interpersonal violence. *Journal of Consulting and Clinical Psychology, 76*, 243–258. doi:10. 1037/0022-006X.76.2.243

Rothbaum, B. O., Astin, M. C., & Marsteller, F. (2005). Prolonged exposure versus eye movement desensitization and reprocessing (EMDR) for PTSD rape victims. *Journal of Traumatic Stress, 18*, 607–616. doi:10.1002/jts.20069

Schnurr, P. P., Friedman, M. J., Engel, C. C., Foa, E. B., Shea, M. T., Chow, B. K., ... Bernardy, N. (2007). Cognitive behavioral therapy for posttraumatic stress disorder in women: A randomized controlled trial. *JAMA: Journal of the American Medical Association, 297*, 820–830. doi:10.1001/jama.297.8.820

Schnurr, P. P., Lunney, C. A., Bovin, M. J., & Marx, B. P. (2009). Posttraumatic stress disorder and quality of life: Extension of findings to veterans of the wars in Iraq and Afghanistan. *Clinical Psychology Review, 29*, 727–735. doi:10.1016/j.cpr.2009.08.006

Shapiro, S. (2001). Enhancing self-belief with EMDR: Developing a sense of mastery in the early phase of treatment. *American Journal of Psychotherapy, 55*, 531–542.

Shapiro, F., & Maxfield L. (2002). Eye movement desensitization and reprocessing (EMDR): Information processing in the treatment of trauma. *Journal of Clinical Psychology, 58*, 933–946. doi:10.1002/jclp.10068

Tarrier, N., & Gregg, L. (2004). Suicide risk in civilian PTSD patients: Predictors of suicidal ideation, planning and attempts. *Social Psychiatry and Psychiatric Epidemiology, 39*, 655–661. doi:10.1007/s00127-004-0799-4

Tarrier, N., Pilgrim, H., Sommerfield, C., Faragher, B., Reynolds, M., Graham, E., & Barrowclough, C. (1999). A randomized trial of cognitive therapy and imaginal exposure in the treatment of chronic posttraumatic stress disorder. *Journal of Consulting and Clinical Psychology*, 67(1), 13–18.

Tarrier, N., Taylor, K., & Gooding, P. (2008). Cognitive-behavioral interventions to reduce suicidal behavior: A systematic review and meta-analysis. *Behavior Modification*, 32, 77–108. doi:10.1177/0145445507304728

Taylor, S., Thordarson, D. S., Maxfield, L., Fedoroff, I. C., Lovell, K., & Ogrodniczuk, J. (2003). Comparative efficacy, speed, and adverse effects of three PTSD treatments: Exposure therapy, EMDR, and relaxation training. *Journal of Consulting and Clinical Psychology*, 71, 330–338. doi:10.1037/0022-006X.71.2.330

van der Kolk, B. A., Spinazzola, J., Blaustein, M. E., Hopper, J. W., Hopper, E. K., Korn, D. L., & Simpson, W. B. (2007). A randomized clinical trial of eye movement desensitization and reprocessing (EMDR), fluoxetine, and pill placebo in the treatment of posttraumatic stress disorder: Treatment effects and long-term maintenance. *Journal of Clinical Psychiatry*, 68, 37–46. doi:10.4088/JCP.v68n0105

Weathers, F. W., Blake, D. D., Schnurr, P. P., Kaloupek, D. G., Marx, B. P., & Keane, T. M. (2013). *The Clinician-Administered PTSD Scale for DSM-5 (CAPS-5)*. Retrieved from www.ptsd.va.gov

Weathers, F. W., Litz, B. T., Herman, D. S., Huska, J. A., & Keane, T. M. (1993, October). *The PTSD Checklist (PCL): Reliability, validity, and diagnostic utility*. Paper presented at the annual meeting of the International Society for Traumatic Stress Studies, San Antonio, TX.

Weathers, F. W., Litz, B. T., Keane, T. M., Palmieri, P. A., Marx, B. P., & Schnurr, P. P. (2013). *The PTSD Checklist for DSM-5 (PCL-5)*. Retrieved from www.ptsd.va.gov

Weathers, F. W., Blake, D. D., Schnurr, P. P., Kaloupek, D. G., Marx, B. P., & Keane, T. M. (2013). *The Clinician-Administered PTSD Scale for DSM-5 (CAPS-5)*. www.ptsd.va.gov.

Yehuda, R., Halligan, S., Golier, J., Grossman, R., & Bierer, L. (2004). Effects of trauma exposure on the cortisol response to dexamethasone administration in posttraumatic stress disorder and major depressive disorder. *Psychoneuroendocrinology*, 29, 389–404. doi:10.1016/S0306-4530(03)00052-0

13

The Collaborative Assessment and Management of Suicidality with Suicidal Service Members

David A. Jobes

Blaire C. Schembari

Keith W. Jennings

Jon is a 22-year-old, Caucasian, unmarried U.S. Army sergeant; he was assigned to a light infantry battalion as an infantry team leader at a military installation in the southeastern United States. During his four years of military service, he deployed once to Iraq (in support of Operation Iraqi Freedom) for a 10-month tour and once to Afghanistan (in support of Operation Enduring Freedom) for an 8-month deployment. He presented to the local behavioral health clinic after his company commander referred him to treatment because the commander was concerned about Jon's suicide risk.

Following his unit's return from Afghanistan approximately six months ago, Jon had begun acting "differently." He had gradually stopped socializing with his friends in the company and spent most of his off-duty time alone in his room in the barracks. His fellow platoon-mates had noticed that Jon, who used to be the one organizing a night out with half the platoon on the weekends, rarely left his room anymore and that he sometimes would not even answer his phone or respond to a knock on his door when they attempted to invite him out with them.

Additionally, Jon's work performance as a Soldier and team leader had taken a dramatic downturn. As recently as several months prior, he was viewed by most of the other Soldiers in his unit as the "go-to" guy when presented with a difficult problem. His chain of command had viewed him as one of the most promising and "squared-away" Soldiers in the company, which had resulted in his being promoted to team leader; he became responsible for the supervision and welfare of himself and three other Soldiers. Jon had considered this to be a great honor and took this responsibility very seriously. He took great pride in being one of the most technically and tactically proficient Soldiers in the company. He also prided himself on being in great physical shape, consistently "maxing out" his scores on the Army's mandated physical training (PT) test. But along with withdrawing socially from his friends, Jon had stopped exercising regularly over the last few months and had begun to have difficulty "keeping up" during his company's daily PT, which was a source of great embarrassment to him.

Over the past several weeks, Jon had begun arriving late to company formations and, in some cases, missing them completely. The first few times, his chain of command had let this slide, but it was seen as very out of character for Jon. After it occurred several times, his squad leader and platoon sergeant gave him several written, disciplinary counselings and eventually recommended Uniform Code of Military Justice (UCMJ) action for his "repeated failures to arrive to training at the right place, at the right time, and with the right gear." Jon was angry and ashamed for "letting" himself get into this situation and now feared that he would lose his position as a team leader and even be demoted in rank with a consequent reduction in pay.

One night soon after Jon was told that he would be undergoing UCMJ action for his "subpar" performance, a fellow Soldier and friend heard Jon yelling and swearing in his barracks room. Concerned, the Soldier knocked on Jon's door several times, and after there was no response (just continued yelling and swearing), the Soldier decided to enter the room to check and make sure Jon was okay. When he entered the room, the Soldier found Jon drunk, with a bottle of bourbon in one hand and a pistol in the other. Jon told the Soldier to leave the room because he was going to kill himself and did not want anyone else around. Jon then dropped the pistol and the bottle and began to cry. Not knowing what else to do, the Soldier called the platoon sergeant to seek guidance. Jon's chain of command was notified, and by the time the commander arrived at the company area, it was nearly time for PT formation. Jon's squad leader then escorted him to the clinic where he was seen as a walk-in appointment.

CHAPTER OVERVIEW

This fictional case vignette is based loosely on actual cases of suicidal service members who were effectively treated by the Collaborative Assessment and Management of Suicidality (CAMS; Jobes, 2006; 2016a). CAMS is an evidence-based therapeutic framework that specifically targets a person's suicide risk through a collaborative clinical engagement. Although CAMS is used across different settings and populations, the military application described in this chapter is especially germane as various empirical investigations have examined the use of CAMS with military personnel, and its application with this population is growing (Jobes, 2012; Jobes & Drozd, 2004; Jobes, Wong, Conrad, Drozd, & Neal-Walden, 2005; also see James, 2012; Jobes, Lento, & Brazaitis, 2012). The case of Jon is revisited throughout this chapter to further illustrate the clinical philosophy of the approach and use of the Suicide Status Form(SSF), which functions as a multipurpose assessment, treatment planning, tracking, and outcome clinical tool. The chapter concludes with a review of 25 years of clinical research that provides empirical support for the CAMS approach.

CASE STUDY CONTINUED: INITIAL ENGAGEMENT

As he sat in the waiting room with his squad leader, Jon could not understand how he had ended up in the clinic, waiting to see a "shrink." As he filled out the myriad of intake forms required by the clinic, he could not stop thinking about how his career was ruined and how all his platoon-mates would not want to go to war with him because he would now be thought of as "weak" and "crazy" for seeing a behavioral health care provider. When the psychologist came to get him, and he followed her to his office, he could not help but think, "I'm way too broke to fix. I don't know what good this is going to do." As the psychologist began talking, Jon couldn't take it anymore and abruptly cut her off. He said, "We both know why I'm here. . . . I'm one of those guys that I used to hear about . . . the crazy ones who just can't cut it anymore and try to kill themselves. . . . I don't know why I'm like this; I don't want to be, but I just can't fix it." The initial conversation follows:

Clinician: Jon, it is clear that you are going through a very difficult time . . . your chain of command is worried about you, which is why they had you come here today. I'd like to learn more about what's brought you to this dark place and see if we can figure out a better way to deal with it. Why do you think that you want to kill yourself?

Jon: (long pause) Well, that's what I'm saying, I just don't know . . . and I don't know if there's anything we can do about it. I just feel so horrible . . . like I don't want to get up in the morning anymore, I don't want to see anybody anymore . . . I just feel like I can't do anything right anymore. I've become this total screw-up.

Clinician: You know, it sounds like you've been feeling this way for a while, but that you haven't always felt this way. It is not unusual for some Soldiers to get into some very dark places—feeling depressed, trapped, hopeless, or desperate. . . feeling this way can often lead to suicidal thoughts.

Jon: Yeah, no kidding. . . . I just don't want to feel like this anymore. But I just don't know what to do.

Clinician: You know, when someone is feeling all of those nasty ways, I can see how suicide might make sense . . . for someone who suffers and feels desperate, thoughts of suicide can and do happen and may even feel comforting. Suicide is just one among many things that occur to some people in emotional trouble as a means of coping with seemingly unbearable pain and suffering. Perhaps not surprisingly, it is not my favorite coping approach! But I do understand the feelings and I appreciate that for some people it gives them a sense of control and power within difficult situations . . . but I have a bias in favor of other means of coping that are not so costly and irreversible as suicide is . . . here in the clinic we have a particular approach to dealing with this kind of pain and suffering. It is called CAMS, and I believe you are an excellent candidate for this particular intervention. If you are willing, and with your permission, I would like to complete an assessment form together that will help us understand more about your pain and suffering and what we can do to perhaps better get your needs met.

CAMS PHILOSOPHY

As seen in the preceding depiction, the clinician skillfully illustrated the core *philosophy* of CAMS, which is central to the CAMS approach. Jon's suicidal thoughts and desires were met with openness and empathy from the clinician; she neither shamed the patient nor judged his suicidal behaviors. The clinician heard the patient's pain and provided an empathic and safe space for the Soldier, thereby opening the door for a strong therapeutic alliance. The CAMS philosophy is marked by understanding that the patient's suicidal desires, intentions, and thoughts reflect a form of coping. The patient's suicidality is invariably rooted in legitimate needs (e.g., the need to end one's pain, or the need to reassert control and power, or the need to stop the voices of psychosis). While CAMS clinicians appreciate the subjective validity of suicide as a coping method for the suicidal patient, they nevertheless assert that there are more effective ways of coping and getting suicidal needs met. It should be noted that empathy for the patient's suicidal state does not equate to endorsing suicidal actions. Indeed, CAMS clinicians are encouraged to remind patients that state law requires them to prevent clear and imminent self-harming behaviors potentially through a voluntary or involuntary psychiatric hospitalization. By leading with empathy and a forthright discussion of the legal implications, the CAMS clinician endeavors to create a meaningful *outpatient* alliance that can be used to develop worthwhile strategies for coping and creating alternative means for satisfying the person's needs.

The CAMS philosophy is to communicate and inspire a belief within patients that they are respectfully understood and being offered a potentially life-saving intervention. This kind of engagement can help stabilize the patient and open the door to development of alternative coping methods. It also enables further assessment, targeting and treatment of the patient's suicidal "drivers"—the problems and issues that make suicide compelling to the patient (Jobes, Comtois, Brenner, Gutierrez, & O'Connor, 2016). CAMS is designed to promote three overarching goals of care: (a) the development of a suicide-specific CAMS Stabilization Plan to keep the patient out of the hospital, (b) the identification and problem-focused treatment of patient-defined suicidal drivers, and (c) the explicit development of existential purpose and meaning.

In order to effectively assess (and thereby treat) a patient's suicidal risk, a thorough assessment of the patient's suicidality throughout the course of CAMS care is critical. Within CAMS this ongoing assessment begins with the first two pages of the SSF. Within CAMS this assessment is done in a collaborative manner through a particular method of interaction between the patient and clinician.

CASE STUDY CONTINUED: INITIAL ASSESSMENT: SECTIONS A AND B OF THE SSF

Jon was taken aback by the psychologist's agreement that it made sense that he had turned to suicide as an option given his pain and emotional suffering. He was also a little shocked that she seemed to have a structured plan for how to address his suffering. Despite himself, Jon found himself wondering if perhaps there was something he could do to relieve some of the misery he was experiencing. Hesitantly, he agreed when she asked for permission to sit next to him to work through an assessment form together. As she moved her desk chair adjacent to him (close but not too close), the clinician handed him a clipboard with the first page of the SSF (Appendix A). The therapeutic engagement continued:

Clinician: Jon, this is an assessment tool called the Suicide Status Form. It is a way to help us figure out exactly what you're experiencing and to develop a focused treatment plan to give you some breathing room from all of this distress. There are no wrong answers to this. The goal here is to help me see it like you see it. We want to explore what you're going through together so that I can understand what it's like for you. Please consider each rating here ("Section A") and rate each construct . . . and after you rate each scale, I then want you to write in descriptions that further clarify what is going on with you. Okay?

Jon: I'm not sure how to do this right. . . . I don't really know what you mean here by "psychological pain." Does that mean feeling like crap?

Clinician: Kind of . . . what we're talking about here is not physical pain, not like a broken bone . . . but what this rating is after is something more akin to emotions or misery . . . a kind of psychological suffering . . . not physical, but more mental.

As depicted in the case study of Jon, the initial SSF assessment is completed by both the clinician and patient, sitting side-by-side in a collaborative fashion. The aim of this process is for both the patient and the clinician to gain a richer understanding of the patient's psychological suffering. As shown in Appendix A, much of Jon's anguish and distress focuses on his poor work performance as a Soldier and team leader. Jon is quite depressed and has withdrawn socially; he seems to believe a critical inner voice that says he is a "total screw-up."

The patient and clinician proceed with the completion of Section A together. The SSF Core Assessment is a sequence of rating scales centering on five theoretical constructs—psychological pain, stress, agitation, hopelessness, and self-hate. Each of these constructs also contains a prompt (e.g., "What I find most painful is . . ."). Following the rating of each construct, patients are further encouraged to respond to the prompts by writing responses in their own words. The Core Assessment concludes with the assessment of the patient's overall risk of suicide. The additional ratings following the Core Assessment include other scales assessing the degree to which the client's suicidal state is related to themselves or to others, a qualitative description of the patient's "Reasons for Living" (RFL) and "Reasons for Dying" (RFD) and additional ratings of the patient's "Wish to Live" versus "Wish to Die," and a final qualitative prompt that assesses the "One thing" that would cause the patient to no longer be suicidal.

Once Section A has been completed, the clinician remains seated next to the patient and takes the clipboard back in order to complete Section B. This section contains various empirically supported risk factors that are closely associated with suicide, including previous suicide attempts, current suicidal ideation and intent, current suicide plan, and so on. As shown in Appendix B, Jon's suicide plan was to drink heavily and shoot himself in the head with his pistol. After the collaborative completion of Section B, Jon's suicidal state has been rigorously evaluated through various quantitative and qualitative assessments, as well as through a detailed evaluation of the key psychological risk factors of suicide, leaving the clinician and Jon with an extremely clear understanding about the nature of Jon's suicide risk.

CAMS INITIAL TREATMENT PLANNING

As previously discussed, the initial session of CAMS is comprised of a particular collaborative interaction between the patient and clinician as they work their way through an SSF-guided evaluation of the patient's suicidal risk. The initial session concludes with the interactive creation of a viable treatment plan (see Section C of the initial session SSF). As noted, one of the main goals of CAMS is to keep a suicidal patient *out* of an inpatient setting. For this goal to be realized, the patient and clinician must work together to devise a feasible outpatient treatment plan. This plan should make *both* parties feel comfortable moving forward on an outpatient basis.

To further the collaborative relationship, the clinician and patient dyad continues to cooperate as "co-authors" of the CAMS treatment plan. Although the clinician completes Section C, the patient (still positioned next to the provider) is encouraged to actively contribute to the development of the treatment plan. The patient's active engagement in treatment planning, where the patient is quite literally the co-author of the treatment plan with the clinician, is designed to seize upon the patient's elevated readiness to change that has formed over the course of the collaborative suicide risk assessment completed during the session (Britton, Conner, & Maisto, 2012; Jobes, 2011; O'Connor et al., 2015). We have observed in our clinical trials of CAMS that patients are very satisfied with this kind of engagement and with having direct involvement in their own care (Comtois et al., 2011).

As shown in Section C (see Appendix B), the dyad is faced with the need to address the patient's "Self-Harm Potential." In CAMS this issue is not negotiable and must be addressed by the treatment goal/objective of outpatient "Safety and Stability." It follows that self-directed violence and the need for outpatient safety and stability are operationalized through the formation of a SSF Stabilization Plan (see Appendix C). This plan acts as a stabilization strategy should the patient's suicidality worsen between sessions. It is designed to help get the patient through any "dark moments" in the course of his or her CAMS care.

The CAMS SSF Stabilization Plan was initially derived from Rudd, Joiner, and Rajab's (2001) Crisis Response Plan and is analogous to Stanley and Brown's (2012) Safety Planning intervention and other coping-oriented stabilization methods (Brown et al., 2005; Chiles & Strosahl, 2005; Linehan, 1993). While there is some variation, the Stabilization Plan typically stresses four fundamental interventions: (a) the removal of the patient's access to lethal means, (b) the construction and utilization of a crisis card, (c) an increased effort to reduce interpersonal isolation, and (d) a commitment to fully engage in CAMS care for a certain duration of time (e.g., one to three months; see Jobes et al., 2011; Jobes et al., 2016).

Following the construction of the Stabilization Plan, the patient is then directed to describe two suicide-specific "drivers" that will be written in as Problem #2 and #3, respectively. Each identified driver should have corresponding treatment goals/objectives and interventions as noted in Section C. Thus the overall goal of the CAMS treatment plan is to create a suicide-specific treatment that is employed during an allotted treatment time frame; both parties commit to pursuing the CAMS Stabilization Plan while the identified suicidal drivers are targeted and treated through a problem-focused approach. When the dyad has developed a mutually satisfactory treatment plan, an outpatient course of care can methodically eliminate the patient's need for suicidal coping.

CASE STUDY CONTINUED: TREATMENT PLANNING: SECTION C

Clinician: (Still sitting adjacent to the client, reviewing SSF assessment Sections A and B) Okay, Jon . . . I think we've got a pretty good idea about what's leading you to consider suicide as an option. I've got some ideas, but given everything we've covered and put down on the SSF over the last 40 minutes or so, I'm curious: What you think is going on?

Jon: Well, I don't know . . . I haven't ever really put it all down like this before. . . . I guess I just really feel like crap about how I've gone from being a "water-walker" in the company, to what feels like the biggest turd ever. I think that's what's making me feel so guilty and ashamed all the time.

Clinician: Yes, that sense of loss over not being the guy that you used to be really seems to keep coming up as we go through all this. It sounds like it's not only this sense of loss, but that there's also a sense of shame because you just don't know what happened . . . it seems like you're not sure how you went from being "that guy" to the "screw-up," as you call it, that you feel you are now.

Jon: Yeah, it's like no matter what I do nowadays, I just mess it all up.

Clinician: And I wonder about this other theme that seems to be emerging here . . . this idea that you feel so alone all the time?

Jon: Well, it's like I used to hang out with people all the time and have a ton of friends, but now, I just feel like everyone would be better off without me. I just feel ashamed over some of the things that happened on my last rotation [in

Afghanistan], and it's like I just don't want to face any of my family this way. Hell, I don't even want to face the guys in my unit, because I feel like if I let them down before, it could happen again ... I'm like dead weight, dragging them down with me.

Clinician: That sounds like a lot of stuff to be carrying around on your shoulders ... between feeling like a screw-up all the time, and then feeling too ashamed to face people close to you because you feel like you let them down ... these are all very heavy and painful issues for you to be enduring ... no wonder you have been feeling suicidal!

Jon: Yeah, I feel like they'd be better off without me, and that if I removed myself from the equation, then ... well ... it'd just be better for everyone ...

Clinician: (softly) Yes ... I understand how you can feel that way ... but in my experience most families and friends are usually not relieved to lose someone to suicide ... (pause) ... but you know, Jon, there may be another way we can deal with all this ... in this approach that we are doing here, there is a way that we can try to help you both feel better in the short term and potentially address some of these larger and painful issues that make you want to kill yourself. While it may sound weird, I would like to suggest that you give me a chance to see if we can effectively treat your suicidal despair over the next few months ... from my perspective you have everything to gain and nothing to lose by trying ... while it sounds provocative, you could always kill yourself later if it is the only thing left to do ... but remember if you are going to try to kill yourself while you are in treatment, by law I have a duty to try to prevent that by having you go to the hospital ...

Jon: (panicking) Whoa, hang on here, I do not want to be hospitalized for this! It's already bad enough that I'm seeing a shrink; I can only imagine what everybody thinks! But if you throw me in the hospital, I'm done ... my career is over. Please don't do that.

Clinician: (gently but firmly) Jon ... there is no need to freak out ... in fact my explicit goal in this approach is to work with you safely and appropriately on an outpatient basis ... I want to find a way to keep you out of the hospital if at all possible! But to do that, we have to turn our attention to developing a viable outpatient treatment plan that justifies outpatient care and gives us both a sense that it is safe to proceed together in this treatment here with me ... okay? I want to work together with you to figure out if there is a way we can get you to a better place, where you can go back to your unit, and be "that guy" that you feel that you used to be.

Jon: (much relieved) Yeah ... that's what I want ... I was feeling like there was no way that could happen before ... but now, I feel like maybe if it could be possible to do something about all this, maybe. All I know is that I don't want to keep feeling like this, and if there's a way to do that, then I want to give it a shot ... how do we do this?

As noted in Section C of the first session SSF (see Appendix B), the dyad has developed a viable Stabilization Plan and recognized two main suicidal drivers to focus on and treat during the process of CAMS care. When both the patient and clinician are satisfied with and are committed to the treatment plan, they confirm this by signing their names in the spaces provided at the bottom of the page. If the dyad is unable to produce an agreed-upon outpatient treatment plan, *then* inpatient care may be necessary.

If both the patient and clinician are able to devise a satisfactory treatment plan within the context of CAMS-guided care there will be subsequent interim CAMS sessions that regularly assess the risk and treat suicidal drivers until the suicidality is basically eliminated. At the end of every CAMS session, patients are always given a copy of the SSF assessment and treatment plan (and their SSF Stabilization Plan developed in session 1). Quite literally, we encourage the patient to fold up their stabilization plan and keep it in the utility pocket of their sleeve for ready access. Alternatively, we have service members take pictures of their CAMS documents for access on their smart phones. After the first session (and after each interim session), the clinician completes a final documentation form (Appendix D). In clinical settings, the SSF is often used as the official medical record progress note in place of standard documentation practices until the patient's suicide risk has fully resolved (i.e., the aim here is not to duplicate documentation but to intentionally make the patient's suicide risk the focus of care). Jobes (2016a) notes that this degree of attention and focus to suicide-specific documentation has the added advantage of notably lessening the possibility of malpractice tort litigation for wrongful death should a fatal outcome occur. In contrast, having the patient sign a "no-suicide" or "no-harm" contract does not reduce suicide risk in any way and does not protect providers from malpractice litigation (Goldsmith, Pellmar, Kleinman, & Bunney, 2002; Lewis, 2007; Rudd, Mandrusiak, & Joiner, 2006).

CASE STUDY CONTINUED: SSF STABILIZATION PLANNING AND INITIAL TREATMENT PLAN

By the end of the first CAMS session, Jon felt much better than when he first came in. He felt like he was engaged in a course of action that might help bring him out of the dark place he was in. He was not sure if it would work, but he just felt much better now that he was doing something about it, rather than just wallowing in his own misery. He felt reassured because the clinician seemed to have clear sense of what to do. He was relieved that her goal was to treat him on an outpatient basis, and he felt she was being honest about this goal. He liked the CAMS Stabilization Plan idea because it gave him a sense of what to do if he got into deep trouble. Even though he initially balked at giving up his guns, once she had explained the reasoning behind this finite intervention, he had to agree that having a friend keep his guns in a safe place was a sensible move. They had also agreed to have the clinician contact Jon's commander to help coordinate Jon's treatment attendance. They agreed to meet later in the week to begin focusing on Jon's suicidal drivers. They also identified friends (including his squad leader) to whom he could reach out to for support. With some reluctance Jon agreed to meet with a clinic psychiatrist to discuss medication options. Finally, Jon agreed to start a "dysfunctional thought record" to track his thoughts, emotions, and behaviors to identify any patterns and themes in his distress. While still overwhelmed and a bit shaky, Jon left the first session with a sense of direction. He felt a tiny glimmer of hope that maybe there was a way to deal with his situation.

CAMS TRACKING/UPDATE INTERIM SESSIONS

One of the key goals in CAMS care is the realization of three consecutive sessions in which the patient has marked reductions in suicidal thoughts, feelings, and behaviors. Three such sessions mark the "resolution" of suicidal risk and the potential end of CAMS care. Jobes (2016a) has previously noted that this operational definition of suicide resolution is both theoretically and empirically rooted. Previous CAMS studies indicate that suicide resolution is generally observed within 10 to 12 sessions (Jobes et al., 2005; Jobes, Rudd, Overholser, & Joiner, 2008); some CAMS patients achieve resolution in as few as six to eight sessions (Comtois et al., 2011). However, prior to achieving suicide resolution, each CAMS session starts by having the patient complete Section A of the SSF Tracking/Update form that is used for all interim sessions (see Appendix D). Moreover, the patient and clinician return to sitting alongside each other and revisit the treatment plan (Section B of the CAMS Tracking/Update form) at the end of each ongoing session.

Updating the CAMS treatment plan each session may require further crafting of the CAMS Stabilization Plan as well as "sharpening" the problem-focused treatment of suicidal drivers. The continual session-by-session re-evaluation and treatment planning updating process is central to effective CAMS care. Like the initial treatment planning session, patients always receives a photocopy of their interim SSF document at the conclusion of each session. Similar to the initial session form, SSF Section C (see Appendix E) is completed after each CAMS session to further supplement the case documentation.

At this juncture it is perhaps important to note that we conceive CAMS as a flexible clinical approach that can be tailored to a wide array of suicidal patients and a variety of clinical environments (Jobes, 2016a). Indeed, Jobes and colleagues (2011; 2016) have explicitly stated that CAMS is *not* a new psychotherapy. CAMS is best understood as a "therapeutic framework" within which a clinician can mold clinical care in ways that are consistent with his or her own psychotherapeutic orientation. When determining what treatment technique to employ, clinicians operating within the CAMS framework are always expected to rely on their own clinical judgment and experience to determine which techniques and methods make the most sense. Our goal is to sustain the flexibility and adaptability of CAMS so that clinicians from various theoretical orientations can utilize this intervention without needing to adopt a whole new clinical lens.

Although some clinicians are highly attracted to the flexibility CAMS offers, others desire more structure when it comes to working with suicidal clients in ongoing outpatient-based care. In order to strike a balance between structure and flexibility, a variety of suicide-focused interventions can be effectively employed within the CAMS therapeutic framework. To this end, we have seen the effective importation of several suicide-specific techniques within the CAMS clinical framework, including "chain analysis" from Dialectical Behavior Therapy (Linehan, 1993), "bibliotherapy" such as *Choosing to Live* (Ellis & Newman, 1996), the creation and utilization of a "Hope Kit" (Wenzel, Brown, & Beck, 2009), and the use of numerous suicide-specific cognitive-behavioral techniques (see Rudd, Joiner, & Rajab, 2001).

The adherence framework that we use within CAMS treatment research is perhaps the best example of our desire to find a balance between flexibility and more structured guidance (Jobes, 2010; Jobes et al., 2011; Jobes et al., 2016). Within our clinical trials, CAMS clinicians are required to adhere to a set of components that we consider essential to effective CAMS care. Ideally, these components should be included in every CAMS session:

1. Collaborative assessment of suicidal risk
2. Collaborative treatment planning
3. Collaborative identification of issues that make the client suicidal (identifying the patient's suicidal "drivers")
4. Collaborative problem-focused treatment of those issues that cause the patient's suicidality (i.e., targeting and treating the suicidal drivers)
5. Collaborative construction and understanding of existential purpose and meaning

CASE STUDY CONTINUED: CAMS TRACKING/UPDATE INTERIM SESSIONS

While there were a few bumps in the road, Jon nevertheless responded well to CAMS-guided care. In general, he found speaking with the clinician remarkably helpful. He appreciated not being shamed by the clinician for feeling suicidal—her empathic approach was quite comforting. He was especially struck by what she called their "collaborative empiricism" and how they explored what could have contributed to his "downward spiral." As they explored things together, they learned that Jon felt responsible for the deaths and injuries of several of his fellow Soldiers during his last deployment, which seemed to contribute greatly to his current state. Jon was not open to the clinician referring him to another provider for trauma-focused therapy, so they decided to address some of this work within his CAMS care because it greatly contributed to his suicidality.

By session four (see Appendix D—CAMS Tracking/Update interim session form), Jon seemed significantly back on track. He was exercising regularly again, engaging socially with his platoon-mates, and performing well as a team leader within his company. He began taking an antidepressant medication prescribed by the psychiatrist and, after a few weeks, felt that it had a positive impact on his mood. He also began to sleep regularly again, which was a major improvement from his sleeping habits prior to the start of therapy. After only a handful of sessions, Jon began noticing, with the encouragement of his psychologist, a genuine sense of hope and direction. In turn, his previous desperate feelings and suicidal thoughts began to fade.

CAMS CLINICAL OUTCOME/DISPOSITION

After three successive interim tracking sessions demonstrate the meaningful reduction or elimination of suicidal thoughts, feelings, and behaviors, CAMS may be concluded by the completion of the CAMS Outcome/Disposition Final Session form (Appendix F). Although complete resolution is the ideal clinical outcome, the Outcome/Disposition form also provides the opportunity to note alternative outcomes such as additional non-CAMS based therapy, clinical referrals, clinical dissolutions, and/or inpatient hospitalization. As previously noted, the CAMS outcome form offers important documentation relating to various treatment results and the general disposition of the case. (Note: The Appendices do *not* contain all the SSF documents that are part of a complete course of care.) Although not every CAMS case is free of hindrances and difficulties during the course of treatment, we have observed significant success for the *majority* of U.S. Air Force service personnel who completed CAMS/SSF-based treatment for suicidal ideations and behaviors (Jobes et al., 2005; Jobes et al., 2009).

CASE STUDY CONTINUED: CLINICAL OUTCOME/DISPOSITION

The case of Jon was a textbook CAMS success story. His slide into suicidal despair was explored and understood; his CAMS treatment effectively targeted and treated those issues that caused his suicidality. Jon reached resolution criteria by Session 8, at which point his CAMS care came to an end. Even though the CAMS phase of his treatment was formally drawn to a close, Jon continued his weekly counseling sessions, focusing on his PTSD symptoms through a course of cognitive processing therapy to help him with his combat-related traumas. Identifying the etiology of his current suicidal distress proved crucial for Jon's treatment. Recognizing that his agony and despair were related to his combat experiences served to reduce his distress. Exploring these issues in depth and working through his sense of responsibility and guilt proved to be a turning point in treatment. Jon entered CAMS care in state of desperation and with significant reluctance. He concluded CAMS care after a total of eight sessions with new skills for coping and a budding sense of hopefulness. Jon rediscovered a sense of competence and excitement about continuing his Army career. He also realized that he had people in his life, particularly in his chain of command, who cared for him and supported him during this difficult time.

REVIEW OF THE SSF/CAMS RESEARCH

CAMS has been used across a variety of clinical settings in both the United States (Conrad et al., 2009; Ellis, Allen, Woodson, Frueh, & Jobes, 2009; Jobes, Bryan, & Neal-Walden, 2009) and throughout the world (e.g., Arkov, Rosenbaum, Christiansen, Jonsson & Munchow, 2008; Schilling, Harbauer, Andreae, & Haas, 2006). To date, the SSF has been translated into six different languages, and the main CAMS reference book (Jobes, 2006) has been translated into Chinese and Korean and a second edition has recently been published (Jobes, 2016b). There is a rapidly growing body of empirical support for the clinical use of the CAMS and the SSF (Jobes, 2012). What follows is a brief review of research support for the use of the SSF and CAMS.

Suicide Status Form Studies

In 1988, the first author initiated the development of a new tool for evaluating and treating suicidal college students at a university-based mental health center. This novel assessment tool ultimately became the SSF (Jobes & Berman, 1993). There have been two psychometric studies of validity and reliability of the SSF "Core Assessment"—which measures, among other variables, psychological pain, stress, agitation, hopelessness, self-hate, and overall suicide risk. Jobes, Jacoby, Cimbolic, and Hustead (1997) found that these variables function quasi-independently and demonstrate good convergent and criterion prediction validity in a sample of suicidal college students. Moreover, the constructs demonstrated good test–retest reliability. Conrad et al. (2009) further established the psychometric validity and reliability of the SSF Core Assessment with a sample of suicidal inpatients. This study reported even stronger results for the validity and reliability of the SSF Core Assessment. Indeed, factor analytic results showed a robust two-factor solution that described 72% of the total variance.

Beyond purely psychometric considerations, other SSF-related research has produced support for its use. Jobes and colleagues (1997) used index session SSF quantitative ratings to distinguish two categories of clinical responders: "acute resolvers" and "chronic nonresolvers." These data were used to expand the clinical theory of suicidal typologies (Jobes, 1995) and laid the groundwork for subsequent SSF-related within-group investigations. One excellent example of a robust within-subject study examined the SSF ratings of suicidal college students using hierarchical linear modeling (Jobes Kahn-Greene, Greene, & Goeke-Morey, 2009). In this study, first-session index SSF quantitative ratings predicted significant and differential reductions in suicidal ideation and the moderation of this effect.

A different line of SSF research has investigated qualitative responses to the SSF. For example, Jobes and Mann (1999) examined patient-generated "Reasons for Living" (RFL) and "Reasons for Dying" (RFD) on the SSF. Results from this investigation revealed that SSF RFL and RFD responses could be reliably coded into distinct and meaningful categories (Jobes & Mann, 1999). When examining the RFL responses, family- and future-oriented responses made up over half of the total reasons for living; this indicates that when treating a suicidal client it may be beneficial to focus treatment on the creation of future plans and goals, as well promoting connection and support among the client's family system. The RFD responses yielded several aspects of escape as chief motivators driving a client's suicidal ideation and behaviors. Jobes and Mann suggested that clinicians should consider focusing treatment on affording the client some form of security or sanctuary for (i.e., providing a "therapeutic escape"). Subsequent research examined SSF qualitative descriptions of psychological pain, stress, agitation, hopelessness, and self-hate (Jobes et al., 2004).

While these early micro-coding studies of qualitative SSF responses were clinically useful, more recent macro coding of SSF qualitative responses have been used to create reliable suicidal typologies that can enhance the study of within-group treatment-related changes over time. For example, Corona et al. (2013) reliably categorized RFL/RFD suicidal patient responses into different suicidal motivations: life-motivated, ambivalent, and death-motivated. These different suicidal motivations were then studied in relation to index and posttreatment SSF core construct ratings in a community study of suicidal outpatients in Denmark. This line of typology-oriented research continues in the form of well-powered randomized clinical trials. Twenty-five years of SSF research clearly demonstrates the additive value of having both quantitative and qualitative suicide-relevant data generated directly by the suicidal patient (Jobes, 2012).

Research on CAMS

Years of SSF-related risk assessment research rather naturally led to the development of CAMS, which in turn morphed into a suicide-specific clinical intervention (Jobes, 2012). As described in this chapter, there are now seven SSF/CAMS-related published studies that show that the use of CAMS is associated with significant reductions in suicidal ideation as well as significant decreases in mental health symptom distress. Critical within this line of research was a nonrandomized case control study of CAMS versus "treatment as usual" (TAU) with a sample of suicidal Air Force service members (Jobes et al., 2005). In this study the results showed that patients receiving CAMS care displayed significantly more rapid resolutions of their suicidal ideation and generally resolved their suicidal ideation and behaviors in approximately 10 to 12 sessions compared to 23 to 25 sessions in the TAU control condition. Additionally, CAMS treatment was associated with statistically significant reductions in the utilization of nonmental health care services (i.e., emergency department and primary care visits) in the six months following the beginning of suicide-focused mental health care when compared to TAU patients. Various post hoc analyses of potential "third variables" that may have accounted for differences between CAMS and TAU showed that CAMS care was consistently superior to TAU care.

While the correlational findings of the Air Force study were encouraging, experimental randomized controlled trials (RCTs) investigating the *causal* effectiveness of CAMS has also made a persuasive case for its utility and effectiveness (Comtois et al., 2011). In this study, urban-dwelling suicidal outpatients ($n = 32$) were randomly assigned to CAMS versus "Enhanced Care As Usual" (E-CAU). Even with a small sample and limited statistical power, CAMS was found to be significantly superior to E-CAU on all primary and secondary measures. Specifically, CAMS resulted in significant reductions in suicidal ideation and overall symptom distress while significantly increasing hope and reasons for living. The significant between-group experimental effects were most robust at the 12-month assessment time point, which perhaps suggests an enduring positive effect from CAMS care. A second RCT (Andreasson et al., 2016) comparing Dialectical Behavior Therapy (DBT) versus CAMS for 108 Danish outpatient suicide attempters surprisingly showed no significant differences between treatments for self-harm and suicide attempts. These results are important as DBT is the gold standard treatment for suicidal behaviors and DBT patients received much more care than CAMS in the RCT. Finally a third unpublished

RCT of CAMS versus enhanced care as usual (E-CAU) with 148 suicidal Soldiers showed that CAMS caused the elimination of suicidal thinking within 3 months (Jobes, 2016b).

Further RCT replication is needed with well-powered designs. As of this writing, there are now three RCTs in various stages of completion within the United States and abroad. Results of these trials will provide valuable data about the causal impact of CAMS across a range of settings with suicidal community-based suicidal outpatients and suicidal college students.

CONCLUSION

Although CAMS was originally developed within a university-based counseling center, it has become an increasingly established as an evidence-based intervention applicable to a variety of suicidal patients in a diverse range of clinical settings. As clinical trial research continues, we are continually cultivating new CAMS adaptations and innovations (e.g., electronic versions of the SSF and CAMS group treatments). Given the nature of the military suicide problem, the need for novel and effective methods of working with suicidal military personnel has become critical. We have seen some notable success using CAMS in military treatment settings, and our current research continues to develop the use of CAMS with this population. The particular emphasis in CAMS on developing a strong clinical alliance and enhancing patient motivation seems especially suitable for military populations. Effectively targeting and treating suicidality can become a therapeutic "mission" that these service members may be particularly good at performing.

FUTURE DIRECTIONS

The field of clinical suicidology is growing rapidly in significance and potential impact. Changes in health care delivery in a post-health care reform era have far-reaching implications for the treatment of suicidal risk (Jobes & Bowers, 2015). Among the key questions facing the field is how do we increase the current standard of care for suicidal risk? As discussed elsewhere (Jobes et al., 2008), too many clinicians when facing a suicidal patient still fail to provide sufficient informed consent, do not thoroughly assess suicidal risk, and do not use evidence-based treatments to effectively manage suicide risk. There is still an overreliance on a medication-only approach, no-harm contracting, and an overreliance on brief inpatient stays that are not suicide-specific. In the years ahead we anticipate an increased need to demonstrate the use of evidence-based, suicide-specific care that is least restrictive and most cost-effective (Jobes & Bowers, 2015).

Given these considerations, we feel that CAMS is well-positioned to provide a sensible response to many of these challenges. For example, we are seeing CAMS adapted to a range of settings, including inpatient care and respite care, and even within consultation liaison psychiatry. CAMS training uses a blended model that includes e-learning, role-playing, and case consultation to increase adherence. In addition, we are embarking on the development of CAMS groups, which may effectively treat more suicidal patients in a cost-effective and safe manner. Integrating CAMS and technology is a major focus of our future research.

When we look back on this era we may reflect on how the military suicide problems following the wars in Iraq and Afghanistan fostered a new generation of suicide research. We hope that this era will be marked by major innovations—such as CAMS—that make a lifesaving difference for both service members and the public at large.

CASE STUDY OF JON: EPILOGUE

After his CAMS resolution session Jon's suicidal thinking never recurred. He nevertheless continued to pursue other mental health treatment focusing on reducing his PTSD symptoms and the development of new coping strategies to improve his overall functioning. As he steadily stabilized and improved, Jon began to excel in his position as an infantry fire team leader, and he was eventually promoted to a new position as an acting squad leader in his platoon. Jon continued to clinically improve and ultimately left mental health care altogether. Within a year after his promotion Jon reconnected with a girl from his hometown whom he dated and subsequently married. Of course not all CAMS cases resolve as rapidly or as successfully as Jon's case, but over the years we have seen hundreds of such cases that do resolve quickly and effectively with CAMS guided care. Jon had once stared into the abyss of suicidal despair, convinced that nothing could make a difference. But with appropriate clinical care, engagement, and management he came out the other side, able to re-embrace his competence and reclaim his life with a hard-earned sense of purpose and meaning.

REFERENCES

Andreasson, K., Krogh, J., Wenneberg, C., Jessen, H. K. L., Krakauer, K., Gluud, C., Thomsen, R. R., Randers, L., & Nordentoft, M. (2016). Effectiveness of dialectical behavior therapy versus collaborative assessment and management of suicidality for reduction of self-harm in adults with borderline personality traits and disorder—A randomized observer-blinded clinical trial. *Depression and Anxiety*, DOI 10.1002/da.22472

Arkov, K., Rosenbaum, B., Christiansen, L., Jonsson, H., & Munchow, M. (2008). Treatment of suicidal patients: The Collaborative Assessment and Management of Suicidality. *Ugeskrift for Laeger*, 170, 149–153.

Britton, P. C., Conner, K. R., & Maisto, S. A. (2012). An open trial of motivational interviewing to address suicidal ideation with hospitalized veterans. *Journal of Clinical Psychology*, 68(9), 961–971. doi:10.1002/jclp.21885

Brown, G. K., Have, T. T., Henriques, G. R., Xie, S. X., Hollander, J. E., & Beck, A. T. (2005). Cognitive therapy for the prevention of suicide attempts. *JAMA: Journal of the American Medical Association*, 294, 563–570.

Chiles, J., & Strosahl, K. (2005). *Clinical manual for assessment and treatment of suicidal patients*. Arlington, VA: American Psychiatric Publishing.

Comtois, K. A., Jobes, D. A., S. O'Connor, S., Atkins, D. C., Janis, K., E. Chessen, C., ... Yuodelis-Flores, C. (2011). Collaborative Assessment and Management of Suicidality (CAMS): Feasibility trial for next-day appointment services. *Depression and Anxiety*, 28(11), 963–972.

Conrad, A. K., Jacoby, A. M., Jobes, D. A., Lineberry, T. Jobes, D., Shea, C., ... Kung, S. (2009). A psychometric investigation of the suicide status form with suicidal inpatients. *Suicide and Life-Threatening Behavior*, 39, 307–320.

Corona, C. D., Jobes, D. A., Nielsen, A. C., Pedersen, C. M., Jennings, K. W., Lento, R. M., & Brazaitis, K. A. (2013). Assessing and treating different suicidal states in a Danish outpatient sample. *Archives of Suicide Research*, 17(3), 302–312. doi:10.1080/13811118.2013.777002

Ellis, T. E., Allen, J. G., Woodson, H., Frueh, B. C., & Jobes, D. A. (2009). Implementing an evidence-based approach to working with suicidal inpatients. *Bulletin of the Menninger Clinic*, 73, 339–354.

Ellis, T. P., & Newman, C. F. (1996). *Choosing to live: How to defeat suicide through cognitive therapy*. Oakland, CA: New Harbinger.

Goldsmith S. K., Pellmar, T. C., Kleinman, A. M., & Bunney, W. E. (2002). Programs for suicide prevention. In Institute of Medicine Committee on Pathophysiology and Prevention of Adolescent and Adult Suicide, *Reducing suicide: A national imperative* (pp. 273–330). Washington, DC: National Academies Press. Retrieved from http://www.ncbi.nlm.nih.gov/books/NBK220931/

James, L. C. (2012). Introduction to special section on suicide prevention. *Military Psychology*, 24(6), 565–567.

Jobes, D. A. (1995). The challenge and the promise of clinical suicidology. *Suicide and Life-Threatening Behavior*, 25, 437–449.

Jobes, D. A. (2006). *Managing suicidal risk: A collaborative approach*. New York: Guilford Press.

Jobes, D. A. (2010). Suicidal patients, the therapeutic alliance, and the Collaborative Assessment and Management of Suicidality. In K. Michel & D. A. Jobes (Eds.), *Building a therapeutic alliance with the suicidal patient* (pp. 205–229). Washington, DC: American Psychological Association.

Jobes, D. A. (2011). Suicidal patients, the therapeutic alliance, and the Collaborative Assessment and Management of Suicidality. In M. Konrad & D. A. Jobes (Eds.), *Building a therapeutic alliance with the suicidal patient* (pp. 205–229). Washington, DC: American Psychological Association. Retrieved from http://dx.doi.org/10.1037/12303-012

Jobes, D. A. (2012). The Collaborative Assessment and Management of Suicidality (CAMS): An evolving evidence-based clinical approach to suicidal risk. *Suicide and Life-Threatening Behavior*, 42(6), 640–653.

Jobes, D. A. (2016a). *Managing suicidal risk: A collaborative approach* (2nd Ed.). New York: Guilford Press.

Jobes, D. A. (2016b, August). *Operation Worth Living: A Randomized Controlled Trial of the Collaborative Assessment and Management of Suicidality (CAMS) vs. Enhanced Care as Usual with Suicidal Soldiers*. Presentation at Military Health System Research Symposium, Orlando, FL.

Jobes, D. A., & Berman, A. L. (1993). Suicide and malpractice liability: Assessing and revising policies, procedures, and practice in outpatient settings. *Professional Psychology: Research and Practice, 24*, 91–99.

Jobes, D. A., & Bowers, M. (2015). Treating suicidal risk in a post-healthcare reform era. *Journal of Aggression, Conflict and Peace Research, 7*, 167–178.

Jobes, D. A., Bryan, C. J., & Neal-Walden, T. A. (2009). Conducting suicide research in naturalistic clinical settings. *Journal of Clinical Psychology, 65*, 382–395.

Jobes, D. A., Comtois, K., Brenner, L., & Gutierrez, P. (2011). Clinical trial feasibility studies of the Collaborative Assessment and Management of Suicidality (CAMS). In R. O'Connor, S. Platt, & J. Gordon (Eds.), *International handbook of suicide prevention: Research, policy, and practice* (pp. 383–400). Chichester, UK: Wiley Blackwell.

Jobes, D. A., Comtois, K. A., Brenner, L. A., Gutierrez, P. M., & O'Connor, S. S. (2016). Lessons learned from clinical trials of the Collaborative Assessment and Management of Suicidality (CAMS) (pp. 431–449). In R. C. O'Connor and J. Pirkis (Eds.), *International handbook of suicide prevention: Research, policy, & practice* (2nd ed.). West Sussex, UK: Wiley-Blackwell.

Jobes, D. A., & Drozd, J. F. (2004). The CAMS approach to working with suicidal patients. *Journal of Contemporary Psychotherapy, 34*, 73–85.

Jobes, D. A., Jacoby, A. M., Cimbolic, P., & Hustead, L. A. (1997). Assessment and treatment of suicidal clients in a university counseling center. *Journal of Counseling Psychology, 44*, 368–377.

Jobes, D. A., Kahn-Greene, E., Greene, J., & Goeke-Morey, M. (2009). Clinical improvements of suicidal outpatients: Examining suicide status form responses as moderators. *Archives of Suicide Research, 13*, 147–159.

Jobes, D. A., Lento, R., & Brazaitis, K. (2012). An evidence-based clinical approach to suicide prevention in the Department of Defense: The Collaborative Assessment and Management of Suicidality (CAMS). *Military Psychology, 24*(6), 604–623.

Jobes, D. A., & Mann, R. E. (1999). Reasons for living versus reasons for dying: Examining the internal debate of suicide. *Suicide and Life-Threatening Behavior, 29*, 97–104.

Jobes, D. A., Nelson, K. N., Peterson, E. M., Pentiuc, D., Downing, V., Francini, K., & Kiernan, A. (2004). Describing suicidality: An investigation of qualitative SSF responses. *Suicide and Life-Threatening Behavior, 34*, 99–112.

Jobes, D. A., Rudd, M. D., Overholser, J. C., & Joiner, T. E. J. (2008). Ethical and competent care of suicidal patients: Contemporary challenges, new developments, and considerations for clinical practice. *Professional Psychology: Research and Practice, 39*, 405–413

Jobes, D. A., Wong, S. A., Conrad, A. K., Drozd, J. F., & Neal-Walden, T. (2005). The Collaborative Assessment and Management of Suicidality versus treatment as usual: A retrospective study with suicidal outpatients. *Suicide and Life-Threatening Behavior, 35*, 483–497.

Lewis, L. M. (2007). No-harm contracts: A review of what we know. *Suicide and Life-Threatening Behavior, 37*(1), 50–57. doi:10.1521/suli.2007.37.1.50

Linehan, M. M. (1993). *Cognitive behavioral treatment of borderline personality disorder.* New York: Guilford Press.

O'Connor, S. S., Comtois, K. A., Wang, J., Russo, J., Peterson, R., Lapping-Carr, L., & Zatzick, D. (2015). The development and implementation of a brief intervention for medically admitted suicide attempt survivors. *General Hospital Psychiatry, 37*(5), 427–433.

Rudd, M. D., Joiner, T. E., & Rajab, M. H. (2001). *Treating suicidal behavior: An effective, time-limited approach.* New York: Guilford Press.

Rudd, M. D., Mandrusiak, M., & Joiner, T. E. (2006). The case against no-suicide contracts: The commitment to treatment statement as a practice alternative. *Journal of Clinical Psychology, 62*(2), 243–251. doi:10.1002/jclp

Schilling, N., Harbauer, G. Andreae, A., & Haas, S. (2006, March). *Suicide risk assessment in inpatient crisis intervention.* Poster presented at the Fourth Aeschi Conference, Aeschi, Switzerland.

Stanley, B., & Brown, G. K. (2012). Safety planning intervention: A brief intervention to mitigate suicide risk. *Cognitive and Behavioral Practice, 19*(2), 256–264.

Wenzel, A., Brown, G. K., & Beck, A. T. (2009). *Cognitive therapy for suicidal patients: Scientific and clinical applications.* Washington, DC: American Psychological Association.

APPENDIX A
CAMS Suicide Status Form (SSF-IV-R) Initial Session

Patient: Jon Jones Clinician: EJ Date: 1/4 Time: 09:00

Section A (Patient):

Rate and fill out each item according to how you feel <u>right now</u>.
Then rank in order of importance 1 to 5 (1=most important to 5=least important).

Rank	Item
2	1) RATE PSYCHOLOGICAL PAIN (*hurt, anguish, or misery in your mind, **not** stress, **not** physical pain*): Low pain: 1 2 3 4 **(5)** :High pain What I find most painful is: I'm a liability - dead weight
4	2) RATE STRESS (*your general feeling of being pressured or overwhelmed*): Low stress: 1 2 3 4 **(5)** :High stress What I find most stressful is: I can't handle anything anymore
5	3) RATE AGITATION (*emotional urgency; feeling that you need to take action; **not** irritation; **not** annoyance*): Low agitation: 1 2 3 **(4)** 5 :High agitation I most need to take action when: I'm alone at night
3	4) RATE HOPELESSNESS (*your expectation that things will not get better no matter what you do*): Low hopelessness: 1 2 3 4 **(5)** :High hopelessness I am most hopeless about: being able to ever get better
1	5) RATE SELF-HATE (*your general feeling of disliking yourself; having no self-esteem; having no self-respect*): Low self-hate: 1 2 3 4 **(5)** :High self-hate What I hate most about myself is: that I'm such a screwup now
N/A	6) RATE OVERALL RISK OF SUICIDE: Extremely low risk: 1 2 3 **(4)** 5 :Extremely high risk (will **not** kill self) (will kill self)

1) How much is being suicidal related to thoughts and feelings about <u>yourself</u>? Not at all: 1 2 3 4 **(5)** : completely
2) How much is being suicidal related to thoughts and feelings about <u>others</u>? Not at all: 1 **(2)** 3 4 5 : completely

Please list your reasons for wanting to live and your reasons for wanting to die. Then rank in order of importance 1 to 5.

Rank	REASONS FOR LIVING	Rank	REASONS FOR DYING
2	maybe get better	1	no hope
3	be happy again	3	I'm just an embarrassment
1	don't want to hurt my mom	2	end the pain
		4	it's easier

I wish to live to the following extent: Not at all: 0 1 2 3 **(4)** 5 6 7 8 : Very much
I wish to die to the following extent: Not at all: 0 1 2 3 **(4)** 5 6 7 8 : Very much
The one thing that would help me no longer feel suicidal would be: to stop screwing everything up and not be so useless

From *Managing Suicidal Risk: A Collaborative Approach*, 2nd ed. by D. A. Jobes (2016). New York: Guilford Press. Copyright 2016 by Guilford Press. Adapted and printed with permission.

APPENDIX B
CAMS Suicide Status Form (SSF-IV-R) Initial Session

Section B (Clinician):

(Y) N Suicide ideation   Describe: "all the time" I'm a screw up
- Frequency   ✓ per day   ✓ per week   ✓ per month dead weight
- Duration   ___ seconds   ___ minutes   ___ hours

(Y) N Suicide plan
When: at night - get drunk and shoot self
Where: barracks
How: handgun   Access to means (Y) N
How: _____   Access to means Y N

Y (N) Suicide preparation   Describe: _____

(Y) N Suicide rehearsal   Describe: loaded gun to head

Y N History of suicidal behaviors
- Single attempt   Describe: n/a
- Multiple attempts   Describe: n/a

(Y) N Impulsivity   Describe: "I just do things without thinking about them"
(Y) N Substance abuse   Describe: etoh
(Y) N Significant loss   Describe: loss of sense of self
(Y) N Relationship problems   Describe: wants a relationship
(Y) N Burden to others   Describe: "dead weight"
Y (N) Health/pain problems   Describe: _____
(Y) N Sleep problems   Describe: some
(Y) N Legal/financial issues   Describe: current UCMJ action
(Y) N Shame   Describe: "for what I let happen in AF"

Section C (Clinician):   TREATMENT PLAN (Refer to Sections A & B)

Problem #	Problem Description	Goals and Objectives	Interventions	Duration
1	Self-Harm Potential	Safety and Stability	Stabilization Plan Completed ✓	3 mths
2	feel like a screw up	to be competent and a "water-walker" again	individual psychotherapy - insight oriented + CBT	3 mths
3	Shame & guilt	to understand why I feel so much shame	ind. psychotherapy insight + CBT	3 mths

YES ✓ NO ___ Patient understands and concurs with treatment plan?
YES ___ NO ✓ Patient at imminent danger of suicide (hospitalization indicated)?

Jon Jones   1/4   Emme Johnson   1/4
Patient Signature   Date   Clinician Signature   Date

APPENDIX C

CAMS Suicide Status Form (SSF-IV-R) STABILIZATION PLAN

Ways to reduce access to lethal means:

1. have squad leader store my gun
2. decrease drinking
3. _____

Things I can do to cope differently when I am in a suicide crisis (consider crisis card):

1. go for a walk
2. play video games with guys in the platoon
3. go to the gym
4. call my mother
5. read a book
6. Life or death emergency contact number: 1-800-273-TALK (Lifeline)

People I can call for help or to decrease my isolation:

1. My squad leader (Tom) 555-1234
2. Mike (good friend in my platoon) 555-2345
3. My mom 555-3456

Attending treatment as scheduled:

Potential Barrier: Solutions I will try:

1. duties at night – talk to squad leader about coming to appointments
2. no car – ask squad leader if I can borrow his car for appointments

THE COLLABORATIVE ASSESSMENT AND MANAGEMENT OF SUICIDALITY

APPENDIX D
CAMS Suicide Status Form (SSF-IV-R) Tracking/Update Interim Session

Patient: Jon Jones   Clinician: EJ   Date: 1/17   Time: 11:00

Section A (*Patient*):

Rate each item according to how you feel right now.

1) RATE PSYCHOLOGICAL PAIN (*hurt, anguish, or misery in your mind, **not** stress, **not** physical pain*):
   Low pain: 1 2 3 **(4)** 5 :High pain

2) RATE STRESS (*your general feeling of being pressured or overwhelmed*):
   Low stress: 1 2 3 **(4)** 5 :High stress

3) RATE AGITATION (*emotional urgency; feeling that you need to take action; **not** irritation; **not** annoyance*):
   Low agitation: 1 2 3 **(4)** 5 :High agitation

4) RATE HOPELESSNESS (*your expectation that things will not get better no matter what you do*):
   Low hopelessness: 1 2 3 **(4)** 5 :High hopelessness

5) RATE SELF-HATE (*your general feeling of disliking yourself; having no self-esteem; having no self-respect*):
   Low self-hate: 1 2 3 4 **(5)** :High self-hate

6) RATE OVERALL RISK OF SUICIDE: Extremely low risk: 1 2 **(3)** 4 5 :Extremely high risk
   (will **not** kill self)   (will kill self)

In the past week: Suicidal Thoughts/Feelings Y ✓ N __   Managed Thoughts/Feelings Y ✓ N __   Suicidal Behavior Y __ N ✓

Section B (*Clinician*):
Resolution of suicidality, if: current overall risk of suicide <3; in past week: no suicidal behavior and effectively managed suicidal thoughts/feelings ☐ 1st session ☐ 2nd session
Complete SSF Outcome Form at 3rd consecutive resolution session

TREATMENT PLAN UPDATE

Patient Status:
☐ Discontinued treatment ☐ No show ☐ Cancelled ☐ Hospitalization ☐ Referred/Other: _____

Problem #	Problem Description	Goals and Objectives	Interventions	Duration
1	Self-Harm Potential	Safety and Stability	Stabilization Plan Updated ✓	11 weeks
2	Self-hate (screw-up)	↓ self hate ↑ ⊕ self regard	insight + CBT psychotherapy	11
3	Shame + guilt	↓ shame + guilt	" "	11

Patient Signature: Jon Jones   Date: 1/17
Clinician Signature: Emme Johnson   Date: 1/17

APPENDIX E

Section C *(Clinician Post-Session Evaluation)*:

MENTAL STATUS EXAM (circle appropriate items):
- ALERTNESS: **ALERT** DROWSY LETHARGIC STUPOROUS
  OTHER: _____
- ORIENTED TO: **PERSON** **PLACE** **TIME** **REASON FOR EVALUATION**
- MOOD: EUTHYMIC ELEVATED **DYSPHORIC** AGITATED ANGRY
- AFFECT: FLAT BLUNTED CONSTRICTED **APPROPRIATE** LABILE
- THOUGHT CONTINUITY: **CLEAR & COHERENT** GOAL-DIRECTED TANGENTIAL CIRCUMSTANTIAL
  OTHER: _____
- THOUGHT CONTENT: **WNL** OBSESSIONS DELUSIONS IDEAS OF REFERENCE BIZARRENESS MORBIDITY
  OTHER: _____
- ABSTRACTION: **WNL** NOTABLY CONCRETE
  OTHER: _____
- SPEECH: **WNL** RAPID SLOW SLURRED IMPOVERISHED INCOHERENT
  OTHER: _____
- MEMORY: **GROSSLY INTACT**
  OTHER: _____
- REALITY TESTING: **WNL**
  OTHER: _____

NOTABLE BEHAVIORAL OBSERVATIONS: _____

DIAGNOSTIC IMPRESSSIONS/DIAGNOSIS (**DSM**/ICD DIAGNOSES):

309.81 PTSD; 296.2 MDD

PATIENT'S OVERALL SUICIDE RISK LEVEL (check one and explain):

- ☐ MILD (WTL/RFL)
- ☑ MODERATE (AMB) — **Explanation:** Suicidal but strongly committed to CAMS tx.
- ☐ HIGH (WTD/RFD)

CASE NOTES: Stabilization plan effective; working "hanging out" with other people again; using stabilization plan

Next Appointment Scheduled: 1 week   Treatment Modality: individual tx

Clinician Signature: *Emma Johnson*   Date: 1/17

APPENDIX F
CAMS Suicide Status Form (SSF-IV-R) Outcome/Disposition Final Session

Patient: Jon Jones Clinician: EJ Date: 3/26 Time: 1300

Section A (*Patient*):

Rate each item according to how you feel right now.

1) RATE PSYCHOLOGICAL PAIN (*hurt, anguish, or misery in your mind, **not** stress, **not** physical pain*):
   Low pain: 1 (2) 3 4 5 :High pain

2) RATE STRESS (*your general feeling of being pressured or overwhelmed*):
   Low stress: (1) 2 3 4 5 :High stress

3) RATE AGITATION (*emotional urgency; feeling that you need to take action; **not** irritation; **not** annoyance*):
   Low agitation: (1) 2 3 4 5 :High agitation

4) RATE HOPELESSNESS (*your expectation that things will not get better no matter what you do*):
   Low hopelessness: 1 (2) 3 4 5 :High hopelessness

5) RATE SELF-HATE (*your general feeling of disliking yourself; having no self-esteem; having no self-respect*):
   Low self-hate: 1 2 (3) 4 5 :High self-hate

6) RATE OVERALL RISK OF SUICIDE: Extremely low risk: 1 (2) 3 4 5   :Extremely high risk
   (will **not** kill self)   (will kill self)

In the past week: Suicidal Thoughts/Feelings Y__ N✓ Managed Thoughts/Feelings Y✓ N__ Suicidal Behavior Y__ N✓

Were there any aspects of your treatment that were particularly helpful to you? If so, please describe these. Be as specific as possible. not everything was my fault - connecting my thoughts to how I was acting with others

What have you learned from your clinical care that could help you if you became suicidal in the future? Stabilization plan - knowing I can come here for help if I need to

Section B (*Clinician*):

Third consecutive session of resolved suicidality: ✓ Yes ___ No (if no, continue CAMS tracking)

**Resolution of suicidality, if for third consecutive week: current overall risk of suicide <3; in past week: no suicidal behavior and effectively managed suicidal thoughts/feelings

OUTCOME/DISPOSITION (Check all that apply):

✓ Continuing outpatient psychotherapy ___ Inpatient hospitalization

___ Mutual termination ___ Patient chooses to discontinued treatment (unilaterally)

___ Referral to:_____

___ Other. Describe:_____

Next Appointment Scheduled (if applicable): 4/2

Jon Jones 3/26   _Emma Johnson_ 3/26
Patient Signature Date   Clinician Signature Date

14

Healing the Hidden Wounds of War
Treating the Combat Veteran with PTSD at Risk for Suicide

Herbert Hendin

Although storytellers and writers going back to Homer described the profound psychological effects of war on those who fought, it was not until World War I that psychiatrists in Austria and Germany began to describe in detail the traumatic reactions to combat. They concluded that combat trauma involved a breaking through of the individual's defense against stimuli (*reitzschutz*). "Traumatic neurosis" was the result of fright (*schreck*)—a condition occurring when one encountered a danger without being adequately prepared. The repetitive nightmares seen in the disorder were considered an attempt to be prepared after the fact, to dissipate by repetition the anxiety generated by the experience (Freud, Ferenczi, Abraham, Simmel, & Jones, 1921).

Abram Kardiner's work with World War I veterans, and subsequent collaboration with John P. Spiegel, who worked with World War II veterans, clearly delineated the symptoms of posttraumatic stress disorder (PTSD; Kardiner & Spiegel, 1947). Although Kardiner acknowledged *reitzschutz* theory as a starting point for his own thinking, he incorporated traumatic stress into an adaptational frame of reference. Kardiner saw trauma as an alteration in the individual's usual environment in which the adaptive maneuvers suitable to previous situations no longer sufficed. With the balance between the individual and his or her adaptive equipment broken, a new adaptation was not possible, and the individual accommodated his shrunken inner resources with the development of symptoms.

Kardiner described the features of what was then called traumatic war neurosis: fixation on the trauma, repetitive nightmares, irritability, exaggerated reactions to unexpected noise (startle reactions), proclivity to explosive aggressive behavior, and a contraction of the general level of functioning, including intellectual ability. He also saw loss of interest in activity as a result of the breakup of organized channels of action that were replaced by periodic outbursts of disorganized aggression. The internal conception of the self became altered, confidence was lost, the world was seen as a hostile place, and the patient lived in perpetual dread of being overwhelmed (Kardiner & Spiegel, 1947). The symptoms he described were subsequently categorized for use in a civilian population in the *Diagnostic and Statistical Manual of Mental Disorders* of the American Psychiatric Association that became the basis for the diagnosis of PTSD.

PTSD is a different disorder in a veteran population, and Department of Veterans Affairs (VA) researchers developed their own instrument to diagnose PTSD and to evaluate its severity. The most recent revision of the diagnostic criteria for PTSD (in the *Diagnostic and Statistical Manual of Mental Disorders*, fifth edition) places PTSD among the various "stress disorders," which makes some sense but the DSM-5 diagnostic criteria do not fully describe the clinical phenonmenom of PTSD in combat veterans. VA researchers and clinicians, whether implicitly or explicitly, use their own conceptual model to assess and treat this disorder.

As early as the end of the nineteenth century, the effect of combat in contributing to suicide was recognized by the French sociologist Émile Durkheim, who reported that suicide rates among European military men were up to 10 times greater than those among male civilians of comparable age (Durkheim, 1951). Only after the war in Vietnam were systematic analyses undertaken in the United States that confirmed a higher rate of suicide among men who had served in the military than among other men of the same age (U.S. House of Representatives, Committee on Veteran Affairs, 1978; Centers for Disease Control and Prevention, 1987). Clinicians who worked closely with men who saw considerable combat in Vietnam noted that suicidal behavior was frequently a manifestation of what came to be known as PTSD (Jury, 1979; Kolb, 1986; Lipkin, Blank, Parson, & Smith, 1982; Stuen & Solberg, 1972). Vietnam veterans with PTSD have been shown to be four times more likely to die by suicide than veterans without PTSD (Bullman & Kang, 1996).

A seven-year research and treatment project begun in 1978 at a Veterans Administration medical center, consisting of combat veterans of the Vietnam War with PTSD who were at risk for suicide, laid the groundwork for the material in this chapter (Hendin & Haas 1984a, 1984b). Comparable research being done at the Michael DeBakey VA Medical Center in Houston, Texas, which focuses on veterans of the wars in Iraq and Afghanistan, is building on this work (Hendin, Al Jurdi, Houck, Hughes, Turner, 2010). This chapter is aimed at integrating this work to create a helpful guideline for the understanding and treatment of the combat veteran with PTSD at risk for suicide.

PTSD AND THE RISK FOR SUICIDE

The study of Vietnam combat veterans with PTSD provided insight into some of the factors associated with suicide among them. One-hundred combat veterans with PTSD completed a comprehensive questionnaire and five semistructured interviews, which we have described in detail elsewhere (Hendin & Haas, 1991). The interviews elicited additional information about the individual's life before and after his service in Vietnam. None of the veterans had made a suicide attempt or been preoccupied with suicide prior to his combat experience.

Persistent severe guilt over combat experiences was found to be the major factor differentiating veterans who had attempted suicide and those who were seriously preoccupied with suicide from those veterans who were neither. Nineteen of the 100 combat veterans with PTSD had attempted suicide at least once since returning from Vietnam. Guilt related to combat actions was significantly marked in all 19 of the suicide attempters but in only 32 of the 66 nonsuicidal veterans ($\chi^2 = 14.24$, $df = 1$, $p < 0.001$). Fifteen, who were not suicidal, had been seriously preoccupied with suicide since they left the service. Guilt was also marked in 12 of these 15 veterans compared to the 66 nonsuicidal veterans ($\chi^2 = 3.71$, $df = 1$, $p = 0.05$).

Although anxiety, survivor guilt, and depression marked those at risk for suicide, combat guilt outperformed the other three predictors when all three were entered into a logistic regression simultaneously.

Although logistic regression analysis did not identify survivor guilt as a significant predictor of suicide attempts, additional one-way analysis provided some evidence of the importance of the concurrent presence of the two types of guilt. Forty of the 100 veterans studied, for example, showed both marked guilt about combat actions and marked survivor guilt. Among the group, 14 (35%) had made a suicide attempt. By contrast, none of the 30 veterans who showed neither marked combat guilt nor survivor guilt had attempted suicide ($\chi^2 = 13.3$, $df = 1$, $p < .001$). The findings suggest that the combination of these two types of guilt play a significant role in determining suicidal risk among patients.

Moreover, among the 17 veterans who had killed civilians while feeling out of control and felt guilty about such actions but were not suicidal, only 2 had survivor guilt. By contrast 9 of the 12 suicide attempters who had killed civilians while feeling out of control experienced survivor guilt in addition to guilt over their combat actions ($\chi^2 = 12.21$, $df = 1$, $p < .001$).

The combat experiences of the suicidal veterans were examined for possible determinants of their guilt. The chaotic nature of guerilla warfare in Vietnam, the uncertainty about who the enemy was, the emphasis on body counts, and the Viet Cong's use of women, children, and the elderly as combatants contributed to combat actions about which veterans felt severe guilt.

The Viet Cong, for example, would strip American soldiers they had killed and hang their naked bodies from a tree with their genitals stuffed into their mouths. Such tactics, designed to frighten soldiers, also tended to infuriate them and contributed to atrocities on both sides.

MEANING OF COMBAT

How each veteran experienced the combat events (i.e., the meaning of the combat experience to the veteran) was integral in determining the nature of the guilt and the risk for suicidal behavior. The term "meaning of combat" refers to the subjective, often unconscious perception of the traumatic event and includes the affective state of the veteran before the event took place, when it took place, and the affects experienced subsequently.

Repetitive nightmares and re-experiencing symptoms are cardinal symptoms of PTSD. Both are valuable tools in determining the meaning of the experience to the veteran. The following case example is illustrative (Hendin & Haas 1984a).

Case 1

Throughout his tour Greg L. thought he would be killed in action. The thought was comforting to him because it would enable him to avoid having his friends, family, and fiancée discover that he had lost control of his anger and killed without reason in Vietnam. During the last two weeks of his tour, when he learned that he was not going to be assigned to any more combat missions, he tried to kill himself with an overdose of drugs.

He had been an artillery spotter in Vietnam. He was preoccupied with a memory of a friendly village that he and his sergeant had helped to destroy in a contest designed to see who could call in the best coordinates. Through his binoculars, Greg had watched with excitement as the shells landed. As the village was being destroyed he saw an old woman with betel nut stains on her teeth running in his direction. She was shaking her arms trying to get him to stop the shelling. As she ran toward him, she was killed by an artillery round.

After he returned to the United States, Greg was tormented by a painful recurring nightmare that expressed his intense guilt over the destruction of the village. In the dream he is captured by South Vietnamese villagers, strung on a pole like a pig carcass, and paraded around the village so that everyone can curse him, spit on him, hit him, and throw stones at him. The old woman with the betel nut-stained teeth is taunting him. The villagers hold him responsible for all the death and destruction in their village. He knows they are going to kill him.

Greg made a second suicide attempt during a re-experiencing event in which he thought he saw the villagers covered in blood. He cut his wrists and described feeling a sense of relief as the blood spurted out. Both the nightmare and the reliving experience express his sense of guilt and need for punishment.

The nightmares of most veterans with PTSD correspond closely with the combat experiences and the terror over being killed that they engender. Veterans who have severe guilt over their combat are more likely to experience nightmares that reflect guilt, are punitive in nature, and indicate risk for suicide. Greg's experience of feeling out of control while in Vietnam was usual among the suicide attempters. Sixteen of the 19 (suicide attempters 82%) had felt out of control as a result of excessive fear or rage during periods of their tours of duty, including the situations in which their anger led to deaths about which they felt guilty (Hendin & Haas, 1991). Veterans like Greg, who feel out of control while in combat, and remained so in civilian life, are the most difficult to involve in treatment, so it was not surprising that Greg turned down the offer of short-term psychotherapy that was available to participants in the research project. During the course of study, three of the veterans who also felt out of control, and did not accept the offer of treatment, did kill themselves.

Although killing the enemy or being in danger of being killed increases the likelihood of being severely distressed by combat, the work described here suggests the importance of looking closely at perceptual and adaptive factors, rather than simply at objective aspects of the combat experience, in seeking to understand veterans' combat experiences. How combat events and situations are perceived, integrated, and acted on bears a primary relationship to the aftereffects of combat on the veteran.

RECOGNIZING THE VETERAN AT RISK FOR SUICIDE

Treating the veteran at risk for suicide requires identifying correctly those veterans who are at risk. In a previous study with patients who were not veterans, detailed data was obtained from therapists of patients who committed suicide while in treatment with them. Written responses to questionnaires and subsequent personal interviews with the therapist were used to determine what patients were feeling and experiencing in their lives immediately before their suicides (Hendin, Maltsberger, Lipschitz, Haas & Kyle, 2001). The data was contrasted with data from the same therapists on comparably depressed patients in treatment with them who were not suicidal. We found that the suicides were preceded by a time-limited state of suicide crisis that was marked by three factors that usually occurred in combinations of two or three in a single patient: a precipitating event, behavioral changes, and intense affective states.

Intense affective states that were intolerable and uncontrollable proved to be the factor most related to suicide (Hendin, Maltsberger, & Szanto, 2007). The uncontrollable nature of the affects engendered fear on the part of the patients that they were fragmenting (i.e., "falling apart"). Nine affects were examined: anxiety, rage, desperation, abandonment, loneliness, hopelessness, self-hatred, guilt, and humiliation. A striking contrast was observed in the patients who went on to suicide and the comparably depressed patients who were not suicidal. Just before death the suicides averaged more than three times the number of intense affects than comparably depressed nonsuicidal patients. These differences remained when controlled for severity of depression, psychiatric disorders, and borderline personality disorders.

That work made it possible to develop the Affective States Questionnaire, which was tested prospectively and successfully for its ability to predict short-term risk for suicidal behavior (three months) among a population of 240 outpatient and inpatient veterans, not selected for the presence of PTSD or the risk for suicide. Recognizing the intense, overwhelming emotional states that leave patients feeling out of control in a crisis period immediately preceding their suicidal behavior is critical in this process (Hendin, Al Jurdi, Houck, Hughes & Turner, 2010).

TREATING THE VETERAN WITH PTSD AT RISK FOR SUICIDE

Earlier concepts of the unconscious that are outdated have been rejected by modern psychodynamics that has recognized its underlying, enduring contributions to our understanding and ability to treat mental illness and, in particular, the role of the unconscious in influencing behavior, the value of dreams, and the nature of the relationship with the therapist conducting the treatment. Even slight differences in the dream and the actual experience can be helpful in understanding the nature of the experience and making treatment possible. Tom B. is an example.

Case 2

Troubled by violent impulses toward his family as well as suicidal thoughts, Tom's entire postcombat life had been pervaded by PTSD. For years he had suppressed the symptoms with drugs, which he began using when he returned from Vietnam. He stopped, because he felt they were destroying his body, but he then became aware of his preoccupation with Vietnam and the disturbing nature of his nightmares. Tom had one recurrent nightmare that he said "scares the hell out of me. It's so real but I don't know if it actually happened." In the dream he is carrying the dead body of a young woman and trying to bury it so no one can find it. Upon waking from this dream he would sense that he had some involvement in the girl's death but would be unable to recall what it was. When asked if he had ever raped any Vietnamese women, Tom replied that he had not. When asked if he had ever witnessed a rape, he said that he had.

His squad had been assigned to secure the entrance to a tunnel complex while four men from another squad went underground to explore the tunnels. His squad was in radio contact with the other squad and learned that they had found a Viet Cong hospital base. A short while later Tom heard shouting and the sounds of grenades exploding. The four men came out of the tunnel dragging a French nurse who was bleeding from arm wounds. Each of the four raped the nurse while Tom's squad watched. When the last man was finished he pulled out his knife and killed the woman. When this happened Tom and his squad departed; he never knew how the men disposed of the nurse's body. He did know that when the four soldiers reported the incident they made no mention of taking anyone alive. Tom claimed to have had no particular reaction to the event. He admitted that he had been sexually excited while watching what had happened, but he had never connected the episode with his nightmare.

Tom was seen for several months of short-term psychotherapy during which time he was helped to explore and feel the emotions connected with his nightmare. Just as in the dream where he was carrying and trying to find a place to bury a woman's dead body, he had tried for years to bury the entire experience. Although he had succeeded on a conscious level, the burden of guilt he was nonetheless carrying is evident in his dream. In therapy he was able to connect it with the rape and killing of the nurse he had witnessed, to recognize that he was a "participant" in her rape, and to experience the emotions connected with it. He stopped having the nightmares, became less angry with his family, no longer had thoughts of suicide, and remained so on follow-up a year later.

Tom had been treated with behavioral therapy and medication without any improvement before he was referred to the research and treatment program. His nightmare, however, had been treated only as a symptom to be suppressed with sleeping medication rather than an opening to unconscious feelings that were troubling him.

Tom is not likely to have had a successful therapy if he had not first stopped his substance abuse. Substance abuse and/or difficulty functioning—at work and in family and social relations—increase the risk for suicide (Hendin et al., 2010). For substance-abusing veterans with PTSD at risk for suicide, enrollment in a treatment program needs to be a requirement for participation in a psychodynamic treatment.

WHAT PROTECTS SOME

Given the nature of combat in Vietnam, it is not difficult to understand the high incidence of PTSD among men who fought in that war. It is more difficult to attempt to explain what protected some who saw intense combat from developing the disorder. Since combat adaptation is one of the factors significantly correlated with PTSD and the form it takes, it was desirable to examine the experience and adaptation of veterans who experienced severe combat but who did not develop the disorder (Hendin & Pollinger Haas 1984a).

Although a large-scale survey (U.S. House of Representative, Committee on Veterans Affairs, 1978) had established that there are Vietnam combat veterans who did not appear to be suffering from posttraumatic stress, no detailed examination of their combat adaptation had been made. We undertook a pilot study of a representative group of 10 veterans, selecting them because, in spite of having had intense combat that included killing enemy soldiers and the possibility or likelihood of being killed, being wounded, or seeing comrades wounded or killed, these veterans did not develop PTSD. They appeared to be dealing with postwar civilian life without evidence of a stress disorder. Each participated in the study on a volunteer basis and completed the same five-session clinical evaluation that we gave to Vietnam veterans with PTSD but did not need the additional seven sessions that the other veterans did.

From comparison of the results of this analysis with data on the 100 veterans with PTSD whose study was completed, several key differences between the combat adaptations of the two groups were evident. Veterans in the group that did not develop PTSD regarded the ability to stay calm under pressure as a good soldier's most important attribute. All seemed to see impulsiveness as a threat to individual and group survival. Consistent with the emphasis these men put on self-control for proper decision-making was their ability to tolerate better than most veterans the unstructured nature of the war. Among combat soldiers there was a widespread sense that the conflict was utterly meaningless and that they had no control over any of it, a sense that seems to have been expressed in the "don't mean nothing" phrase frequently used by combat soldiers in Vietnam to describe whatever was going on. The combat veterans who did not develop PTSD were better able than most veterans to see the war in terms of the limited objectives of each day's mission. The following case is illustrative.

Case 3

Paul B., a 34-year-old. tall, thin veteran, had been married for 11 years and owned a garage in a small town in New York State. He had a calm, relaxed manner and related the events of his arduous combat tour in Vietnam as though these experiences were part of an unpleasant but distant past.

Paul had grown up with three siblings in a working-class family in the same town in which he currently lived. From his father, who was an auto mechanic, he learned how to repair cars, and after graduation from high school he obtained full-time employment in this field. He was involved in a steady relationship with a young woman when he was drafted.

During his first months in Vietnam, his company had a commanding officer whose poor judgment, Paul felt, resulted in several unnecessary deaths. Paul's sense that he too would not survive was heightened when the company was overrun by enemy soldiers, and in a fierce battle two-thirds of the 100-man company was killed.

Shortly after the episode, while on a search-and-destroy mission, Paul's squad set up an ambush and 12 enemy soldiers walked into it. After the unit's machine gun jammed, Paul grabbed grenades from the other men and circled the area in which the enemy was trapped, throwing grenades. He described this as the only time during his tour in which he felt angry, something he was unable to explain. Although he later received a Bronze Star for that action, he felt the way he had exposed himself to danger had been reckless. Two days later, in a similar ambush, he fired at the enemy but was careful not to expose himself in the same way.

Paul received a second Bronze Star in another close encounter with the enemy, in which he was fired on from only a few feet away after he had thrown a grenade into a North Vietnamese bunker. Although his hand was nicked by a bullet, he got away by crawling on his knees; he "had the shakes" for several days after the incident. In time, however, he said his fear in combat became less as he became more experienced.

One of his recollections of Vietnam was an action in which women and children were unnecessarily killed. His company had been told that anyone they saw in a particular area would be unfriendly. When they came on a group of people walking through a field, the lead platoon fired and only later realized that they had killed women and children. Paul's platoon was in the rear that day and he considered it fortunate that he was not directly involved in the killing.

Paul shared the commonly held view that the war was pointless, saying he even felt bad about having to kill enemy soldiers, since "You killed people in one place and then had to do it again in the same place or someplace nearby. It wasn't that you took territory and then moved on to the next area." He considered, however, that much of what happened to anyone in Vietnam was a function of leadership. He had become a squad leader, and part of being a good leader, he felt, was explaining to the men what their particular mission was for each day. He thought it was demoralizing to be "slogging away feeling it don't mean nothing." He was proud that none of the men in the squad he led were killed during his tour.

When Paul returned from Vietnam he went back to work as a mechanic and resumed the relationship with his girlfriend, whom he married one year later. He described his wife as a thoughtful, caring person, and from the beginning he was happy in his marriage. After several years he bought his own garage, which he ran with the assistance of one of his brothers. His father also helped out a few hours a day. His parents were in good health and lived in the same town as he did, and he had remained close to them as well as to his siblings.

Although Paul never developed PTSD, he had had some residua of his combat experience. There had been times, soon after he returned from Vietnam, when he would drive around alone late at night thinking of the things he had experienced during the war. In the early years he had occasional dreams about his company being overrun and being fired at from close range by North Vietnamese soldiers, but these had completely stopped. Rather than feeling emotionally numb or socially withdrawn since coming home from Vietnam, he said he had generally felt more self-confident and more talkative, adding that his wife had also noticed this difference.

Paul had occasional startle reactions but did not suffer from insomnia, had experienced no difficulties with his concentration, and had no combat-related guilt. Neither did he report any significant episodes of anxiety, explosiveness, or depression. Indeed, there was no evidence that his combat experiences had had a lasting negative effect on his life.

In looking for the explanation of how Paul and others emerged basically intact from the war, one is struck by the number of fortuitous occurrences that seemed to have protected them both physically and psychologically. Paul could easily have been severely wounded or killed by the enemy soldier who had fired at him from close range. Despite his skill as a squad leader, chance determined that no one in the squad was killed, particularly in the operation in which his company was decimated. Paul could also have been in the platoon that mistakenly killed women and children. In either case he would have had to deal with the increased potential for resulting distress. Although luck often played a role in the combat histories of these veterans, each of them brought to combat a way of thinking, feeling, and acting that appeared to constitute a more systematic form of protection.

Sometimes the exercise of judgment required these men to take responsibility for countermanding orders from their superior officers. One veteran described an incident in which his squad was out on ambush and about 60 Viet Cong came by. Realizing that the Claymore mines they had set would have killed only about half of the enemy, which would have left his squad at the mercy of another 30 soldiers, he went against standing orders and directed his men not to activate the mines. After the Viet Cong passed by, he called in to request artillery fire in the direction they were heading. Another veteran told of a comparable incident in which his squad had been instructed to make a body count at night in an unsafe area. Considering it foolish to go out before daybreak, he lied and told his officer that the count had been completed. In no case did such disobedience appear to express defiance or a need simply to challenge authority. Rather, these men trusted their own values and judgment and made choices that were consistent with both effectiveness and survival.

Acceptance of fear in themselves and others was also characteristic of this group. Paul had "the shakes" for a few days after he was fired at from point-blank range. He accepted his fear as an appropriate reaction to what had happened, expected that it would subside, and was not ashamed of what he felt. He shared with other veterans in the group the feeling that experience increased the ability to distinguish dangerous versus safe situations so he was not in constant state of anxiety. As a group, they also accepted signs of fear in their comrades.

Many combat soldiers report excitement and a sense of triumph during engagements in which they killed the enemy. Some were stimulated by rage or hatred of the enemy to develop a lust for killing and for more engagement. None of the veterans who did not develop PTSD reported triumphal killing; they tended to regard killing enemy soldiers as a disturbing but unfortunate necessity. Although rage occasionally led some to perform heroic acts, sustained rage was more likely to lead to behavior that was self-destructive both to the individual and his comrades. As a group, the veterans who did not develop PTSD believed that rage clouded judgment and led to dangerous mistakes. One veteran in this group told of a man in his squad who spent hours sharpening his long hunting knife and talking about how he was planning to sneak up on unsuspecting Viet Cong and slit their throats. Feeling this man was a danger to the unit, the veteran had arranged with his squad leader to have him transferred out.

Consistent with their maintaining a high degree of emotional and intellectual control during combat and not being stimulated by violence was the absence of actions over which they felt or needed to deny guilt. None of these veterans had engaged in nonmilitary killings of civilians, prisoners, or other Americans; in sexual abuse; or in mutilation of enemy dead. Since such actions—and consequent guilt—were seen in a significant proportion of the veterans who developed PTSD, their absence among the non-PTSD group was striking.

The veterans in this group did not dehumanize the enemy in their attitudes, speech, or behavior even when others in their units were doing so. Several in this group related incidents that suggested not only the absence of rage and dehumanization but also a strong sense of humanity and compassion. One man, for example, had been with another soldier when they saw at some distance people who appeared to be part of a Viet Cong hospital unit carrying litters. Although they felt they would have been able to kill at least some of the group from their concealed position behind the tree line, they decided not to fire.

The cluster of traits seen among veterans who did not develop PTSD—calmness under pressure, intellectual control, ability to create structure, acceptance of their own and others' emotions and limitations, and lack of excessively violent or guilt-arousing behavior—constituted an adaptation that was uniquely suitable for the preservation of emotional stability in situations that were often unstructured and unstable. These veterans experienced combat in Vietnam as a dangerous challenge to be met effectively while attempting to stay alive. They did not perceive combat as a test of their own worth as men, as an opportunity to express anger or vengeance, or as a situation in which they were powerless victims.

Certainly, the combat adaptations of these men reflected their prewar character structure and emotional stability. At the same time, the fact that under the intense pressure of combat other men whose prewar personalities seemed stable and intact evidenced destructive behavior argues against viewing combat functioning simply as an expression of pre-existing strengths or vulnerabilities. In addition, although most of the men in this group came from supportive families who gave them a sound beginning in life and played a significant role in their postwar readjustment, three were virtually forced to raise themselves. Interestingly, all three were functioning well at the time they were drafted. It appeared throughout our work that how well a veteran was functioning immediately prior to going to the service, although not infallible, was a good measure of how he would function in combat.

Large-scale controlled studies of what protects soldiers exposed to combat from subsequently developing PTSD are indicated. The anonymous surveys that have been done focus more generally on the degree of combat exposure (Sareen et al., 2012) and multiple deployments (Kuehn, 2010) as contributing to "mental disorders." An exception is a study that looked at the interaction between combat exposure and unit cohesion in predicting suicidal ideation among postdeployment soldiers. The results indicated that soldiers who had greater combat exposure but also had high levels of unit cohesion had lower levels of suicide-related ideation. Those who had higher levels of combat exposure and lower unit cohesion were most at risk for suicide-related ideation (Mitchell, Gallaway, Millikan, & Bell, 2012). This finding is consistent with what was reported by the veterans who did not develop PTSD as well as established experience that social support protects against suicidal behavior.

Although killing the enemy or being in danger of being killed increases the likelihood of being severely distressed by combat, the work described here suggests the importance of looking closely at perceptual and adaptive factors rather than simply at objective aspects of the combat experience in seeking to understand veterans' combat experiences. How combat events and situations are perceived, integrated, and acted on bears a primary relationship to the aftereffects of combat on the veteran.

TREATING COMBAT VETERANS OF THE WARS IN IRAQ AND AFGHANISTAN WITH PTSD AND AT RISK FOR SUICIDE

There is a significant difference in the population that served in Vietnam War and the wars in Iraq and Afghanistan. The veterans of the wars in Vietnam were drafted, their average age was 20, and they rarely had histories of depression or suicidal behavior prior to the war. Veterans of the wars in Iraq and Afghanistan were volunteers, their average age was 28, and they frequently had histories of precombat mental illness including suicidal behavior (LeardMann et al., 2013). In cases we have seen, their enlistment was often a way of trying to provide structure to their lives, which left them vulnerable when it did not work.

The combat experiences of Vietnam veterans also differed significantly from the experiences of veterans of the wars in Iraq and Afghanistan, where improvised explosive devices were a principal cause of traumatic brain injury (TBI). Veterans with TBI are also more likely to die by suicide than those without TBI (Brenner, Ignacio, & Blow, 2011). Guilt over the killing of noncombatants is less likely to play a role in their suicide than it is with Vietnam veterans who experienced the chaotic combat firefights and sustained guerilla warfare of the war in Vietnam. Iraq veterans with experiences roughly comparable to those of Vietnam veterans usually fought in battles, like those in Fallujah, Ramadi, and Nasiriya, in which sustained firefights (over months and years) in cities and within buildings led to actions in which women and children were killed and situations where soldiers felt guilt afterward. Multiple deployments, however, that characterized the wars in Iraq and Afghanistan have been shown to contribute to veterans' physical and mental health problems (Kline et al., 2010) and may also contribute to suicide independent of combat exposure.

Many therapists who are treating combat veterans of the wars in Iraq with PTSD are fearful about treating suicidal patients. The primary goal of the psychodynamic therapy is to treat veterans with PTSD who are at risk for suicide and are disturbed (guilty) over their behavior in combat. This involves understanding the experience and relieving the guilt associated with it. Yet anxiety often causes therapists to avoid raising or discussing the veteran's combat experience. They fear what the veteran's reaction will be, including the possibility of being blamed if the veteran becomes suicidal. These therapists need to be helped to overcome this anxiety to a degree that they cannot only ask the right questions but are also comfortable in doing so. Otherwise the inquiry is apt to result in the reaction they fear. At first they may need to assume a calm they do not yet feel.

Another problem for therapists stems from the fact that the nightmares of veterans who are guilty over their behavior in combat need to be elicited and understood to fully understand their guilt, and most therapists are not used to utilizing dreams in psychotherapy. Here the problem is eased by the fact that the combat nightmares of veterans are not that disguised—are less symbolic—and not difficult to understand by therapists with no prior experience in dealing with dreams.

Therapy often flounders when the veteran has shared at least some of the disturbing specifics of his combat tour. The therapist may inadvertently respond with revulsion, anger, or fear. More frequently, the therapist's discomfort is communicated in the need to convey understanding before he or she is in a position to do so. When this happens, it is the therapist's discomfort rather than what is specifically said that the veteran responds to, only increasing his distress.

It is better for the therapist to accept and respect the veteran's guilt, to acknowledge the pain of the experience, to indicate that he has already punished himself enough, and to work to help him not let the event continue to define his life. Telling a veteran who feels appropriately guilty about his behavior in combat "These things happen in war" is counterproductive.

Another principle of psychodynamic therapy applicable in treating nonveterans as well as veterans, and of use in eliciting the combat experiences of veterans, is not to accept all statements made by patients at face value.

Examples

One patient seen in Boston in his initial session after saying hello to his therapist said, "I want you to know I am not a Boston Red Sox fan." Most people probably know that Boston's passion and devotion to the Red Sox in good times and bad are unmatched in baseball. That he was letting his therapist know that he was a contrarian became more evident in his therapy.

Another patient dreamed of saving his child from drowning. He interprets the dream as saving his child from danger, saying he worries that something bad will happen to her. In his sessions he had been expressing his anger with his child for the strain he felt at the child's impact on his relationship with his wife. He is the author of the dream, and he has put the child in a life-threatening situation. In his dream he is expressing his anger and denying it at the same time. Nor does the dream mean he actually wants the child to die. The discussion of the dream made it possible to work with him to resolve the difficulties he was having in being a father.

Knowledge of this principle will help therapists deal with one of the most common problems in eliciting combat experiences of veterans. Veterans will often relate a somewhat troublesome experience to avoid discussing a far more troublesome experience. That is either a plea bargain and/or a test of what the therapist's response will be. A common one used by veterans of the war in Iraq is of shooting and killing drivers who do not stop at check points—learning later from others in the car that the driver could not see or read the check point sign.

An example is Jack B, a veteran of the war in Iraq. He related an incident in which he shot and killed an old man who did not stop at a check point sign. From other passengers in the car he learned that the man had not seen the sign. The veteran described this as his most disturbing experience. It was pointed out to him that while he had killed someone he had not needed to kill, he had done so inadvertently while following military procedures. He was asked if he was more disturbed by killing that he had done when he was not following military procedures. He revealed that he had killed a captured prisoner of war whom he was guarding.

Therapists learn more when they ask about the veteran's nightmares. A veteran of the Iraq war had killed a captured prisoner but said it did not bother him because "everybody did it." He had a recurrent nightmare, however, in which he is captured and killed by Iraqi soldiers. When this contradiction was pointed out, it enabled the therapist to help him deal with his guilt.

Another factor increasing the need for punishment and complicating treatment occurs when veterans who have lost emotional control in combat remain out of control in civilian life (Hendin & Pollinger Haas, 1991). This condition can be the result of neurochemical or physical changes in the brain or epigenetic changes caused by the stress of combat. To what degree psychological treatments can result in beneficial epigenetic changes has yet to be determined. For now, mood-stabilizing medication may be needed to make psychological treatment possible.

The relationship between the veteran and the therapist plays a key role in the healing process of veterans who have PTSD, and this is particularly true for those who have severe combat guilt and are at risk of suicide. The veteran needs to forgive himself for the behavior that triggered his guilt and the self-punitive way it is expressed. When the veteran feels relief at having shared the experience with a trusted therapist, the therapist is in a position to give him "permission" to forgive himself, to resolve problems that have developed in the course of the illness, and to go on with his life. Guilt is an emotion that can be harmful when it is self-punitive, but it can be a powerful force for changing the direction of one's life.

DISCUSSION

Both the VA and the Defense Department have expressed concerns that treatments currently in use are not slowing the rate of suicide among the active military and combat veterans with PTSD (Frances, 2013; Shinseki, 2010). Cognitive Behavioral Therapy (CBT), and Dialectical Behavioral Therapy (DBT) are the treatments of choice of the VA for veterans with PTSD. Although they have shown the ability to reduce some PTSD symptoms, they have not yet shown the ability to prevent suicidal behavior among this population. Prolonged Exposure Therapy is being used to treat civilian and military personnel with PTSD, but has not been tested for its ability to prevent suicidal behavior among military personnel or veterans.

Several factors are likely to explain why the current treatments used by the military and the VA are so far not proving effective in preventing suicide among those with PTSD who are at risk. Not determining the often unconscious, subjective, emotional meaning of the traumatic combat experiences of the veterans is only part of the problem. Of equal importance is not adequately recognizing the ways in which the relationship between the veteran and the therapist can be used to enable the veteran to give up self-destructive behavior. The belief that human behavior can be understood without reference to unconscious processes runs counter to advances in neuroscience that see the mind as operating largely by unconscious processes taking place in the brain (Kandel, 2013).

A less constrictive approach has been incorporated in short-term psychodynamic therapy that also utilizes techniques of interpersonal therapy, dialectical behavioral therapy, and cognitive behavioral therapy. There is reason to believe that these therapies would be more successful in treating veterans with PTSD at risk for suicide if they incorporated some simple psychodynamic principles.

No psychological approach is the whole answer to treating veterans with PTSD who are at risk for suicide. Many who are severely depressed need medication in order to be amenable to any psychotherapy. A large number of those most at risk, who were emotionally out of control during the combat experience over which they feel guilty, are continuing to suffer from that same loss of control in civilian life. Mood stabilizers may need to be employed to make psychotherapy with them possible.

CONCLUSION

A unique short-term (12 sessions) treatment approach has been presented that targets the guilt from combat-related experiences that underlies suicidal behavior in combat veterans with PTSD who are most at risk. The risk is intensified if they also have survivor guilt. It has shown promise of being able to prevent suicidal behavior with veterans of the wars in Vietnam and Iraq and Afghanistan who have experienced chaotic firefights resulting in out-of-control behavior that aroused guilt. The section describing the adaptation of veterans who were exposed to severely traumatic combat experiences but did not develop PTSD provides a perspective on what appears to help some to avoid developing the disorder.

The treatment's essential components define the meaning of combat to the veteran with the aid of the veteran's nightmares and address relief of guilt. The next step is to test the treatment with a control group large enough to determine its effectiveness. We are hopeful that practitioners of other therapies will incorporate psychodynamics into their practice and research. Their doing so might increase the possibility that more veterans will receive the treatment they need.

The quality of care provided by the VA has improved dramatically in the past 30 years, but the VA is underfunded and understaffed in relation to the increased need for its services. The VA and the Department of Defense have been criticized for failing to implement properly and evaluate treatments employed with combat veterans at risk for suicide; they are now working together to change that.

Reducing suicide among military personnel and veterans is a challenge that needs to be met. Although the large majority of patients are not suicidal, those who are suffering from guilt over combat experiences are an important subgroup responsible for a disproportionate number of suicides.

Mental health professionals in the VA are ready to learn, develop, and test treatment approaches to PTSD that will work. The public and Congress are currently in a mood to support their treatment. That support tends to weaken as the years after veterans return go by. There will need to be an ongoing effort to sustain public awareness of the problem so that Congress provides adequate funding. The need for help does not fade, nor does the danger of suicide abate, in a disorder that is rightly described as "posttraumatic."

REFERENCES

Brenner, L. A., Ignacio, R. V., & Blow, F. C. (2011). Suicide and traumatic brain injury among individuals seeking Veterans Health Administration services. *The Journal of Head Trauma Rehabilitation*, 26, 257–264.

Bullman, T. A., & Kang, H. K. (1996). The risk of suicide among wounded Vietnam veterans. *American Journal of Public Health*, 86, 662–667.

Centers for Disease Control and Prevention. (1987). Postservice mortality among Vietnam veterans. *JAMA: Journal of the American Medical Association*, 257, 790–795.

Durkheim, É. (1951). *Suicide*. Glencoe, IL: Free Press.

Frances, A. (2013, March 27). The epidemic of suicide in the military. *The Huffington Post Healthy Living*.

Freud, S., Ferenczi, S., Abraham, K., Simmel, E., & Jones, E. (1921). *Psychoanalysis and the war neuroses*. London: Psychoanalytic Press.

Hendin, H., Al Jurdi, R. K., Houck, P. R., Hughes, S., & Turner, J. B. (2010). Role of intense affects in predicting short-term risk for suicidal behavior: A prospective study. *The Journal of Nervous and Mental Disease*, 198, 220–225.

Hendin, H., & Haas, A. P. (1991). Suicide and guilt as manifestations of PTSD in Vietnam combat veterans. *The American Journal of Psychiatry*, 148, 586–591.

Hendin, H., Maltsberger, J. T., Lipschitz, A., Haas, A. P., & Kyle, J. (2001). Recognizing and responding to a suicide crisis. *Suicide and Life-Threatening Behavior*, 34, 115–128.

Hendin, H., Maltsberger, J. T., & Szanto, K. (2007). The role of intense affects in signaling a suicide crisis. *The Journal of Nervous and Mental Disease*, 195, 363–368.

Hendin, H., & Pollinger Haas, A. (1984a). Combat adaptations of Vietnam veterans without posttraumatic stress disorders. *The American Journal of Psychiatry*, 141, 956–960.

Hendin, H., & Pollinger Haas, A. (1984b). *Wounds of war: The psychological aftermath of combat in Vietnam*. New York: Basic Books.

Jury, D. (1979). The forgotten warriors: New concern for the Vietnam veteran. *Journal of Behavioral Medicine*, 6, 38–41.

Kandel, E. R. (2013). From nerve cells to cognition. In E. R. Kandel, J. H. Schwartz, T. M. Jessell, S. A. Siegelbaum, & A. J. Hudspeth (Eds.), *Principles of neuroscience* (5th ed., pp. 370–391). New York: McGraw-Hill.

Kardiner, A., & Spiegel, H. (1947). *War stress and neurotic illness*. New York: Hoeber.

Kline, J., Falca-Dodsin, M., Sussner, B., Ciccone. D. S., Chandler, H., Calllahan, L., & Losonczy, M. (2010). Effects of repeated deployment to Iraq and Afghanistan on the health of New Jersey Army National Guard Troops: Implications for military readiness. *Journal of Public Health*, 100, 276–283.

Kolb, L. C. (1986). Post-traumatic stress disorder in Vietnam veterans (editorial). *The New England Journal of Medicine*, 3, 641–642.

Kuehn, B. M. (2010). Military probes epidemic of suicide: Mental health issues remain prevalent. *JAMA: Journal of the American Medical Association*, 304, 1427–1430.

LeardMann, C., Powell, T. M., Smith, T. C., Bell, M. R., Smith, B., Boyko, E. J., . . . Hoge, C. W. (2013). Risk factors associated with suicide in current and former US military personnel. *JAMA: Journal of the American Medical Association*, 310, 406–505.

Lipkin, J. O., Blank, A. S., Parson, E. R., & Smith, J. (1982). Vietnam veterans and posttraumatic stress disorder. *Hospital & Community Psychiatry*, 33, 908–912.

Mitchell, M. M., Gallaway, M. S., Millikan, A. M., & Bell, M. (2012). Interaction of combat exposure and unit cohesion in predicting suicide-related ideation among post-deployment soldiers. *Suicide and Life-Threatening Behavior*, 42, 486–494.

Sareen, J., Cox, B. J., Afifi, T. O., Stein, M. B., Belik, S., Meadows, G., & Asmundsen, G. J. G. (2012). Combat and peacekeeping operations in relation to prevalence of mental disorders and perceived need for mental health care. *Archives of General Psychiatry*, 64, 843–852.

Shinseki, E. (2010, November 12). *Suicide rates soar among US vets*. Retrieved from military.com.

Stuen, M. R., & Solberg, K. B. (1972). The Vietnam veteran: Characteristics and needs. In L. I. Sherman & E. M. Caffey (Eds.), *The Vietnam Veteran in Contemporary Society*. Washington, DC: Veterans Administration.

US House of Representatives, Committee on Veterans Affairs. (1978). *Presidential review memorandum on Vietnam-era veterans: House Report 38*. Washington, DC: US Government Printing Office.

15

Understanding Traumatic Brain Injury and Suicide Through the Lens of Executive Dysfunction

Beeta Y. Homaifar

Melodi Billera

Sean M. Barnes

Nazanin Bahraini

Lisa A. Brenner

INTRODUCTION

This chapter presents background information regarding traumatic brain injury (TBI; i.e., definition, severity classifications, epidemiology, assessment, and common postinjury sequelae/psychiatric disorders) are presented to provide context for a discussion of the complicated relationships between brain injury and suicidal thoughts and behaviors. The potential contribution of executive dysfunction (e.g., impairment in reasoning and/or decision-making) is reviewed. In addition, the idea that propensity toward or against engaging in risky behavior can be used to increase understanding regarding the relationship between TBI and suicidal ideation and behaviors is discussed. Last, clinical challenges and future research directions are presented.

DEFINITION OF TBI/CLASSIFICATION OF SEVERITY

Although many definitions of TBI exist, the one endorsed by the Department of Defense and the Department of Veterans Affairs in 2009 is as follows: A TBI is a

> traumatically induced structural injury and/or physiological disruption of brain function as a result of an external force that is indicated by new onset or worsening of at least one of the following clinical signs, immediately following the event:
>
> - Any alteration in mental state at the time of the injury (confusion, disorientation, slowed thinking, etc.) (Alteration of consciousness/mental state [AOC])
> - Any period of loss of or a decreased level of consciousness (LOC)
> - Any loss of memory for events immediately before or after the injury (post-traumatic amnesia [PTA])
> - Neurological deficits (weakness, loss of balance, change in vision, praxis, paresis/plegia, sensory loss, aphasia, etc.) that may or may not be transient
> - Intracranial lesion. (p 16)

TABLE 15.1 Departments of Defense and Veterans Affairs Consensus Based Classification of Closed TBI Severity

	Mild TBI/Concussion	Moderate TBI	Severe TBI
Structural imaging results	Normal	Normal or abnormal	Normal or abnormal
AOC (time)	A moment to 24 hours	Greater than 24 hours	Greater than 24 hours
LOC (time)	0–30 minutes	> 30 minutes but < 24 hours	> 24 hours
PTA (time)	0–1 day	> 1 but < 7 days	> 7 days

Note: TBI = traumatic brain injury; AOC = alteration of consciousness/mental state; LOC = level of consciousness; PTA = posttraumatic amnesia.
Source: Department of Veterans Affairs & Department of Defense (2009).

The three levels of TBI severity are mild, moderate, and severe. These determinations pertain to disruption of functioning at the time of injury (e.g., AOC, LOC, PTA) versus symptom severity post-TBI. The presence of neuroimaging findings is also considered (Table 15.1). It should be noted that diagnosing moderate or severe TBI is often less complicated, as individuals who sustain more severe injuries generally seek medical attention and positive physiologic markers are frequently present (e.g., imaging; Brenner & Homaifar, 2009). Conversely, assessment of mild TBI (mTBI) is often more difficult in part because the majority of these individuals are not hospitalized (Ivins et al., 2003) or do not seek other medical attention (National Center for Injury Prevention and Control, 2003). Moreover, physiological markers for mTBI that could be used in routine clinical practice have not yet been identified.

EPIDEMIOLOGY OF TBI

TBIs occur in civilian, military, and veteran populations. With regard to civilians, the Centers for Disease Control and Prevention (CDC) estimates that approximately 1.7 million people sustain TBIs annually (CDC, 2012). Of those, almost 75% are mTBIs, which are sometimes referred to as concussions. Falls are the leading cause of injury, followed by unknown/other causes, motor-vehicle/traffic accidents, and assault. TBI rates are higher for males than for females. The age groups at greatest risk for sustaining TBIs are children (ages zero to 4 years), older adolescents (ages 15 to 19 years), and adults (ages 65 and older; CDC, 2012).

There has been much discussion in both the popular press and academic journals about TBIs sustained by military personnel serving in Iraq and Afghanistan. Rates reported widely vary, from 8% to 23%, and are contingent on a number of factors such as time served and military occupational specialty (Terrio et al., 2009; Vasterling et al., 2006). As with civilians, the majority of injuries sustained are mild in nature (Terrio et al., 2009). The Defense and Veterans Brain Injury Center, the group tracking medical diagnoses of TBI among U.S. military personnel, reported the incidence of TBI between 2000 and 2015 (Quarter 1) was 327,299, with 269,580 (82.4%) mild, 27,728 (8.5%), moderate, and 3,422 (1.0%) severe cases. An additional 4,865 (1.5%) were classified as penetrating, and 21,704 (6.6%) were not classifiable (Military Health System Office of Strategic Communications, 2015).

Among those who sustained TBIs in Iraq or Afghanistan, the most frequent mechanism of injury was blast exposure associated with the detonation of improvised explosive devices (Terrio et al., 2009). A number of studies suggest that blast-induced injuries due to barotrauma (i.e., direct effect of blast overpressure on tissue) can be deleterious to the brain (Cernak, Savic, Ignjatovic, & Jevtic, 1999; Cernak, Wang, Jiang, Bian, & Savic, 2001). In addition to barotrauma (primary blast injury), military personnel can sustain secondary TBIs from shrapnel or objects being thrown off explosives (secondary injury) and tertiary injuries from bodily displacement (CDC, 2006).

Brenner et al. (2013) noted high rates of TBI among veterans seeking mental health services (mean age = 53.2, range = 21 to 86). In their sample of 316 veterans who completed a structured interview regarding lifetime history of TBI, the mean number of lifetime injuries reported was 2.5, with the most frequent causes being assault, transportation, and falls. More than a quarter of the most severe lifetime injuries reported by the participants were moderate to severe in nature.

TBI Assessment

In order to assess for a history of TBI, clinicians must obtain information regarding the injury event and associated alteration in consciousness (e.g., AOC, LOC, PTA). If they exist, imaging results can provide additional useful information. When possible, information about the injury should be obtained from medical records, collateral sources, and/or eyewitness accounts. In cases where medical attention is not sought and collateral information is unavailable, assessment of TBI is based on the individual's account of the injury event and self-report of AOC, LOC, and/or PTA. Such recollection can be complicated by the fact that events both preceding and following a TBI may be impacted by several factors, including limited awareness, poor recall, stigma associated with the injury (Corrigan & Bogner, 2007a), or psychological distress. As noted by Ruff et al. (2009), fear or stress at the time of the injury can also result in an AOC.

Despite these challenges, the current gold standard for TBI diagnosis, particularly mTBI, is structured clinical interview (Corrigan & Bogner, 2007a). Formal clinical interviews such as the Ohio State University TBI Identification Method (OSU TBI-ID; Corrigan & Bogner, 2007b) are available and, with practice, relatively easy to administer. Recent work by Brenner et al. (2013) suggests that the use of a four-item screen may be helpful in determining which patients need more extensive evaluation (such as the OSU TBI-ID). The four screening items are: (a) "Have you ever been hospitalized or treated in an emergency room following a head or neck injury?"; (b) "Have you ever been knocked out or unconscious following an accident or injury?"; (c) "Have you ever injured your head or neck in a car accident or from some other moving vehicle accident?"; and (d) "Have you ever injured your head or neck in a fight or fall?"

TBI Sequelae

An additional barrier to accurate diagnosis is the frequent confusion of history of TBI (injury event with subsequent AOC) with the symptoms of TBI (e.g., headaches, cognitive impairment, and sleep difficulties). The severity of reported sequelae (e.g., headache) should not be used to determine whether such a historical event occurred. Moreover, the severity of symptoms should not be used to determine the severity of the TBI. In fact, research has shown that those with mTBI often endorse significantly more symptoms than those with moderate to severe TBI (Gordon, Haddad, Brown, Hibbard, & Sliwinski, 2000).

Sequelae experienced after an mTBI are often referred to as postconcussive symptoms and include a range of nonspecific physical/somatic (e.g., dizziness, headaches, photophobia, fatigue), cognitive (e.g., poor concentration, decreased memory), and emotional (e.g., depression, anxiety, irritability) complaints (Vasterling, Verfaelli, & Sullivan, 2009). In the context of these milder injuries, most symptoms resolve over time. Although most individuals return to baseline functioning after a TBI within three months, some experience persistent sequelae (Ryan & Warden, 2003). The nonspecific nature of postconcussive symptoms and their large overlap with a number of psychological and minor medical conditions have raised questions about the etiology of prolonged postconcussive symptoms in those with mTBI. Although evidence concerning the maintenance of postconcussive symptoms is still inconclusive, the consensus in the literature is that persistent symptoms following mTBI are likely influenced by a variety of neuropathological and psychosocial factors (Whitaker, Kemp, & House, 2007). Factors that may contribute to the onset and maintenance of these sequelae underscore the importance of thorough clinical assessment for premorbid and comorbid psychological disorders, as well as preinjury variables (e.g., history of learning problems). More severe injuries are typically associated with sustained cognitive, mood, behavioral, and physical impairments and associated functional challenges (Table 15.2). In addition to the physical symptoms listed in Table 15.2, more severe injuries can also be associated with seizures and paraplegia.

TBI AND PSYCHIATRIC DISORDERS

Post-TBI mental health complaints are common for all severity levels, with symptoms associated with preinjury psychiatric problems, damage to areas of the brain, and postinjury adjustment. With regard to mood disorders, persisting symptoms of depression have been noted (Hibbard, Uysal, Kepler, Bogdany, & Silver, 1998), with prevalence rates ranging from 33% to 42% (Jorge et al. 2004; Kreutzer, Seel, & Gourley,

TABLE 15.2 Potential Long-Term Impairments Associated with TBI

Cognition	Mood	Behavior	Physical
Attention	Apathy	Poor initiation	Headache
Processing speed	Depression	Disinhibition	Dizziness
Communication	Hopelessness	Impulsivity	Paresis/plegia
Learning/memory	Irritability	Agitation	Balance problems
Executive dysfunction	Emotional lability	Aggression	Fatigue

Note: TBI = traumatic brain injury.

2001). In a recent study of depression post-TBI, minor depression, defined as having two to four symptoms of depression, with at least one cardinal symptom (i.e., depressed mood or loss of interest), was as common as major depressive disorder (Hart et al., 2011). Many of these studies focus on depression immediately postinjury or within the first few years of injury. However, long-term follow-up studies and cross-sectional data indicate that persisting psychological problems including depression and social isolation can occur up to 30 years post-TBI (Brooks, Campsie, Symington, Beattie, & McKinlay, 1986, 1987; Hoofien, Gilboa, Vakil, & Donovick, 2001). Studies have attempted to parse out the differences in severity of TBI with regard to the extent to which individuals experience depressive symptomatology. In one study of individuals who were two years postinjury, those with severe TBIs reported lower levels of depressive symptoms than those with mild injuries (Malec, Testa, Rush, Brown, &Moessner, 2007). The authors suggested that the difference between those with mild versus severe injuries was due, in part, to self-awareness. In other words, those with mTBI may have had more self-awareness than those with more severe injuries, thus making them more likely to report symptoms of depression.

With regard to anxiety disorders, a recent study showed that individuals with mTBI were significantly more likely than those without mTBI to develop social phobia, panic disorder, agoraphobia, and posttraumatic stress disorder (Bryant et al., 2010). Bryant et al. also noted that in the first year post-mTBI, 23% of patients had developed new psychiatric conditions. The authors suggested that these psychiatric disorders may have developed secondary to brain damage that increased vulnerability among these patients. Individuals with more severe TBIs are also at risk for developing postinjury anxiety-related conditions (Alway, Gould, McKay, Johnston, & Ponsford, 2015).

TBI AND SUICIDE

In addition to increasing vulnerability for developing psychiatric disorders, TBI is associated with an increased risk for suicidal ideation and suicidal behavior in both civilian (Rodger, Wood, & Lewis, 2010; Tsaousides, Cantor, & Gordon, 2011) and military/ veteran populations (Brenner, Ignacio, & Blow, 2011; Bryan, Clemans, Hernandez, & Rudd, 2013). Rodger et al. observed a high rate of suicidal ideation (33%) in a TBI population and noted that feelings of worthlessness were the greatest predictor for suicidal ideation. In a similar study, Tsaousides et al. found suicidal ideation in 23.8% of adult TBI patients. A lack of social support or appropriate medical/psychiatric treatment, combined with postinjury depression or anxiety, may increase risk for suicidal ideation.

The New Haven National Institute of Mental Health epidemiological catchment area study found that civilians with a lifetime history of TBI were 2.39 times more likely (95% confidence interval [CI]: 1.32–4.31) to have had a lifetime history of a suicide attempt (SA) than those without such a history, even after adjusting for demographics, quality of life, alcohol abuse, and psychiatric comorbidity (Silver, Kramer, Greenwald, & Weissman, 2001). In an important study conducted in Denmark, Teasdale and Engberg (2001) examined TBI and suicide rates in all civilian patients diagnosed with TBI over a 15-year period and found that people with a history of TBI had 2.7 to 4.1 times higher rates of death by suicide than the general population (standardized mortality ratio 3.0, 2.7, and 4.1 for those with concussions, cranial fractures, or cerebral contusions, respectively). Although some smaller civilian studies have failed to find a significant association between TBI and suicidality (e.g., Lewin, Marshall, & Roberts, 1979; Shavelle, Strauss, Whyte, Day, & Yu, 2001), the bulk of the research indicates that civilians with TBI are at increased risk of suicide.

Recent research also substantiates the link between TBI and suicidality in military and veteran populations. In a study of individuals with and without TBI, researchers (Bryan et al., 2013) found that patients with a TBI were more likely to endorse suicidal ideation/behaviors than their counterparts. Of the patients without TBI, none endorsed suicidal ideation or behaviors whereas 16% of patients with mTBI reported suicidal ideation/behavior (13.1% ideation, 1.4% plan, 1.5% previous attempt). Brenner, Ignacio, and Blow (2011) examined medical records of all Veterans Health Administration users from 2001 to 2006 and found that those with a history of TBI were 1.55 times more likely to have died by suicide than veterans without a history of TBI. This relationship persisted even after accounting for other psychiatric and demographic factors. It should be noted that risk was present across the TBI severity continuum. Compared to those without TBI, individuals with a concussion/cranial fracture were 1.98 times more likely (95% CI: 1.39–2.82) to die by suicide, and patients with a cerebral contusion/traumatic intracranial hemorrhage were 1.34 times more likely (95% CI: 1.09–1.64) to die by suicide. In discussing the increased risk across the continuum of injury severity, Brenner et al. (2011) noted that "further work is required to clarify whether those with concussion/cranial fracture versus cerebral contusion/traumatic intracranial hemorrhage are unique populations. It is likely that factors associated with increased risk vary depending on the severity of injury sustained. It may also be that preexisting factors contribute to a greater degree for a subset of the population (e.g., those with concussion)" (p. 263).

In an active duty military population, Skopp, Trofimovich, Grimes, Oetjen-Gerdes, and Gahm (2012) did not find a relationship between TBI and suicide. The authors noted that their findings should be interpreted in light of potential differences in the underlying populations of their sample as compared with others who have found a relationship between TBI and suicide. Skopp et al. suggested that mTBIs sustained in the general civilian population may be associated with risky behaviors (e.g., fighting, alcohol abuse) as well as with psychopathology (Kim, 2002; Bjork & Grant, 2009; Simpson & Tate, 2007), whereas mTBIs sustained in military populations may reflect injuries that commonly occur during training or combat. Therefore, individuals who sustain mTBIs in the line of duty may represent a different cohort than individuals who engage in a pattern of high-risk behaviors (e.g., perhaps due to impaired decision-making ability or other pre-existing factors) that are likely to result in a TBI (e.g., substance abuse, driving recklessly, physical fights). What remains unknown is whether sustaining a mTBI in the military places one at subsequent risk for future unsafe behaviors (e.g., future TBIs, suicide).

Risk likely waxes and wanes post-TBI, and there is no clear indication regarding when risk may be heightened for this patient population. For example, in a study of veterans with TBI and a history of at least one SA (Group 1) and those with TBI and no history of SA (Group 2), the postinjury attempts for those in Group 1 varied widely (months to nearly 30 years; Homaifar, Brenner, Forster, & Nagamoto, 2012). Data from other civilian, military, and veteran studies support the notion of elevated suicide risk persisting from immediately postinjury to several decades following a TBI (Anstey et al., 2004; Homaifar et al., 2012; Mainio et al., 2007; Teasdale & Engberg, 2001).

It is also important to be aware of differences in suicide risk that are associated with different levels of TBI severity. Moderate and severe TBIs are more likely than mTBIs to be associated with persistent TBI-related deficits, such as cognitive and motor disturbances, increased impulsivity, personality changes, and postinjury changes in mental and physical capacity (Hoofien et al., 2001). In turn, these deficits can lead to changes in work status, income, quality of life, and psychiatric problems (Kuipers & Lancaster, 2000; Reeves & Laizer, 2012). This subpopulation may experience increased suicidal ideation/behavior directly related to the impact of TBI sequelae and functioning. Indeed, feelings of loss due to role changes, difficulties in cognitive functioning (including executive dysfunction), and emotional and psychiatric difficulties are common precipitating factors for suicidality among veterans with TBI (Brenner, Homaifar, Wolfman, Kemp, & Adler, 2009).

In contrast, elevated suicide risk among individuals who experience mTBIs may be primarily driven by pre-existing psychosocial and psychiatric vulnerabilities (Brenner & Homaifar, 2009; Simpson & Tate, 2007). Bryan and colleagues (2013) noted that premorbid functioning likely plays a large role in determining the risk of developing a psychiatric disorder or becoming suicidal after a TBI. It is difficult to determine to what extent antecedent factors (e.g., psychopathology, aggressiveness, substance abuse) contribute to suicide risk relative to the TBI itself (Reeves & Laizer, 2012). Findings from several studies suggest that among individuals who sustain mTBIs, comorbid psychopathology may be more indicative of suicide risk than the TBI per se. For example, Barnes, Walter, and Chard (2012) compared suicide risk factors in two groups of veterans seeking treatment for PTSD. Differences in the assessed risk factors were small and suggested that if PTSD and mTBI were associated with increased suicide risk relative to PTSD alone, the added risk was likely mediated or confounded by PTSD symptom severity. Bryan and colleagues found that the highest suicide risk in a population of military personnel who had sustained mTBIs were service members who had reported high levels of depressive symptoms or depressive and posttraumatic stress symptoms around the time of their injury. Similarly, Simpson and Tate (2005) found that individuals with a history of TBI who also had a comorbid psychiatric disorder and substance abuse postinjury were 21 times more likely to attempt suicide than individuals with no such history.

Taken together, these studies suggest multiple pathways through which a history of TBI can impact suicide risk. Making things even more complicated is the fact that the pathways themselves are influenced by a range of pre-existing and postinjury psychological, neurobiological, and psychosocial factors. In addition to examining the wide range of factors that can impact the trajectory of recovery and suicide risk following TBI, a number of recent studies have attempted to increase understanding of specific neurobiological/neuropsychological mechanisms that contribute to increased risk for suicide in those with a history of TBI. Yurgelun-Todd et al. (2011) studied veterans with mTBI and found that significant reductions in frontal white matter tracts were associated with impulsivity and suicidality, suggesting a neurobiological vulnerability to suicide risk. Additionally, research suggests that executive functioning as measured by commonly employed and novel neuropsychological instruments may be a particularly important factor to consider in evaluating suicide risk when TBI is present (Homaifar et al., 2012; Mann, Waternaux, Haas, & Malone, 1999; Oquendo et al., 2004, Yurgelun-Todd et al., 2011).

EXECUTIVE DYSFUNCTION AND SUICIDE

Executive functioning refers to a set of higher-order mental activities primarily governed by the frontal lobes including reasoning, decision-making, and insight, and executive dysfunction refers to impairments in these areas (Berger & Posner, 2000; Lezak, Howieson, & Loring, 2004; McDonald, Flashman, & Saykin, 2002). Executive dysfunction is a characteristic feature of a number of psychiatric disorders that are associated with increased risk for suicide, including depression (McLennan & Mathias, 2010), bipolar disorder (Dixon, Kravariti, Frith, Murray, & McGuire, 2004), and schizophrenia (Reed, Harrow, Herbener, & Martin, 2002).

A number of recent studies among individuals with mental health conditions suggest that the nature of neuropsychological impairments contributing to both suicidal ideation and behavior may reflect deficits beyond those typically attributable to psychiatric illness alone (Jollant et al., 2005; Jollant et al. 2010; Keilp et al., 2001; Marzuk, Hartwell, Leon, & Portera, 2005). That is, executive dysfunction may play a role in increasing risk for suicidal behavior among those with no known neurological injuries (e.g., TBI). In a study comparing depressed patients with and without suicidal ideation, those with suicidal ideation displayed poorer performance on measures of reasoning and mental flexibility (Marzuk et al, 2011). Similarly, Keilp et al. found that deficits in certain executive functions, namely abstract reasoning and concept formation, distinguished between depressed patients who had a history of high-lethality SAs and depressed patients with no such history.

There is some evidence suggesting that even among individuals with a history of psychiatric illness and suicidal behavior, deficits in executive function may only be salient under certain conditions or mood states. Westheide and colleagues (2008) found that among depressed patients who had recently attempted suicide, only those with suicidal ideation at the time of assessment showed deficits in executive functioning, with impaired decision-making being the most salient. Such findings are consistent with work by Jollant et al. (2010) demonstrating that under conditions of uncertainty, individuals with a history of SAs had difficulties evaluating the potential risk associated with certain response options, which contributed to disadvantageous decisions. In addition, those with a history of SAs showed decreased activity in the orbitofrontal and occipital cortices when making risky, relative to safe, choices, whereas the opposite pattern (i.e., increased activity during risky, relative to safe, choices) was found in healthy controls performing the same task (Jollant et al., 2010). Translating these findings to suicidal behavior, the authors suggested that difficulties in learning to appropriately evaluate risk associated with different response options, under conditions of uncertainty, may contribute to high-risk behavior (e.g., SA), which, despite its long-term negative consequences, produces an immediate solution (e.g., reduction of pain). Such deficits in risk evaluation as depicted on both neuropsychological testing and neuroimaging reflect potential neurobiological processes that increase one's vulnerability to suicidal behavior (Courtet, Gottesman, Jollant, & Gould, 2011).

In addition to disadvantageous decision-making, disinhibition and aggression are other components of executive dysfunction that have been linked to suicidal behavior. In a study comparing individuals who recently attempted suicide to those who had suicidal ideation and no history of suicidal behavior, disinhibition was significantly associated with the SA group, controlling for other relevant demographic, clinical, and neuropsychological variables (Burton, Vella, Weller, & Twamley, 2011).

TBI, SUICIDE, AND EXECUTIVE FUNCTIONING

Both qualitative and quantitative studies have implicated executive dysfunction as a risk factor for suicidal ideation/behavior in the TBI population. Brenner, Homaifar, et al. (2009) found that for a group of individuals with mild, moderate, or severe TBI, self-described executive dysfunction precipitated suicidal ideation or behavior. In a sample of veterans with various TBI severity levels, with and without histories of SA, Homaifar et al. (2012) found that perseveration, as measured by a task of executive functioning, distinguished between the two groups. This suggests that if perseveration is present, adaptive coping may be impeded because it may be more difficult to break out of the repetitive thoughts of suicide. This is supported by Brenner, Homaifar, et al. (2009) wherein participants often noted that suicide was seen as the "only way out."

In a psychiatric population of individuals with and without a history of SAs, Mann et al. (1999) found that, independent of psychiatric diagnosis, individuals with a past history of SAs were more likely to have had a TBI (operationalized as TBI that led to LOC) and had higher scores on measures of impulsivity and lifetime aggression than those without a history of attempts. This suggests that impulsivity may underlie suicidal and aggressive behavior. Lending further support to the link between executive dysfunction and suicide in those with mTBI, Oquendo et al. (2004) found increased suicidal behavior in those with elevated lifetime aggression. They also found a bidirectional relationship between aggression and TBI. The presence of elevated lifetime aggression was a risk factor for sustaining a TBI, and those with TBI endorsed higher levels of aggression post-injury. This finding suggested that, in their sample, postinjury aggression may have been in part related to TBI. Finally, Brenner and colleagues (2015) examined the relationship between executive dysfunction and SA history to explore the potential impact of having a history of moderate to severe TBI. When compared to those without TBI and SA, those with TBI only, and those with SA only, those with both TBI and SA displayed difficulty learning on the Iowa Gambling Test. The authors suggested that the finding highlight specific decision-making difficulties faced by those with histories of TBI and SA.

CONCEPTUALIZING SUICIDE RISK IN TBI: INCORPORATING EXECUTIVE DYSFUNCTION

There is a dearth of literature that could provide a conceptual framework with which to understand an individual's subjective experience of suicidal ideation or behavior, let alone an individual with a history of TBI. Merely having access to lists of risk factors and warning signs does not help clinicians organize and incorporate client-specific factors into conceptualizations of suicide risk that facilitate further connection to and collaboration with a suicidal individual.

To augment traditional suicide risk assessment and aid in conceptualizing the added risk of TBI, it can be helpful to organize risk factors in a conceptual manner.

Simpson and Brenner (2011) suggested that for a subset of those with TBI and suicidal ideation/attempts, the concept of risky behavior may provide a unifying framework for understanding the complicated interrelationship between these two phenomena. Risky behavior has been defined as "sensation seeking" or as a behavioral propensity with impulsive qualities that includes both a tendency to enjoy and pursue exciting activities and the ability to remain open to new experiences that could be dangerous (Whiteside & Lynam, 2001). Research suggests that the propensity toward risk-seeking behavior is largely influenced by both individual differences in neurocognitive functioning and contextual factors (Paulsen, Carter, Platt, Huettel, & Brannon, 2012).

It can be argued that, compared to those who are risk-averse, individuals who are risk-seeking may be more likely to engage in behaviors (e.g., driving over the speed limit, drug use, extreme and/or contact sports) that increases their chance of sustaining a TBI. Similarly, individuals who have a propensity toward risk-taking behavior can be viewed as sharing similar traits (e.g., executive dysfunction manifested as impulsivity, aggression, or poor decision-making) that place them at increased risk for other negative outcomes, such as suicidal behavior.

However, in the context of risk-taking behavior, the relationship between pre-existing risk factors, TBI, and suicide is complicated and is best conceptualized over a life course. For example, it may be the case that for some individuals, TBI and suicidal behavior are two different manifestations of risk-taking behavior that are largely influenced by longstanding personality traits and neurocognitive risk factors such as aggression and poor decision-making. In such circumstances, an individual's risk for suicidal behavior is high regardless of TBI history, and the impact of the actual TBI on overall suicide risk may not be as pronounced when compared to other, more salient risk factors (e.g., history of substance abuse, mood disorder, and multiple SAs predating an mTBI). In other cases, however, sustaining a TBI may place individuals at risk for the intensification of pre-existing risk factors (e.g., impulsivity, aggression) that when coupled with potential sequelae secondary to TBI (e.g., poor sleep, pain, depression), significantly heighten an individual's risk for suicidal behavior.

Further complicating the matter is the fact that risk-taking behavior is largely influenced by contextual factors. For example, some individuals may be more prone to risk-taking when it comes to economic or financial decisions but may be risk-averse with respect to recreational or health-related decisions, whereas the opposite pattern may be true for others. The emotional context of a situation is also important to consider. Specifically, some individuals may exhibit risky behavior under emotional distress due to difficulties integrating affective and cognitive information when making decisions, yet such decision-making difficulties may not be apparent in affectively neutral situations. In the former instance, heightened emotions and impulsivity might work together to increase risk-taking by promoting action before the potential negative consequences of those actions have been fully measured (Paulsen et al., 2012).

In sum, a subset of individuals with a history of risk-taking behavior may be more likely to sustain TBIs. This risk-taking behavior may be in part rooted in executive dysfunction. If such individuals sustain TBIs, their executive dysfunction may be compounded by potential enduring cognitive sequelae of TBI (i.e., more executive dysfunction), thus placing them at greater risk for continued risky behavior, the consequences of which may culminate over time leading to suicidal behavior.

CONCLUSION AND FUTURE DIRECTIONS

The intersection of TBI and suicide presents challenges to both clinicians and researchers. It is important for clinicians to be aware that while some research has found that more severe TBIs have been associated with higher suicide risk than mTBIs (Simpson & Tate, 2007), an individual who has experienced a mTBI is not necessarily at lower risk of suicide than someone with a more severe injury. TBIs occur within the context of complex clinical realities that can markedly impact the associated suicide risk. A mTBI might confer a greater risk for suicide in an individual with poor premorbid functioning and substance abuse postinjury than a moderate/severe TBI experienced by an individual with good premorbid functioning and a stable social support system. Therefore, it is incumbent upon clinicians to do a comprehensive assessment when determining how a patient's TBI might be related to his or her risk of suicide, as opposed to only focusing on the most obvious (i.e., severe) cases.

With regard to executive dysfunction, clinicians should be aware that impairment in this cognitive domain is often not confined to impairment in one area (e.g., deficits only in impulsivity but not decision-making; Lezak et al., 2004). Therefore, providers will likely have to assess and work collaboratively with patients to address multiple areas of executive dysfunction as they relate to suicide risk (Homaifar et al., 2012). Additionally, as discussed earlier, under some circumstances, executive dysfunction may facilitate suicidal ideation/behavior in some individuals but not others (Westheide, et al., 2008), therefore the relationship of executive dysfunction to suicide risk is best conceptualized in the context of other factors that may increase an individual's propensity to engage in risky behaviors.

Future research is needed to develop and optimize treatments for individuals with histories of TBI and suicidal thoughts and/or behaviors. Thankfully, moderate/severe TBI and suicide are fairly low baserate events. However, this makes participant recruitment challenging. Furthermore, the many risk factors and complex histories of this heterogenous population present formidable challenges to drawing clear conclusions regarding causality and generalizing from the results of small studies. Researchers are working hard to overcome these challenges and to develop evidence-based treatments for this population. There is currently only one evidence-based suicide prevention treatment for those with TBI. A recent phase II randomized controlled trial examined the efficacy of a treatment program for hopelessness, which is a risk factor for suicidal behavior after severe TBI (Simpson et al., 2011). Compared to waitlist controls, treatment group participants reported a significant decrease in hopelessness and a trend toward reduced levels of suicidal ideation. Research is currently underway to replicate these findings (Matarazzo et al., 2014).

A crucial area of investigation is the continued development and refinement of treatments focused on facilitating recovery and reducing the accumulation of risk and adversity over time. Research on executive dysfunction may provide information to optimize such treatments and develop effective prevention programs. For example, longitudinal research that focuses on integrating neuroimaging and laboratory tasks with the assessment of real-life risky behavior and decision-making during daily activities has the potential to highlight causal mechanisms and clarify the relationships among the many common risk factors of TBI and suicide. In particular, researchers should work to clarify the interplay of TBI and suicide risk factors through the lens of executive dysfunction to provide clinicians with evidenced-based avenues for prevention/intervention of both TBI and suicide.

REFERENCES

Alway, Y. J., Gould, K., McKay, A., Johnston, L., & Ponsford, J. (2015). The evolution of posttraumatic stress disorder following moderate to severe traumatic brain injury. *Journal of Neurotrauma*, 33(9), 825–831.

Anstey, K. J., Butterworth, P., Jorm, A. F., Christensen, H., Rodgers, B., & Windsor, T. D. (2004). A population survey found an association between self-reports of traumatic brain injury and increased psychiatric symptoms. *Journal of Clinical Epidemiology*, 57(11), 1202–1209.

Barnes, S. M., Walter, K. H., & Chard, K. M. (2012). Does a history of mild traumatic brain injury increase suicide risk in veterans with PTSD? *Rehabilitation Psychology*, 57(1), 18–26. doi: 10.1037/a0027007

Berger, A., & Posner, M. I. (2000). Pathologies of bran attentional networks. *Neuroscience and Biobehavioral Review*, 24(1), 3–5.

Bjork, J. M., & Grant S. J. (2009) Does traumatic brain injury increase risk for substance abuse? *Journal of Neurotrauma*, 26(7), 1077–1082.

Brenner, L. A., & Homaifar, B. Y. (2009). Deployment acquired TBI and suicidality: Risk and assessment. In L. Sher & A. Vilens (Eds.), *War and suicide* (pp. 189–202). New York: Nova Science.

Brenner, L. A., Homaifar, B. Y., Wolfman, J. H., Kemp, J., & Adler, L. E. (2009). Suicidality and Veterans with a History of Traumatic Brain Injury: Precipitating Events, Protective Factors, and Prevention Strategies. *Rehabilitation Psychology*, 54(4), 390–397.

Brenner, L. A., Homaifar, B. Y., Olson-Madden, J. H., Nagamoto, H., Huggins, J., Schneider, A. L., . . . Corrigan, J.D. (2013). Prevalence and screening of traumatic brain injury among veterans seeking mental health services. *Journal of Head Trauma Rehabilitation, 28*(1), 21–30.

Brenner, L. A., Ignacio, R. V., & Blow, F. C. (2011). Suicide and traumatic brain injury among individuals seeking Veterans Health Administration services. *Journal of Head Trauma Rehabilitation, 26*(4), 257–264.

Brooks, N., Campsie, L., Symington, C., Beattie, A., & McKinlay, W. (1986). The five year outcome of severe blunt head injury: A relative's view. *Journal of Neurology, Neurosurgery & Psychiatry, 49*, 764–770.

Brooks, N., Campsie, L., Symington, C., Beattie, A., & McKinlay, W. (1987). The effects of severe head injury on patient and relative within seven years of injury. *Journal of Neurology, Neurosurgery & Psychiatry, 2*, 1–13.

Bryan, C. J., Clemans, T. A., Hernandez, A. M., & Rudd, M. D. (2013). Loss of consciousness, depression, posttraumatic stress disorder, and suicide risk among deployed military personnel with mild traumatic brain injury. *Journal of Head Trauma Rehabilitation, 28*(1), 13–20.

Bryant, R. A., O'Donnel, M. L., Creamer, M., Mcfarlane, A. C., Clark, C. R., & Silove, D. (2010). The psychiatric sequelae of traumatic injury. *The American Journal of Psychiatry 167*, 312–320.

Burton, C. Z., Vella, L., Weller, J. A., & Twamley, E. W. (2011). Differential effects of executive functioning on suicide attempts. *Journal of Neuropsychiatry and Clinical Neurosciences, 23*(2), 173–179.

Centers for Disease Control and Prevention. (2006, June 14). *Explosions and blast injuries: A primer for clinicians.* Retrieved from http://www.bt.cdc.gov/masscasualties/explosions.asp

Centers for Disease Control and Prevention. (2012). *Get the stats on traumatic brain injury in the United States.* Retrieved from http://www.cdc.gov/traumaticbraininjury/pdf/BlueBook_factsheet-a.pdf.

Cernak, I., Savic, J., Ignjatovic, D., & Jevtic, M. (1999). Blast injury from explosive munitions. *Journal of Trauma, 47*(1), 96–103.

Cernak, I., Wang, Z., Jiang J., Bian, X., & Savic, J. (2001). Cognitive deficits following blast injury-induced neurotrauma: Possible involvement of nitric oxide. *Brain Injury, 15*(7), 593–612.

Corrigan, J. D., & Bogner, J. (2007a). Initial reliability and validity of the Ohio State University TBI identification. *Journal of Head Trauma Rehabilitation, 22*(6), 318–329.

Corrigan, J. D., & Bogner, J. (2007b). Screening and identification of TBI. *Journal of Head Trauma Rehabilitation, 22*(6), 315–317.

Courtet, P., Gottesman, I. I. Jollant, F., & Gould, T. D. (2011). The neuroscience of suicidal behaviors: What can we expect from endophenotype strategies? *Translational Psychiatry, 1*, e7.

Department of Veterans Affairs, & Department of Defense. (2009). *VA/DOD clinical practice guidelines.* Retrieved from http://www.healthquality.va.gov/

Dixon, T., Kravariti, E., Frith, C., Murray, R. M., & McGuire, P. K. (2004). Effect of symptoms on executive function in bipolar illness. *Psychological Medicine, 34*(5), 811–821.

Gordon, W. A., Haddad, L., Brown, M., Hibbard, M. R., & Sliwinski, M. (2000). The sensitivity and specificity of self-reported symptoms in individuals with traumatic brain injury. *Brain Injury, 14*, 21–33.

Hart, T., Brenner, L., Clark, A., Bogner, J., Novack, T., Chervoneva, I., . . . Arango-Lasprilla, J. (2011). Major and minor depression after traumatic brain injury. *Archives of Physical Medicine and Rehabilitation, 92*, 1211–1219.

Hibbard, M. R., Uysal, S., Kepler, K., Bogdany, J., & Silver, J. (1998). Axis I psychopathology in individuals with traumatic brain injury. *Journal of Head Trauma Rehabilitation, 13*, 24–29.

Homaifar, B.Y., Bahraini, N., Silverman, M., & Brenner, L. A. (2012). Executive functioning as a component of suicide risk assessment: Clarifying its role in standard clinical applications. *Journal of Mental Health Counseling, 34*(2), 110–120.

Homaifar, B. Y., Brenner, L. A., Forster, J. E., & Nagamoto, H. (2012). Traumatic brain injury, executive functioning, and suicidal behavior: A brief report. *Rehabilitation Psychology, 57*(4), 337–341.

Hoofien, D., Gilboa, A., Vakil, E., & Donovick, P.J. (2001). Traumatic brain injury (TBI) 10-20 years later: A comprehensive outcome study of psychiatric symptomatology, cognitive abilities and psychosocial functioning. *Brain Injury, 15*(3), 189–209.

Ivins, B. J., Schwab, K. A., Warden, D., Harvey, L. T., Hoilien, M. A., Powell, C. O., . . . Salazar, A. M. (2003). Traumatic brain injury in US Army paratroopers: prevalence and character. *Journal of Trauma, 55*, 617–621.

Jorge, R. E., Robinson, R. G., Moser, D., Tateno, A., Facorro, B., & Arndt, S. (2004). Major depression following traumatic brain injury. *Archives of General Psychiatry, 61*, 42–50.

Keilp, J. G., Sackeim, H. A., Brodsky, B. S., Oquendo, M. A., Malone, K. M., & Mann, J. J. (2001). Neuropsychological dysfunction in depressed suicide attempters. *The American Journal of Psychiatry, 158*, 735–741.

Kim, E. (2002). Agitation, aggression, and disinhibition syndromes after traumatic brain injury. *NeuroRehabilitation, 17*(4), 297–310.

Kreutzer, J. S., Seel, R. T., & Gourley, E. (2001). The prevalence and symptom rates of depression after traumatic brain injury: A comprehensive examination. *Brain Injury, 15,* 563–576.

Jollant, F., Bellivier, F., Leboyer, M., Astruc, B., Torres, S., & Verdier, R. (2005). Impaired decision making in suicide attempters. *The American Journal of Psychiatry, 162*(2), 304–310.

Jollant, F., Lawrence, N. S., Olie, E., O'Daly, O., Malafosse, A., & Courtet, P. (2010). Decreased activation of lateral orbitofrontal cortex during risky choices under uncertainty is associated with disadvantageous decision-making and suicidal behavior. *Neuroimage, 51*(3), 1275–1281.

Keilp, J. G., Sackeim, H. A., Brodsky, B. S., Oquendo, M. A., Malone, K. M., & Mann, J. J. (2001). Neuropsychological dysfunction in depressed suicide attempters. *The American Journal of Psychiatry, 158,* 735–741. doi:10.1176/appi.ajp.158.5.735

Kuipers, P., & Lancaster, A. (2000). Developing a suicide prevention strategy based on the perspectives of people with brain injuries. *Journal of Head Trauma and Rehabilitation, 15*(6), 1275–1284.

Lewin, W., Marshall, T. F., & Roberts, A. H. (1979). Long-term outcome after severe head injury. *BMJ, 77,* 1533–1538.

Lezak, M. D., Howieson, D. B., & Loring, D. W. (2004). *Neuropsychological assessment* (4th ed.). New York: Oxford University Press.

Mainio, A., Kyllönen, T., Viilo, K., Hakko, H., Särkioja, T., & Räsänen, P. (2007). Traumatic brain injury, psychiatric disorders and suicide: A population-based study of suicide victims during the years 1988–2004 in Northern Finland. *Brain Injury, 21*(8), 851–855.

Malec, J. F., Testa, J. A., Rush, B. K., Brown, A. W., & Moessner, A. M. (2007). Self-assessment of impairment, impaired self-awareness, and depression after traumatic brain injury. *Journal of Head Trauma Rehabilitation, 22,* 156–166.

Mann, J. J., Waternaux, C., Haas, G. L., & Malone, K. M. (1999). Toward a clinical model of suicidal behavior in psychiatric patients. *The American Journal of Psychiatry, 156*(2), 181–189.

Marzuk, P. M., Hartwell, N., Leon, A. C., & Portera, L. (2005). Executive functioning in depressed patients with suicidal ideation. *Acta Psychiatrica Scandinavica, 112*(4), 294–301. doi:10.1111/j.1600-0447.2005.00585.x

Matarazzo, B. B., Hoffberg, A. S., Clemans, T. A., Signoracci, G. M., Simpson, G. K., & Brenner, L. A. (2014). Cross-cultural adaptation of the Window to Hope: A psychological intervention to reduce hopelessness among US veterans with traumatic brain injury. *Brain Injury, 28*(10), 1238–1247.

McDonald, B. C., Flashman, L. A., & Saykin, A. J. (2002). Executive dysfunction following traumatic brain injury: Neural substrates and treatment strategies. *NeuroRehabilitation, 17*(4), 333–344.

McLennan, S. N., & Mathias, J. L. (2010). The depression-executive dysfunction (DED) syndrome and response to antidepressants: A meta-analytic review. *International Journal of Geriatric Psychiatry, 25*(10), 933–944.

Military Health System Office of Strategic Communications. (2015). *DoD numbers for traumatic brain injury worldwide.* Retrieved from http://dvbic.dcoe.mil/dod-worldwide-numbers-tbi

National Center for Injury Prevention and Control. (2003). *Report to Congress on mild traumatic brain injury in the United States: Steps to prevent a serious public health problem.* Atlanta, GA: Author.

Oquendo, M. A., Friedman, J. H., Grunebaum, M. F., Burke, A., Silver, J. M., & Mann, J. J. (2004). Suicidal behavior and mild traumatic brain injury in major depression. *Journal of Nervous and Mental Disease, 192*(6), 430–434.

Paulsen, D. J., Carter, R. M., Platt, M. L., Huettel, S. A., & Brannon, E. M. (2012). Neurocognitive development of risk aversion from early childhood to adulthood. *Frontiers in Human Neuroscience, 5,* 178–195. doi:10.3389/fnhum.2011.00178

Reed, R. A., Harrow, M., Herbener, E. S., & Martin, E. M. (2002). Executive function in schizophrenia: Is it linked to psychosis and poor life functioning? *Journal of Nervous Mental Disease, 190*(11), 725–732.

Reeves R. R., & Laizer J. T. (2012). Traumatic brain injury and suicide. *Journal of Psychosocial Nursing and Mental Health Services, 50*(3), 32–38. doi:10.3928/02793695-20120207-02.

Rodger, L. L., Wood, C. W., & Lewis, R. (2010). Role of alexithymia in suicide ideation after traumatic brain injury. *Journal of the International Neuropsychological Society, 16,* 1108–1114.

Ruff, R. M., Iverson, G. L., Barthe, J. T., Shane, S., Bush, S. S., Broseke, D. K., & NAN Policy and Planning Committee. (2009). Recommendations for diagnosing a mild traumatic brain injury: A National Academy of Neuropsychology education paper. *Archives of Clinical Neuropsychology, 24*(1), 3–10.

Ryan, L. M., & Warden, D. L. (2003). Post concussion syndrome. *International Review of Psychiatry*, 15(4), 310–316.

Shavelle, R. M., Strauss, D., Whyte, J., Day, S. M., & Yu, Y. L. (2001). Long-term causes of death after traumatic brain injury. *American Journal of Physical Medicine and Rehabilitation*, 80, 510–516.

Silver, J. M., Kramer, R., Greenwald, S., & Weissman, M. (2001) The association between head injuries and psychiatric disorders: Findings from the New Haven NIMH Epidemiological Catchment Area Study. *Brain Injury*, 15(11), 935–945.

Simpson, G. K., & Brenner, L. A. (2011). Perspectives on suicide and traumatic brain injury. *Journal of Head Trauma Rehabilitation*, 26(4), 241–243.

Simpson, G., & Tate, R. (2005). Clinical features of suicide attempts after traumatic brain injury. *The Journal of Nervous and Mental Disease*, 193(10), 680–685.

Simpson, G., & Tate, R. (2007). Suicidality in people surviving a traumatic brain injury: Prevalence, risk factors and implications for clinical management. *Brain Injury*, 21, 1335–1351. doi:10.1080/02699050701785542

Simpson, G. K., Tate, R. L., Whiting, D. L., & Cotter, R. E. (2011). Suicide prevention after traumatic brain injury: A randomized controlled trial of a program for the psychological treatment of hopelessness. *Journal of Head Trauma Rehabilitation*, 26(4), 290–300. doi:10.1097/HTR.0b013e3182225250

Skopp, N. A., Trofimovich, L., Grimes, J., Oetjen-Gerdes, L., & Gahm, G. A. (2012). Relations between suicide and traumatic brain injury, psychiatric diagnoses, and relationship problems, active component, US Armed Forces, 2001–2009. *Medical Surveillance Monthly Report*, 19(2), 7–11.

Teasdale, T. W., & Engberg, A. W. (2001). Suicide after traumatic brain injury: A population study. *Journal of Neurology, Neurosurgery & Psychiatry*, 71(4), 436–440.

Terrio, H., Brenner, L. A., Ivins, B. J., Cho, J. M., Helmick, K., & Schwab, K. (2009). Traumatic brain injury screening: Preliminary findings in a US Army brigade combat team. *Journal of Head Trauma Rehabilitation*, 24(1), 14–23.

Tsaousides, T., Cantor, J. B., & Gordon, W. A. (2011). Suicidal ideation following traumatic brain injury: Prevalence rates and correlates in adults living in the community. *Journal of Head Trauma Rehabilitation*, 26(4), 265–275.

Vasterling, J. J., Proctor, S. P., Amoroso, P., Kane, R., Heeren, T., & White, R. F. (2006). Neuropsychological outcomes of army personnel following deployment to the Iraq war. *JAMA: Journal of the American Medical Association*, 296(5), 519–529. doi: 10.1001/jama.296.5.519.

Vasterling, J. J., Verfaelli, M., & Sullivan, K. D. (2009). Mild traumatic brain injury and posttraumatic stress disorder in returning veterans: Perspectives from cognitive neuroscience. *Clinical Psychology Review*, 29, 674–684.

Westheide, J., Quednow, B. B., Kuhn, K. U., Hoppe, C., Cooper-Mahkorn, D., Hawellek, B. (2008). Executive performance of depressed suicide attempters: The role of suicidal ideation. *European Archives of Psychiatry and Clinical Neuroscience*, 258(7), 414–421.

Whiteside, S. P., & Lynam, D. R. (2001). The five factor model and impulsivity: Using a structural model of personality to understand impulsivity. *Personality and Individual Differences*, 30(4), 669–689.

Whittaker, R., Kemp, S., & House, A. (2007). Illness perceptions and outcome in mild head injury: A longitudinal study. *Journal of Neurology, Neurosurgery & Psychiatry*, 78(6), 644–646.

Yurgelun-Todd, D. A., Bueler, C. E., McGlade, E. C., Churchwell, J. C., Brenner, L. A., & Lopez-Larson, M. P. (2011). Neuroimaging correlates of traumatic brain injury and suicidal behavior. *Journal of Head Trauma Rehabilitation*, 26, 276–289. doi:10.1097/HTR.0b013e31822251dc00001199-201107000-00005[pii]

16

The Problem of Suicide in the United States Special Operations Forces

Bruce Bongar
Kate Maslowski
Catherine Hausman
Danielle Spangler
Tracy Vargo

U.S. SPECIAL OPERATIONS FORCES

The U.S. Special Operations Command (SOCOM) was established in 1986 (Feickert, 2015; Hoffman, 2006). Often operating in small teams on high-risk missions, Special Operations Forces (SOF) can also be found working in conjunction with Army, Navy, and Air Force regular forces (Hoffman, 2006). In the years since September 11, 2001, the military has adapted to the irregular nature of today's military conflicts. Hoffman (2006) argued that this adjustment requires the military to take a less direct and targeted approach to warfare. Hoffman further noted that the SOF have proven essential in recent conflicts such as Operation Enduring Freedom in Afghanistan and in counterterrorism efforts around the world. Due to high demand in recent years, the number of SOF members has increased by thousands in recent years (Hoffman, 2006). Currently, SOCOM is comprised of about 66,000 personnel (Feickert, 2015). Since 2004, SOF responsibilities have been extended to counterterrorism efforts (Feickert, 2015). While it is important to note that SOF are utilized by many nations aside from the United States (Natolochnaya & Cherkasov, 2014), the literature reviewed herein focuses primarily on issues concerning the SOF of the United States.

THE PROBLEM OF SUICIDE

A recent report from the U.S. Congress (2015) noted a sharp increase in the suicide rate within SOF since 2012. From 2011 to 2014, the demands of this career path have contributed to 49 special operators dying by suicide (Shanker & Oppel, 2014). Although these individuals make up a small portion of the total armed forces, these losses have had a disproportionate impact on the highly cohesive SOF community. In addition to the potential impact on fighting effectiveness, these deaths by suicide also represent the loss of a significant monetary investment by the government, due to the extensive specialized training required for SOF members (U.S. Government Accountability Office, 2015). There is an urgent need to prevent the loss of these individuals to suicide.

UNIQUE ASPECTS OF THE SOF

Some insist that the SOF represents a unique risk group in terms of mental health challenges (Hanwella & Silva, 2012; U.S. Special Operations Command, 2015). These individuals are often responsible for tasks and missions that require particular training or equipment, are carried out in dangerous or hostile regions, and are conducted under conditions that are time-sensitive and covert (Feickert, 2015). One study indicates that the SOF experience uniquely prolonged high-stress situations (Taverniers, Ruysseveldt, Smeets, & Grumbkow, 2010). According to Strobel (2014), some SOF members have faced multiyear deployments during the recent Middle East conflicts. Further, others note an increase in frequency of deployments for special operators in the past decade (U.S. Government Accountability Office, 2015). Multiple and lengthy deployments have been associated with higher rates of meeting the criteria for a PTSD diagnosis and with increased problems within the service members' families (Barker & Berry, 2009; MacGregor, Han, Dougherty, & Galarneau, 2012). An important component of this finding is that length of dwell time (duration in which a service member spends between deployments) is negatively correlated with PTSD and other mental health problems (such as substance use) following deployment (MacGregor et al., 2012). Thus a key concern with SOF members is that increasing demand for their unique capabilities overseas has led to increased pressure to decrease dwell times in order to maximize the utility of these individuals.

Other researchers have indicated that SOF members may be hesitant to disclose psychological symptoms or problems due to fears that such ailments would end their careers (Strobel, 2014). Another study examined alcohol misuse in U.S. SOF members (Skipper, Forsten, Kim, Wilk, & Hoge, 2014). In this sample, 15% were assessed as having problems with alcohol. The results suggested that certain fighting experiences such as being attacked, firing directly at the enemy, and hand-to-hand combat were associated with alcohol misuse (Skipper et al., 2014). Because alcohol dependence is one of the primary mental health problems associated with suicide, it follows that a group with high-frequency exposure to fighting experiences associated with alcohol misuse would also have elevated suicide risk (Canapary et al., 2002). While Bryan, Stephenson, Morrow, Staal, and Haskell (2014) note that increasing a SOF member's belief that they have achieved or accomplished something important might protect against mental health problems, far more research is needed to address the problem of suicide in this unconventional group of military service members.

RESILIENCE AND ARMED FORCES

Military deployment involves stressors such as combat and being apart from family and friends (Casey, 2011). The concept of resilience can be defined as the soldier's ability to maintain psychological well-being in the face of hardships. Until recent years, little emphasis was placed on the importance of psychological fitness relative to physical fitness by the military (Lester, McBride, & Cornum, 2013). However, resilience, specifically pertains to military families, is now being examined extensively in research (Lester, Taylor, Hawkins, & Landry, 2015).

Social relationships play an important role in a service member's resilience to psychological strain (King, King, Fairbank, Keane, & Adams, 1998). For example, research on veterans has found social support to be negatively associated with posttraumatic stress disorder (PTSD) and depression symptoms (Pietrzak, Johnson, Goldstein, Malley, & Southwick, 2009). Some researchers targeted the social aspect and found that social resilience training (which increased social skills such as empathy and perspective-taking) was shown to decrease the service members' feelings of loneliness (Cacioppo et al., 2015). Interestingly, not all family problems bode poorly for resilience. One study found that some military families comprised of individuals with histories of adverse early childhood experiences may in fact exhibit enhanced family resilience to psychological strain (Oshri et al., 2015).

A recent attempt to improve psychological resilience came in the form of the Comprehensive Soldier Fitness (CSF) program instituted by the U.S. Army (Casey, 2011; Lester et al., 2013). CSF aims to serve as a preventative intervention for healthy service members rather than as a treatment for those already struggling with psychological problems. The CSF model involves tailored training approaches that target service members' individual needs, which are initially assessed through a self-report questionnaire. As part of this program, the U.S. Army incorporated resilience training into its professional leadership education (Lester et al., 2013).

Few researchers have addressed resilience in SOF service members specifically. One study found that graduates from a SOF training course tended to have greater dispositional hardiness to stress than those who did not complete the course (Bartone, Roland, Picano, & Williams, 2008). The notion that SOF service members are typically superior in their resistance to psychological strain is at odds with the trend of increasing suicide in this population and may partially help to explain why limited research exists on improving resilience for these individuals. Increasing a Special Operations who believe strongly that they have accomplished something important during their deployments report fewer postdeployment health problems (Bryan, Stephenson, Morrow, Staal, & Haskell, 2014). Instilling a sense of achievement represents one potential avenue for enhancing SOF resilience. Finally, studies on service members (both special forces and other branches) have shown that resilience may be associated with distinct patterns of brain activity, such as in relation to avoidance and in anticipation of rewards (Lin et al., 2014; Vythilingam et al., 2009). Thus another area for further research is greater exploration of the biological underpinnings of resilience. Advances in this knowledge area may lead to advances in pharmacological treatments or psychotherapeutic interventions.

RATIONALE OF POTFF

Because SOF service members have had a spike in suicide rates since 2010, an examination of the various stressors that could possibly explain this phenomenon has become a priority. A Joint Special Operations University (JSOU) team found that, in addition to the demanding, dangerous, and potentially life-altering challenges SOF members face while deployed, they also face a variety of daunting tasks when they return home and are between tours (JSOU, 2014)., These tasks involve familial, social, spiritual, psychological, and physical demands.

One particular task that each SOF service member must face involves bridging the gap between military culture and civilian life (JSOU, 2014). Civilian culture is different from military culture, and this dichotomy creates a barrier to the temporary immersion of SOF service members into civilian life. Another task involves maintaining preparedness for combat, while honoring family commitments, when members are at home. This balance is one with which many SOF service members and their families have struggled. Similar to this struggle is the task of being apart from their families for months at a time and the frequent familial problems and divorces associated with multiple deployments. An additional task that many SOF service members have encountered when deployed is a disconnection from their spirituality, perhaps due to distance from spiritual leaders during deployment. A final task that has been a particular area of concern involves psychological evaluations. SOF members are required to undergo psychological evaluations to assess the effects of their experiences, and these evaluations can impact future assignments and careers. Stigma surrounding the results of psychological evaluations can create an immense amount of shame for SOF service members who have been impacted by their experiences during deployment. Through the examination of stressors during and outside of deployment, it became clear that SOF service members and their families faced unique demands and challenges.

OBJECTIVE AND DEVELOPMENT

SOCOM initiated a study of the SOF community in 2010 and concluded that the force was unraveling due to the tremendous demands and significant stressors faced by SOF service members and their families (National Military Family Association, 2014). Admiral William McRaven, commander of SOCOM at the time, ventured to address the issue by developing a holistic program for the SOF community called the Preservation of the Force and Family (POTFF). This program focuses on ways to ameliorate some of the stress and to improve the over well-being of the SOF community.

POTFF was created with the ultimate goal of addressing the pressures placed on the SOF community. It aims to maintain operational effectiveness by improving the readiness and overall well-being of SOF service members (U.S. Special Operations Command, 2015). The primary objective of the POTFF is to increase the availability of support services to SOF service members and their families and to extend SOF service careers and improve family well-being.

POTFF has various priorities that it aims to uphold by collaborating with each service component, in order to implement programs that are tailored to the needs of the service type (U.S. Special Operations Command, 2013). Some priorities involve building force resiliency, optimizing SOF performance, and enhancing the readiness of SOF service members. Other priorities involve advocating for policy change, finding solutions to difficult problems, and evaluating the efficacy of the POTFF programs.

A final priority of POTFF is to integrate support to the SOF community by coordinating and collaborating with services. This is done by increasing the proximity of services or by embedding services within the SOF. These integrated services include sleep hygiene, cognitive processing, advanced rehabilitation, strength and conditioning, holistic preventative care, postdeployment and periodic assessment, stress-inoculation training, proximate care and consultation, teaching skills, goal-setting, education, identification and referral for tertiary care, integration of family into all aspects of care, SOF-unique family programs, and specialized pastoral coaching (U.S. Special Operations Command, 2013). Additionally, with the prioritization of providing services to the SOF community, POTFF also prioritized finding ways to reducing stigma associated with help-seeking behaviors.

POTFF FOUR DOMAINS

Because of their stoic appearance, special operators' physical, psychological, social, and spiritual needs have historically been overlooked. As a result the POTFF Task Force (POTFF-TF) was established to examine the unique needs of special operators and their families. The POTFF-TF consulted with experts in each of these domains as well as academics, government agencies, and industry to identify techniques to optimize overall SOF performance (U.S. Special Operations Command, 2013). With this information, the POTFF-TF developed a holistic, multidimensional, and proactive approach to bolstering resiliency within SOF. Specifically, this strength-based movement aims to improve mission readiness, operational effectiveness, and short-term and long-term well-being of SOF members by addressing four key domains: physical, psychological, social, and spiritual well-being. Rather than approaching this goal from the perspective of addressing potential areas of weakness for operators, this movement seeks to enhance resiliency and readiness in order to maximize operators' competitive advantage over future adversaries (U.S. Special Operations Command, 2013).

POTFF features collaboration with not only special operators but also with their family members in order to identify challenges, provide innovative solutions, evaluate existing programs, advocate for policy changes, and reduce stigma associated with seeking help (U.S. Special Operations Command, 2013). This movement incorporates a multidisciplinary response with a team consisting of physical therapists, strength and conditioning coaches, athletic trainers, dieticians, psychologists, licensed clinical social workers, nurse case managers, chaplains, and family readiness coordinators. Further, specific programs are tailored to the unique requirements of each component of SOCOM, including Air Force Special Operations Command, Army Special Operations Command, Joint Special Operations Command, Marine Corps Forces Special Operations Command, and Naval Special Warfare Command. However, POTFF is based on the same four domains across all components of SOF.

Physical Performance

Special force operators "perform hazardous work in unforgiving environments at an unrelenting pace" (U.S. Special Operations Command, 2013). Extensive physical conditioning and training are critical to prevent debilitating injuries or death (U.S. Special Operations Command, 2013). In addition, operators have unique nutritional needs due to the rigorous physical demands of their roles. The nutrition of an operator can greatly affect muscle performance, body composition, and cognitive performance (Ford & Glymour, 2014). Physical injuries have also been found to be associated with posttraumatic stress reactions and poorer general mental health (Dyster-Aas et al., 2012). Thus it is important to prevent physical injuries in order to improve the overall effectiveness of SOF and to avoid other negative mental health outcomes for individual operators. In addition, physical exercise is associated with positive health outcomes and improved cognitive functioning, which can serve as protective factors against mental problems (Deslandes et al., 2009; Hötting & Röder, 2013). To help develop physically well-rounded operators, the POTFF-TF created the Human Performance Program (HPP).

The purpose of HPP is to meet the physical needs of the special force operator in order to reduce the likelihood of debilitating injuries or death and to improve overall performance and resiliency (U.S. Special Operations Command, 2013). This prehabilitative training program incorporates techniques from exercise physiology, nutrition, kinesiology, and sports psychology (U.S. Special Operations Command, 2013). Because the majority of special operators are already physically active, this program instead focuses on improving operators' efficiency and safety while exercising and facilitating quick recovery from injuries. Specifically, strength and conditioning coaches are available to offer specialized training based on unique physical demands of operators, and athletic trainers are available to supervise physical exercise (U.S. Special Operations Command, 2013). Common injuries seen among special operators are similar to those of elite athletes. In addition, injury among elite athletes has been found to have a strong negative effect on social functioning, vitality, and general mental health (McAllister, Motamedi, Hame, Shapiro, & Dorey, 2001). Thus the strength and conditioning coaching, physical therapy rehabilitation techniques, and sports psychology techniques utilized by elite athletes have been applied to treat special operators. Specifically, physical therapists are available to implement rehabilitation programs following injuries and to monitor progress over time. In addition, POTFF dieticians are educating operators and their families about proper nutrition during training periods and daily life. HPP also seeks to maintain a competitive advantage over adversaries by improving overall operator performance and resiliency (U.S. Special Operations Command, 2013). Utilizing techniques from sports psychology, POTFF providers teach operators performance enhancing mental skills training to maintain optimal cognitive capabilities while performing in stressful environments.

Psychological Performance

Special operators must endure the psychological strain of continuous operations, unpredictable deployments, and prolonged separation from family members (U.S. Special Operations Command, 2013). Although mental health utilization in the military has risen over the past decade, stigma around seeking psychological help is still a prevalent problem (Quartana et al., 2014). However, leaders can influence the norms and values of the group by the policies and priorities they establish, directives they give, advice they provide, and actions they take (Weick & Roberts, 1993). Thus SOF leadership can reduce stigma by supporting, utilizing, and speaking openly about mental health treatment (Britt, Wright, & Moore, 2012; Clark-Hitt, Smith, & Broderick, 2012). Once an operator receives mental health treatment, they are more likely to seek help again (Brown, Creel, Engel, Herrell, & Hoge, 2011). To address these psychological factors, the POTFF–TF established the Psychological Performance Program (PPP). By including this domain, POTFF aims to improve cognitive and behavioral performance in order to ultimately bolster resilience of SOF and their families (U.S. Special Operations Command, 2013).

The purpose of the PPP is to not only provide psychological support for service members and their families but also to reduce the stigma surrounding seeking psychological help by engaging leadership at all levels (U.S. Special Operations Command, 2013). This includes SOF leaders encouraging utilization of mental health services and speaking openly about their experiences with mental health treatment (Britt et al., 2012; Clark-Hitt et al., 2012). PPP also includes embedded mental health professionals within each unit to accommodate unpredictable training or deployment schedules and to adjust to the unique needs of the given unit (U.S. Special Operations Command, 2013). A mental health professional specially trained in the needs of the special operations community provides operators with access to someone they can trust and who understands them (Miles, 2013). In addition, POTFF providers educate operators and their families about sleep hygiene, teach stress-inoculation strategies, facilitate cognitive processing, and assist with goal-setting. Postdeployment and periodic psychological assessments normalize the assessment process while providing operators and leaders feedback about their psychological well-being.

Social Performance

Due to multiple, unpredictable deployments, special operators experience extended separations from family members that impact both the operators themselves and their family members (U.S. Special Operations Command, 2013). Military families face challenges not only during times of separation but also during episodes of reunion and reintegration.

Strain from repeated deployments is associated with increased rates of marital conflict, domestic violence, parental maltreatment, spousal depression and anxiety, vicarious posttraumatic stress among family members, and emotional and behavioral problems among military children (Saltzman et al., 2011). In addition, repeated deployments often disrupt family organization, which may contribute to family stress and can hinder child adjustment to inevitable changes (Saltzman et al., 2011). For SOF in particular, family member stress during deployment may be heightened by operators' involvement in high-risk, secretive missions. Further, emotional estrangement and impaired communication following return from deployment may hinder the reintegration process and restoration of family closeness (Saltzman et al., 2011). However, healthy social support from family members and family cohesion can mitigate the effects of stress on emotional well-being and overall performance (Harms, Krasikova, Vanhove, Herian, & Lester, 2013). Further, families can gain individual and family resilience in the face of stress and adversity with adequate resources (Saltzman et al., 2011). Operators need stable and reliable social support networks in order to focus on their missions (U.S. Special Operations Command, 2013). The POTFF social performance domain focuses on the connections among operator resilience and performance, the quality of social relationships, and family functioning (Saltzman et al., 2011).

The purpose of the social performance domain is to strengthen social relationships of special operators and their family members, foster family cohesion, and reduce outside stress associated with social relationships through a variety of programs (U.S. Special Operations Command, 2013). SOCOM Military and Family Life Consultants (MFLC) offer nonclinical counseling services to address a variety of potential issues including deployment and reintegration challenges, marriage and relationship problems, parenting and family issues, stress and anxiety management, and communication problems (U.S. Special Operations Command, 2013). MFLC also provides support for coping with sadness, grief, loss, and the daily life challenges that are unique to special operators (U.S. Special Operations Command, 2013). In addition, SOCOM hopes to build trust between operators, their families, and the command by treating family members as part of the SOF team and integrating family members into all aspects of care. This includes scheduling time for communication among SOF leadership, operators, and family members to address the evolving needs of operators and their families. SOCOM has also developed family advocacy programs and sought to involve family members in the deployment process as part of the SOF team. Although the details of the mission often cannot be disclosed, SOCOM attempts to explain to family members how their sacrifices are contributing to the overall success of the mission (U.S. Special Operations Command, 2013). Research has shown that family belief in the mission was a strong predictor of healthy coping and adaptation among children of military families (Palmer, 2008). In addition, POTFF providers teach ways to stay connected during deployment and utilize advanced social media to assist with ongoing communication during periods of separation when possible. Other initiatives within this domain include sponsored day trips for youth of deployed service members, spouse support groups, family resources guides, child care resources, antibullying classes, and resiliency classes for spouses and family members (U.S. Special Operations Command, 2013).

In addition, JSOU and the Center for Special Operations Studies and Research (CSOSR) have prioritized conducting research to further improve POTFF initiatives in the social performance domain. In 2015, JSOU and CSOSR focused research efforts on the following topics: regulating family commitment between deployments, evaluating family accompaniment policies (i.e., how family accompaniment may affect mission performance and POTFF initiatives, how advanced social media has impacted stress associated with separation from family), analyzing rotation rates (i.e., optimal time spent at home between deployments to allow for mental recuperation, physical recovery, and family bonding as well as identifying rotation rates that are sustainable for the force and families over the long term), and examining differences between the civilian culture and special operations culture (and how these differences affect POTFF initiatives).

Spiritual Performance

Spirituality refers to an internal, personal, subjective, and private experience (Reutter & Bigatti, 2014). Although spirituality does not require subscription to a particular religion, core spiritual components can serve to strengthen the individual's ability to cope with challenges when combined with his or her religious beliefs (U.S. Special Operations Command, 2013). In addition, religiosity (i.e., the degree to which a person adheres to religious values or beliefs) and spirituality are seen as effective resiliency resources (Reutter & Bigatti, 2014). The spiritual domain was established to foster these core spiritual components.

The purpose of the spiritual performance program is to enhance core spiritual beliefs, values, awareness, relationships, and experiences as a basis of healthy living (U.S. Special Operations Command, 2013). Spiritual initiatives include chaplain-led or chaplain-supported programs for service members and their families, specialized pastoral coaching, and parent-child retreats. Again, JSOU and CSOSR have prioritized research involving the assessment of the utility of spiritual support (i.e., changes required to facilitate interaction between operators and chaplains, decision for chaplains or spiritual advisors to be deployed with SOF units). Therefore, POTFF aims to continue to improve the resources offered through the spiritual domain.

FUTURE DIRECTIONS

Evaluating Effectiveness

As with any new program or intervention, data is needed on how effective POTFF is in building resiliency and achieving its mission. This program is unique in that the population it serves is small and hard to access by researchers and clinicians outside of SOF. Currently, there is little data on the effectiveness of POTFF. The integration of tailored POTFF programs to each service of SOCOM is in its infancy. However, a crucial aspect of the POTFF is the POTFF Needs Assessment Survey. Every SOF member and spouse is encouraged to complete the survey, in order to provide information regarding the need for resources and efforts (Zachary, 2015). Surveys have been administered at least three times since 2010. Unfortunately, only about 16% of SOF and spouses have participated in the Needs Assessment Survey (Caserta & Neff, 2014).

In addition to the information being gathered by the POTFF Needs Assessment Survey, service branches have begun to evaluate the efficacy of POTFF. The Naval Inspector General conducted a review of the Commander, Naval Special Warfare Command (CNSWC) in 2014. In the Executive Summary by the CNSWC an evaluation of POTFF took place, in which the efficacy of this task force was evaluated. They found that the POTFF initiative reduced destructive behaviors among sailors, helped sailors to cope with the stress of repeated deployments and combat operations, facilitated a "rest" period as they prepared for future deployments, improved combat readiness through the physical readiness program (HPP), and provided counselors and social workers to support to the families of the sailors. In short, the evaluation determined that POTFF was successful. The official recommendation from that Navy was to continuing funding the POTFF initiative because they believe POTFF to be necessary for preserving the Naval Special Warfare community.

Now that the POTFF program has been in place for a few years, leadership needs to develop a way to measure its actual effectiveness in preventing suicide and building resilience. Currently, the only measures taken and published are those surveying the SOF community to get a sense of perceived effectiveness and availability. The main question remains: Has there been a decrease is suicides within in the SOF community, and is the decrease due to the POTFF or to other factors? Critically and scientifically analyzing this data could further promote the effectiveness of the POTFF programs within its own community and the Department of Defense community at large. Additionally, publishing any positive evidenced-based results could possibly promote more participation from operators and their families in the interventions.

As the initiative of POTFF continues to expand, and increased data is published, it will be necessary to evaluate the efficacy of POTFF as well as the programs within each of the pillars. Only then will positive change be generated in order to for POTFF to truly be an initiative that supports the SOF community.

Promoting from the Top

While POTFF has taken a holistic view to build resilience within SOF, studies show the program may not be utilized to its fullest potential (Caserta & Neff, 2014). In 2013 more than half of the respondents to the Needs Assessment Survey were not aware of the programs provided by POTFF, and about 23% of those who did reply believed resources provided by POTFF were only somewhat beneficial or not beneficial at all (Caserta & Neff, 2014). This lack of awareness of POTFF could be seen as a communication and leadership problem.

In order to fully utilize the multidimensional POTFF model more effort should be focused on developing leaders and leadership to encourage and promote the use of this program. Furthermore, each domain should be given an equal amount of importance and attention in order to exploit the full capacity of the program. Currently, most of the emphasis by leadership is placed on physical and social domains (Wiggins, 2015). Although these domains are most relevant within SOF as they relate directly to job requirements, the other two domains of spirituality and psychological health also play an essential role.

Members of the SOF often underreport cognitive or emotional problems, such as posttraumatic stress, or do not identify symptoms such as sleep disturbance, irritability, and low mood as abnormal (Brown, Creel, Engel, Herrell, & Hoge, 2011). For example, only 7% of respondents to the 2013 Needs Assessment Survey reported high levels of posttraumatic stress, and 86% reported low posttraumatic stress (Caserta & Neff, 2014). Part of the challenge of assessing the need for psychological services in this population is that so many SOF operators have poor insight into their own psychological functioning.

Although having an operational psychologist embedded within a SOF command is intended to help SOF members to develop relationships and become comfortable talking to mental health professional, encouraging service members to consult with operational psychologists in order to develop better cognitive skills and improve performance is not the same as encouraging them to seek help for their psychological symptomology (Wiggins, 2015). Embedding is a great way of establishing familiarity and a relationship with the psychologist, but it does not necessarily help fight the stigma of receiving mental health help. There is a continued stigma against mental health within the military and the SOF community (Britt et al., 2012; Brown et al., 2011). Military leaders can greatly affect the view of mental within their command with their openness to speak about seeking help and leading by example (Britt et al., 2012).

Without the guidance from leadership within the SOF community the POTFF model will not live up to its potential. It is hard to gain a good measure on how effective a program is if it is not being utilized to the fullest. Only a fraction of the SOF community is aware of the multiple opportunities available through POTFF. Promotion by leaders openly discussing each of the four domains will allow the POTFF programs to expand and make the changes they intend to do.

Branch Specific Cultural Influence

There is a cultural component to building resilience and preventing suicide (Chu, Goldblum, Floyd, & Bongar, 2010). The military can be defined as an independent culture. Within the military, each branch has its own culture that influences and motivates its personnel. These branch and rate specific cultural differences are what lead to the development of POTFF. If each SOF military branch enhanced the POTFF multidimensional model with specific factors that are tailored toward their culturally important issues, it would further the effectiveness of the program. The U.S. Army Special Operations Command has begun its initiative Ready, Resilient, Preservation of the Force and Family program to do this very thing. The aim of the POTFF program tailored for this service branch is to ensure the integration of the Army Ready and Resilient Campaign and SOCOM POTFF resources. They aim to provide commanders with conditions that will allow them gain a range of wellness and resiliency capabilities (Department of the Army, 2015).

CONCLUSION

The POTFF multidimensional model looks at the special needs of the SOF community to develop resilience within its service members as well as their families. Taking a holistic look at the SOF enables SOCOM to be proactive in its efforts to increase readiness with in its forces. Unfortunately, prior to the POTFF inception the neglect of the multidimensional aspects of well-being among operators in the past decade led to a greater prevalence of suicide within SOF. SOCOM has taken great strides in combating the unprecedented prevalence of suicide within its community, but further work needs to be done. The POTFF model is still in its infancy, and further data needs to be collected in order to enhance and tailor the program to areas that need further attention. Collecting data in various ways is a way to identify the shortfalls within the domains. Based on initial response, emphasis needs to be placed on promoting the programs offered through POTFF to the members and their families. Additionally, leadership should continue to lead by example when promoting mental health.

REFERENCES

Barker, L. H., & Berry, K. D. (2009). Developmental issues impacting military families with young children during single and multiple deployments. *Mil Med, 174*(10), 1033–1040. doi: 10.7205/milmed-d-04-1108.

Bartone, P. T., Roland, R. R., Picano, J. J., & Williams, T. J. (2008). Psychological hardiness predicts success in US Army Special Forces candidates. *International Journal of Selection and Assessment, 16*(1), 78–81.

Britt, T. W., Wright, K. M., & Moore, D. (2012). Leadership as a predictor of stigma and practical barriers toward receiving mental health treatment: A multilevel approach. *Psychological Services, 9*(1), 26–37.

Brown, M. C., Creel, A. H., Engel, C. C., Herrell, R. K., & Hoge, C. W. (2011). Factors associated with interest in receiving help for mental health problems in combat veterans returning from deployment to Iraq. *Journal of Nervous and Mental Disease, 199*(10), 797–801.

Bryan, C. J., Stephenson, J. A., Morrow, C. E., Staal, M., & Haskell, J. (2014). Posttraumatic stress symptoms and work-related accomplishment as predictors of general health and medical utilization among special operations forces personnel. *Journal of Nervous and Mental Disease, 202*(2), 105–110.

Cacioppo, J. T., Lester, P. B., McGurk, D., Adler, A. B., Thomas, J. L., Chen, H. Y., & Caccioppo, S. (2015). Building social resilience in soldiers: A double dissociative randomized controlled study. *Journal of Personality and Social Psychology, 109*(1), 90–105.

Canapary, D., Bongar, B., & Cleary, K. M. (2002). Assessing risk for completed suicide in patients with alcohol dependence: Clinicians' views of critical factors. *Professional Psychology: Research and Practice, 33*(5), 464–469.

Caserta, R., & Neff, R. (2014) USSOCOM Enterprise report: *Preservation of the Force and Families Wave II needs assessment*. MacDill AFB, FL: USSOCOM.

Casey, G. W. (2011). Comprehensive soldier fitness: A vision for psychological resilience in the US Army. *American Psychologist, 66*(1), 1–3.

Chu, J. P., Goldblum, P., Floyd, R., & Bongar, B. (2010). A cultural theory and model of suicide. *Applied and Preventive Psychology, 14*, 25–40.

Clark-Hitt, R., Smith, S. W., & Broderick, J. S. (2012). Help a buddy take a knee: Creating persuasive messages for military service members to encourage others to seek mental health help. *Health Communication, 27*(5), 429–438. doi:10.1080/10410236.2011.606525

Department of the Army. (2015). *ARSOF NEXT: A return to first principles*. Retrieved from http://www.soc.mil/swcs/SWmag/archive/ARSOF_Next/ARSOF%20Next.pdf

Deslandes, A., Moraes, H., Ferreira, C., Veiga, H., Silveira, H., Mouta, R., ... Laks, J. (2009). Exercise and mental health: many reasons to move. *Neuropsychobiology, 59*(4), 191–198.

Dyster-Aas, J., Arnberg, F. K., Lindam, A., Johannesson, K. B., Lundin, T., & Michel, P. (2012). Impact of physical injury on mental health after the 2004 Southeast Asia tsunami. *Nordic Journal of Psychiatry, 66*(3), 203–208.

Feickert, A. (2015). *U.S. Special Operations Forces (SOF): Background and Issues for Congress*. Washington, D.C.: Congressional Research Service.

Ford, K., & Glymour, C. (2014). The enhanced warfighter. *Bulletin of the Atomic Scientists, 70*(1), 43–53.

Hanwella, R., & Silva, V. (2012). Mental health of Special Forces personnel deployed in battle. *Social Psychiatry and Psychiatric Epidemiology, 47*(8), 1343–1351.

Harms, P. D., Herian, M. N., Krasikova, D. V., Vanhove, A., & Lester, P. B. (2013). The comprehensive soldier fitness program evaluation. Report # 4: Evaluation of resilience training and mental and behavioral health outcomes. Retrieved from http://www.dtic.mil

Hoffman, F. G. (2006). Complex irregular warfare: The next revolution in military affairs. *Orbis, 50*(3), 395–411.

Hötting, K., & Röder, B. (2013). Beneficial effects of physical exercise on neuroplasticity and cognition. *Neuroscience and Biobehavioral Reviews, 37*(9, Part B), 2243–2257.

Joint Special Operations University. (2014). *Special Operations: Research topics 2015*. Retrieved from http://jsou.socom.mil/PubsPages/2015_SpecialOperationsResearchTopics_final.pdf

King, L. A., King, D. W., Fairbank, J. A., Keane, T. M., & Adams, G. A. (1998). Resilience-recovery factors in post-traumatic stress disorder among female and male Vietnam veterans: Hardiness, postwar social support and additional stressful life events. *Journal of Personality and Social Psychology, 74*(2), 420–434.

Lester, P. B., McBride, S., & Cornum, R. L. (2013). *Comprehensive soldier fitness: Underscoring the facts, dismantling the fiction.* Retrieved from http://psychnet.apa.org/books/14190/009

Lester, P. B., Taylor, L. C., Hawkins, S. A., & Landry, L. (2015). Current directions in military healthcare provider resilience. *Military Mental Health, 17*(6), 1–7.

Lin, T., Vaisvaser, S., Fruchter, E., Admon, R., Wald, I., Pine, D. S.,. . . . Hendler, T. (2014). A neurobiological account for individual difference in resilience to chronic military stress. *Psychological Medicine, 45*(5), 1011–1023.

MacGregor, A. J., Han, P. P., Dougherty, A. L., & Galarneau, M. R. (2012). Effect of dwell time on the mental health of US military personnel with multiple combat tours. *American Journal of Public Health,* Suppl1:S55–9. doi: 10.2105/AJPH.2011.300341

McAllister, D. R., Motamedi, A. R., Hame, S. L., Shapiro, M. S., & Dorey, F. J. (2001). Quality of life assessment in elite collegiate athletes. *The American Journal of Sports Medicine, 29*(6), 806–810.

Miles, D. (2013, June 14). *SOCOM strives to boost operators' resilience, readiness.* Retrieved from http://www.defense.gov/news

National Military Family Association. (2014). *Special Operations Forces: A war weary community needs support.* Retrieved from http://blog.militaryfamily.org/2014/03/06/budget-woes-a-special-operations-forces-spouses-perspective/

Natolochnaya, O. V., & Cherkasov, A. A. (2014). The analysis of special operations forces activities in the world. *Voennyi Sbornik, 4*(2), 134–140.

Naval Inspector General. (2014). *Naval Inspector General Command Inspection of Commander, Naval Special Warfare Command 12 To 16 May 2014.* Retrieved from http://www.secnav.navy.mil/ig/FOIA%20Reading%20Room/NAVINSGEN%20Command%20Inspection%20of%20Naval%20Special%20Warfare%20Command%20(SPECWAR)%205%20Aug%202014.pdf

Oshri, A., Lucier-Greer, M., O'Neal, C. W., Arnold, A. L., Mancini, J. A., & Ford, J. L. (2015). Adverse childhood experiences, family functioning, and resilience in military families: A pattern-based approach. *Family Relations, 64*(1), 44–63.

Palmer, C. (2008). A theory of risk and resilience factors in military families. *Military Psychology, 20*(3), 205–217.

Pietrzak, R. H., Johnson, D. C., Goldstein, M. B., Malley, J. C., & Southwick, S. M. (2009). Psychological resilience and postdeployment social support protect against traumatic stress and depressive symptoms in soldiers returning from Operations Enduring Freedom and Iraqi Freedom. *Depression and Anxiety, 28*(8), 745–751.

Quartana, P. J., Wilk, J. E., Thomas, J. L., Bray, R. M., Rae Olmsted, K. L., Brown, J. M., . . . Hoge, C. W. (2014). Trends in mental health services utilization and stigma in US soldiers from 2002 to 2011. *American Journal of Public Health, 104*(9), 1671–1679.

Reutter, K. K., & Bigatti, S. M. (2014). Religiosity and spirituality as resiliency resources: Moderation, mediation, or moderated mediation? *Journal for the Scientific Study of Religion, 53*(1), 56–72.

Saltzman, W. R., Lester, P., Beardslee, W. R., Layne, C. M., Woodward, K., & Nash, W. P. (2011). Mechanisms of risk and resilience in military families: Theoretical and empirical basis of a family-focused resilience enhancement program. *Clinical Child and Family Psychology Review, 14*(3), 213–230.

Shanker, T., & Oppel, R. A. (2014, June 6). War's elite tough guys, hesitant to seek healing. *The New York Times,* p. A1.

Skipper, L. D., Forsten, R. D., Kim, E. H., Wilk, J. D., & Hoge, C. W. (2014) Relationship of combat experiences and alcohol misuse among US Special Operations soldiers. *Military Medicine, 179*(3), 301–308.

Strobel, W. (2014, April 17). *US special forces struggle with record suicides: Admiral.* Retrieved from http://www.reuters.com/article/us-usa-military-suicides-idUSBREA3G2EK20140417

Taverniers, J., Ruysseveldt, J. V., Smeets, T., & Grumbkow J. V. (2010). High-intensity stress elicits robust cortisol increases, and impairs working memory and visuo-spatial declarative memory in Special Forces candidates: A field experiment. *Informa Healthcare, 13*(4), 324–334.

US Congress (2015). *Committee reports 113th Congress (2013–2014): House report 113-473.* Retrieved from http://thomas.loc.gov/cgi-bin/cpquery/?&dbname=cp113&sid=cp113HLTLZ&refer=&r_n=hr473.113&item=&&&sel=TOC_119947&

US Government Accountability Office. (2015). Special Operations Forces: Opportunities exist to improve transparency of funding and assess potential to lessen some deployments. GAO-15-571. Retrieved from http://www.gao.gov/products/GAO-15-571

US Special Operations Command. (2013). *Preservation of the force and family* [Brochure]. Tampa, FL: Author.

US Special Operations Command. (2015). *POTFF directors emphasis*. Retrieved from http://www.socom.mil/POTFF/Shared%20Documents/POTFF_DirectorsEmphasis.aspx

Vythilingam, M., Nelson, E. E., Scaramozza, M., Waldeck, T., Hazlett, G., Southwick, S. M., & Ernst, M. (2009). Reward circuitry in resilience to severe trauma: An fMRI investigation of resilient special forces soldiers. *Psychiatry Research, 172*(1), 75–77.

Weick, K. E., & Roberts, K. H. (1993). Collective mind in organizations: Heedful interrelating on flight decks. *Administrative Science Quarterly, 38*, 357–381.

Wiggins, W. J. (2015). *Generational resilience in support of the global SOF network*. Carlisle, PA: US Army War College.

Zachary, S. (2015, June 23). POTFF helps SOF airmen, families stay resilient. *The Citizen*. Retrieved from http://www.stuttgartcitizen.com/announcements-news/socom-potff-survey-potff-helps-sof-airmen-families-stay-resilient/

17

Managing Suicide in the Older Veteran

Bavna B. Vyas
Lisa M. Brown
David Dosa
Diane L. Elmore

Although the potential for suicide is of concern with all patients, suicide among older adults, defined by Medicare as age >65, and Veterans is increasingly recognized as a serious and rapidly growing public health problem. In the United States, more older adults are living longer than ever before. Most notable is the growth in the population of individuals age 85 and older who are at highest risk for suicide. For more than two decades, older white males have composed the group with the highest risk for death (Hoyert, Kochanek, & Murphy, 1999; McIntosh, Santos, Hubbard, & Overholser, 1994). This trend continues as recent data shows no significant differences in age-adjusted death rates from 2010 to 2011 for intentional self-harm (suicide).

Despite sustained research efforts and the introduction of evidence-based practices into clinical practice, older individuals' risk for suicide remains difficult to predict and treat, even with thorough assessment and competent intervention. Because of these significant challenges, suicide prevention remains a top priority for the U.S. Department of Veterans Affairs (VA) health system, the Veterans Health Administration (VHA) which provides health services to approximately 5.5 million Veterans (one-fifth of all Veterans) each year (McCarthy et al., 2012).

In the past, an important issue encountered by VHA and other large health care systems that hindered efforts was a lack of universally accepted definitions for the range of suicidal behaviors (i.e., deaths, attempts, ideation) (Department of Veterans Affairs, 2008).

Differences in reporting and lack of clarity in data capture and reporting resulted in public and clinical misunderstandings about the past and current scope of the suicide risk for all Veterans as well as various subgroups of Veterans (Department of Veterans Affairs, 2008). Recent endeavors to clarify data reporting include the development of a new nomenclature, designed by the Centers for Disease Control and Prevention, to provide a unified system for collecting and reporting suicide data from various agencies. (http://www.cdc.gov/violenceprevention/suicide/definitions.html.) Together with the Department of Defense, the VHA has leveraged considerable resources to combat Veteran suicide by implementing both public health and clinical intervention strategies. To address the issues of detection of suicide risk, intervention, and treatment of older Veterans this chapter uses the definitions recommended by the 2012 Food and Drug Administration (FDA) industry update on Suicidal Ideation and Behavior (U.S. Department of Health and Human Services, U.S. Food and Drug Administration, & Center for Drug Evaluation and Research, 2012).

SUICIDE RATES

Unclear Rates Among Veterans

Rates of suicide among Veterans are difficult to determine. There exists an ongoing debate regarding both the nature of the data available on suicide in Veterans and methodological approaches for calculating estimates of risk associated with Veteran status (Knox & Bossarte, 2012). A prominent study conducted before Operation Enduring Freedom reported that Veterans in a community sample were twice as likely as non-Veterans to die by suicide (Kaplan, Huguet, McFarland, & Newsom, 2007). Most Veterans were middle-aged or older white men, a reflection of the demographic composition of Veterans during the study period (1986–1997), and most had separated from the military several years prior to enrolling in the study (Miller et al., 2009). However, subsequent work across various studies report conflicting observations on Veteran suicide risk with most reporting similar rates of suicide for Veterans and for age, sex, and race-matched members of the general U.S. population (Bullman & Kang, 1996; Kang & Bullman, 2001; Kang & Bullman, 2008; Ketchum & Michalek, 2005; Michalek, Ketchem, & Akhtar, 1998; Thompson et al., 2002). In 2008, then-Secretary of Veterans Affairs James B. Peak chartered a Blue Ribbon Work Group on Suicide Prevention in the Veteran Population to inform and advise improvements relevant to the assessment and prevention of suicide in the Veteran population. The work group used the "healthy warrior effect" as a possible explanation for lower suicide risk among younger Veterans but could not explain similar or increased suicide risk in Veterans over non-Veterans with aging. As a follow-up to the inconsistent picture of risk of suicide among older Veterans identified by the workgroup, Miller et al. (2009) conducted a large-scale study on middle-aged and older males and concluded that the risk of death from suicide was independent of Veteran status. They subsequently suggested that policies to prevent Veteran suicide should focus on factors that may heighten suicide risk rather than on Veteran status (Miller et al., 2009). This finding was also observed in a later study examining an older age groups' risk for suicide (Kaplan, McFarland, Huguet, & Valenstein, 2012).

General Rates of Suicide in the United States

According to the Centers for Disease Control and Prevention ([CDC] 2013), suicide is the 10th leading cause of death in the United States, resulting in death of more than 41,000 people each year. In 2010, it was estimated that 8,618 older adults died by suicide. Men are seven times more likely to take their lives than women and represent 79.4% of all suicides (National Center for Injury Prevention and Control & Centers for Disease Control and Prevention, 2007). Older Americans represent 13% of the U.S. population and yet account for 18% of all suicide deaths (Arias, Anderson, Kung, Murphy, & Kichanek, 2003). Suicide rates for males are highest among those aged 75 and older (rate 35.7 per 100,000; National Center for Injury Prevention and Control & Centers for Disease Control and Prevention, 2007). The highest rates of completed suicide are among those 65 years or older (14.7/100,000 vs. 11/100,000 for the general population; Centers for Disease Control and Prevention & National Association of Chronic Disease Directors, 2008). Older adults are more likely to complete suicide than any other adult age group, due to more lethal methods of suicide attempt. In older adults, firearms are the most common method (67%) followed by poisoning (14%) and suffocation (12%; Substance Abuse and Mental Health Services Administration & Administration on Aging, 2012).

As noted earlier, the group at highest risk for death by suicide is white men over the age of 85, with a rate of suicide 2.5 times higher than the nation's rate for men (American Association for Suicidology, 2008). In 1999, the U.S. Office of the Surgeon General convened a group to develop the *National Strategy for Suicide Prevention: Goals and Objectives for Action* with the aim of addressing the high risk of suicide in the growing older adult population. These goals and objectives were recently updated in 2012 (U.S. Department of Health and Human Services Office of the Surgeon General & National Action Alliance for Suicide Prevention, 1999, 2012).

RISK FACTORS FOR SUICIDE IN OLDER ADULTS

Identification of risk factors plays a key role in proactive implementation of interventions to prevent suicide behavior. Detection of potential for late-life suicide is complicated by assumptions by clinicians and the public that risk factors and symptoms are part of the normal aging process or components of chronic disease (Brown, Bongar, & Cleary, 2004). Additionally, older adults are less likely to endorse suicidal ideation than younger counterparts (Duberstein et al., 1999).

Proactive identification of suicide risk is imperative in the older Veteran population, which is less likely to report suicidal ideation (Lish et al., 1996).

Veterans

It is widely accepted that the VHA serves a population who possess a constellation of suicide risk factors. An earlier study identified risk factors common to VA patients including male gender, older age, diminished social environment support (exemplified by homelessness and unmarried status), availability and knowledge of firearms, and the prevalence of medical and psychiatric conditions associated with suicide (Lambert & Fowler, 1997). A more recent study specified depression and substantial physical morbidity including substance abuse issues and mental illness as important risk factors in addition to old age and male sex (McCarthy et al., 2012). A review of the literature suggests other links between Veteran suicide and sleep disorders (Pigeon, Britton, Ilgen, Chapman, & Conner, 2012), substance abuse (Ilgen et al., 2012), and dementia (Ayalon, Mackin, Arean, Chen, & McDonel Herr, 2007), although these factors are not consistently ascribed to older adults (McIntosh et al., 1994). Kaplan et al. (2012) attempted to evaluate risk of suicide among male veterans relative to non-Veteran men by age and found Veterans at higher risk for suicide compared with non-Veterans in all age groups except the oldest (>65).

Mental Health

Suicidal ideation may be more prevalent and more likely to be reported among older adults with mental disorders, those taking anxiolytic medications and/or neuroleptic medications, and those with history of cardiac disease, peptic ulcer disease, and three or more physical disorders (Skoog et al., 1996). Psychological autopsy studies reveal that diagnosable major psychiatric illness was present in 71% to more than 90% of older adults who died by suicide (Conwell, 2001; Conwell & Brent, 1995). Several studies and reports have consistently linked suicide risk to clinical depression in older adult Veterans and older adults in the general population (Britton et al., 2012; Duberstein & Conwell, 2000; Hoyert et al., 1999). Older Veterans were more likely than younger Veterans to be suspected of being depressed at the time of death (Kaplan et al., 2012). One study reported a single episode of nonpsychotic, unipolar major depression as the most common psychiatric syndrome of older suicide decedents (Conwell, 2001). Cognitive impairment is another area of concern when discussing late life suicide and mental health risk factors. In one study of older Veterans receiving services in VA primary care clinics, poor cognitive function was reported as a contributor to suicidal ideation (Ayalon et al., 2007). In contrast, Garand et al. (2006) did not detect significant differences between suicidal subjects and controls when reviewing studies of individuals who had a diagnosis of dementia. In the general population, suicide risk in older adult patients with dementia is linked to male gender, high levels of education, and preserved insight but poor performance on tests of executive functioning, attention, and memory and depressive symptoms that need not necessarily meet criteria for major depressive disorder (Lim, Rubin, Coats, & Morris, 2005). Last, psychiatric illness other than depression was also associated with suicidal ideation and behavior. In a 1996 review, primary psychotic illnesses (e.g., schizophrenia, schizoaffective disorder, and delusional disorder), personality disorders, and anxiety disorders was found to play a relatively small role on increased suicide in older adults (Conwell et al., 1996). A 2002 study by the same authors reported affective illness as an important and predominant psychopathology involved in late-life suicide (Conwell, Duberstein, & Caine, 2002).

Transition Periods

Recent studies have focused on transition periods as a time for high suicidal risk. Knox et al. (2012) describe the postdischarge period after acute psychiatric services either in hospital or in the emergency department as a high-risk period for suicide. Similar findings were identified for Veterans during a 12-week period after psychiatric hospitalization (Valenstein et al., 2009). Early work in this area noted the importance of implementing suicide prevention strategies during the transition from inpatient psychiatric settings to home environments (Motto & Bostrom, 2001).

Firearms

Numerous studies have cited firearm ownership as an important risk factor for suicide. According to the CDC (2003) report on the behavioral risk factor surveillance system survey data, Veterans are substantially more likely to own guns than are non-Veterans (45.7% vs. 32.3%; $p < 0.001$) with risk of suicide increasing by nearly five times in homes with guns (Kellerman et al., 1992). Kaplan et al. (2007) found a higher probability that Veterans used firearms as a means of suicide compared with non-Veterans.

A recent follow-up study by the same authors found most male Veterans used firearms for death by suicide, and nearly all older Veterans did so (Kaplan et al., 2012). Common reasons attributed to the high rate of Veteran death by firearms include easy access to firearms and knowledge of how to use them. These findings have resulted in VHA recommendations that health care providers inquire about access to firearms among all depressed or suicidal older patients. The VA's Suicide Prevention Program provides free cable gunlocks at all VA medical centers for Veterans and their family members. Providers are encouraged to guide family members (or other responsible parties) to safely store weapons on behalf of suicidal or unstable Veterans—instead of removing firearms, which may lead to conflict.

Substance Abuse, Including Alcohol

A substance use disorder, particularly alcohol abuse or dependence, is the second most common psychiatric disorder associated with completed elder suicide (Blow, Brockmann, & Lawton, 2004). Substance use disorders are also associated with elder suicide attempts (Blixen, McDougall, & Suen, 1997; Frierson, 1991). Vietnam-era Veterans have been identified as an at-risk group for substance abuse and other psychiatric conditions (most commonly posttraumatic stress disorder; Boehmer, Flanders, McGeehin, Boyl, & Barrett, 2004; Brooks, Laditka, & Laditka, 2008; Cherpitel, Borges, & Wilcox, 2004). The association between alcohol use and later-life suicide is more complicated with probable underestimation of at-risk and problem alcohol use in geriatric suicide. Accurately assessing alcohol use among older adults is challenging due to problems with standard criteria for the diagnoses of alcohol abuse and dependence as they are applied to older persons (Blow, 1998). Additionally, several sources suggest that older adults are more sensitive to the effects of alcohol with aging. In their study calculating age-specific suicide rates for Veterans from the National Violent Death Reporting System (2003–2008), Kaplan et al. (2012) tested suicide decedents' blood for presence of alcohol. They found evidence of acute alcohol use (BAC = 0.08) was present at the time of death in less than 10% of older Veterans, compared with one-third of younger Veterans (Kaplan et al., 2012). In contrast, earlier studies reported greater percentages of older decedents as having consumed alcohol before suicidal attempt (Conwell, Rotenberg, & Caine, 1990; Frierson, 1991).

Sleep Disorders

According to several recent research studies, sleep disturbances may contribute to suicidal ideation and behavior, although there appears to be conflicting evidence regarding the confounding effect of depression (Bernert, Joiner, Cukrowicz, Schmidt, & Krakow, 2005). Because sleep disturbances may precede depression, there exists an early opportunity to monitor for depression and suicidal ideation and behavior in patients who report such symptoms (Baglioni et al., 2011). To examine the role of sleep disturbance in time to suicide since the last treatment visit among Veterans receiving VHA services, Pigeon et al. (2012) studied 423 Veteran suicide decedents. They found that the group with recorded sleep disturbance died more quickly after their last visit and suggest sleep interventions as a possible adjuvant to suicide prevention interventions (Pigeon et al., 2012).

Chronic Illness/Functional Decline/Social Isolation

In contrast to younger cohorts, older adults represent a population where comorbid medical conditions play an important role in suicide risk (Rurvey et al., 2002). Early studies on late-life suicide suggest that social isolation and loneliness were important risk factors for suicidal ideation and behavior (Barraclough, 1971). Higher lethality for suicide attempts among older adults stemmed from declines in physical condition and social isolation making survival and rescue opportunities less likely (Conwell, 2001). Interviews conducted with older suicide decedents' relatives found unemployment, financial disablement, and unmarried status as important risk factors (Duberstein, Conwell, Conner, Eberly, & Caine, 2004). Homeless Veterans represent a particularly vulnerable group with recent estimates that more than 20% of homeless Veterans are aged 55 and older (Khadduri, Culhane, & Cortes, 2010). In a study examining the frequency of recent suicidal behavior in a large sample of older homeless Veterans admitted to a transitional housing intervention program, Veterans with increased suicidal behavior were also more likely to have a history of psychiatric disorders and/or substance abuse (Schinka, Schinka, Casey, Kasprow, & Bossarte, 2012).

Family or Personal History of Suicidal Ideation or Behavior

There is convincing evidence that individuals bereaved by suicide (also referred to as "survivors of suicide loss") may have an increased risk for suicide completion themselves (Jordan & McIntosh, 2011). Several national organizations, including the CDC (2012), consider family or personal history of a suicide attempt as an important risk factor for future suicidal behavior or ideation. History of suicide attempts and family history of suicide represent two of the eight most critical factors in assessing suicide risk in patients with major depression (Peruzzi & Bongar, 1999). However, there are challenges associated with suicide prevention efforts for this priority group due barriers such as resistance to seek help, decreased access to help, and propensity toward complicated grief (McMenamy, Jordan, & Mitchell, 2008). Therefore, national organizations including the American Foundation for Suicide Prevention, Suicide Awareness Voices of Education, and the American Association of Suicidology have increased their efforts to provide help and comfort to those bereaved by suicide (U.S. Department of Health and Human Services Office of the Surgeon General & National Action Alliance for Suicide Prevention, 2012).

SUICIDE ASSESSMENT STRATEGIES IN THE ELDERLY

Despite a robust body of literature with demographic, psychosocial, psychiatric, and health risk factors for suicidal ideation and behavior in older adults, no known assessment tools or strategies are proven to proactively identify suicidal tendency without running the risk of false positives. Assessing for suicide in late life is complex. However, there are a growing number of evidence-based practices for use with older adults that address individual elements in an effort to increase detection of suicidal ideation and behaviors.

Modifiable Risk Factors

Detection and treatment of suicidal behaviors requires identifying *modifiable* risk factors that can be targeted. Some examples of modifiable risk factors include depression, anxiety, panic attacks, psychosis, sleep disorders, substance abuse, command hallucinations, impulsivity, agitation, physical illness, difficult situations (e.g., family, work), and lethal means (e.g., guns, drugs; Simon, 2004).

Role of Personality Traits

A study by Conner et al. (2001) identifies *personality traits* to target in prevention and treatment efforts. The authors describe five constructs typically associated with suicide as follows: (a) impulsive aggression or reactive aggression; (b) social inhibition, behavioral inhibition, introversion, low openness; (c) hopelessness; (d) anxiety; and (e) depression.

Prospective Assessment

A recent update by the FDA on the prospective assessment of suicidal ideation and behavior in clinical trials recommends use of the categories described in the Columbia Suicide Severity Rating Scale, which defines five subtypes of suicidal ideation and behavior as follows (Center for Suicide Risk Assessment at Columbia University Medical Center, n.d.; U.S. Department of Health and Human Services et al., 2012).

- Suicidal ideation (where the definition of *plan* includes intent)
  1. Passive
  2. Active: Nonspecific (no method, intent, or plan)
  3. Active: Method, but no intent or plan
  4. Active: Method and intent, but no plan
  5. Active: Method, intent and plan

- Suicidal behavior
  1. Completed suicide
  2. Suicide attempt
  3. Interrupted attempt
  4. Aborted attempt
  5. Preparatory actions toward imminent suicidal behaviors
- Self-injurious behavior, no suicidal intent

Health Care Decision-Making Capacity

In older adults, an essential assessment element in the management of suicidal behavior is that of cognition and health care decision-making capacity. A report from the American Bar Association Commission on Law & Aging and American Psychological Association Assessment of Capacity in Older Adults Project Working Group (2008) provides a model conceptualizing the components of legal capacity. In the model clinical judgment is a balance between the elements and existing legal standards, which vary by state. A clinician's awareness of state-specific policy is important as it influences assessment and management approaches used with older adults with cognitive impairment and suicidal ideation or behaviors.

PREVENTION/TREATMENT STRATEGIES FOR SUICIDAL BEHAVIORS IN OLDER ADULTS

The following section describes strategies to address suicidal ideation and prevent suicidal behaviors among older adults.

Means Restriction

Empirical evidence exists for "means restriction," defined as modifying the environment and decreasing general access to suicide means, as a suicide prevention strategy (Daigle, 2005; Mann, Apter, & Bertolote, 2005). In the older Veteran population, access to firearms has been cited in several studies as a common risk factor for suicide and an important priority for means restriction. The VA's Suicide Prevention Program promotes firearm safety by providing free cable gunlocks to Veterans and their family members at all VA medical centers. Primary care providers are encouraged to enlist family member support in the effort to safely store weapons on behalf of Veterans at risk for suicidal ideation and behavior.

Provider Education

Despite its high prevalence, studies have reported a 30% to 50% failure rate on the part of primary care physicians following usual care to diagnose depression (Simon & VonKorff, 1995). Additionally, the importance of primary care providers in the prevention of suicidal ideation and behavior is supported by studies that reveal that older adults and women who die by suicide are likely to have seen a primary care provider in the year or even within a month before death (Luoma, Martin, & Pearson, 2002). In a large systematic review on suicide prevention strategies, Mann et al. (2005) concluded that physician education in depression recognition and treatment is an important program component to reduce overall suicide rates. However, they argued that provider education should be made available not only to physicians but to interdisciplinary teams that include social workers, nurses, psychologists, and occupational therapists from a variety of care settings, including outpatient and inpatient medical, outpatient and inpatient mental health, specialty clinics, home, and community. Their findings were also consistent with the results of a recent study that examined the benefits of trained teams and found improvement in the overall quality of case notes, greater ability to recognize important conceptual suicide risk categories, and reported heightened awareness of the importance of late-life suicide (Huh et al., 2012).

Despite the low base rate, suicide ideation and behaviors are the most common clinical emergency faced by mental health professionals (Shein, 1976). Psychologists treat an average of five suicidal patients per month. An estimated 20% to 30% of these practitioners will care for a patient who ultimately dies by suicide (Chemtob, Hamada, Bauer, Kinney, & Torigoe, 1988; Greaney, 1995). Adequate education and training in identifying and treating individuals who are at increased risk for suicide is critical for all mental health care providers. Older Americans underutilize mental health services for a variety of reasons, including inadequate insurance coverage; a shortage of trained geriatric mental health providers;

lack of coordination among primary care, mental health and aging service providers; stigma surrounding mental health and its treatment; denial of problems; and access barriers such as transportation. Geropsychology is an emerging field within psychology devoted to the study of aging and the provision of clinical services for at-risk older adults.

Psychotherapy and Older Veterans

Psychotherapy has been explored in several studies as a viable therapeutic intervention for suicidal ideation and behavior. Studies have shown that cognitive behavioral therapy targeted at individual vulnerability factors associated with suicide ideation and acute hopelessness may reduce rates of repeated suicide attempts by 50% during the following year. Cognitive therapy helped suicide attempters consider alternative actions when thoughts of self-harm arose (Brown et al., 2005). In people with borderline personality disorder, dialectical behavior therapy reduced suicide attempts by half, compared with other types of therapy (Linehan et al., 2006). Several sources recommend not only individual therapy but also family therapy, with emphasis on increased social support from the patient's family and friends when appropriate support is available.

Improvement of Access to Mental Health Services

The mid-1990s was a time period of major transformation in the VA health care system, shifting from its previous focus on inpatient care to the delivery of outpatient and community-based services (Desai, Rosenheck, & Desai, 2008). Access to care has emerged as an important theme when considering the care of older Veterans. The VHA has implemented several interventions to overcome access issues that include limited transportation, rural location, and compliance to care. Through the Home-Based Primary Care program, psychologists provide home visits to older Veterans to conduct an evaluation that includes suicide prevention plans and gun safety. Telemedicine technology has made it possible to remotely implement therapy programs for Veterans with mental health conditions.

The VHA has also played a leadership role in developing and implementing integrated models of care that support a patient-centered approach and promote collaboration among members of the interprofessional health care team. One such initiative is the Patient Aligned Care Team (PACT), which is a patient-centered medical home model being implemented at VHA primary care sites. These teams are managed by primary care providers with the active involvement of other clinical (including mental health professionals) and nonclinical staff. The PACT is designed to increase accessibility, coordination, comprehensiveness, and patient-centered care. According to the VA this model allows Veterans to play a more active role in their care and includes benefits such as increased quality improvement, increased patient satisfaction, and a decrease in hospital costs due to fewer hospital visits and readmissions (U.S. Department of Veterans Affairs, 2011).

Recognition of Pharmacotherapy and Adverse Effects

Given the high prevalence of psychiatric disorders in at least 90% of suicides coupled with treatment failure rates of more than 80% at the time of death (Henriksson, Boethius, & Isacsson, 2001; Lönnqvist et al., 1995), it is imperative that suicide prevention programs include treatment of mood and other psychiatric disorders. According to Mann et al. (2005), evidence is inconclusive for the use of selective serotonin reuptake inhibitors to decrease suicide rates, although their efficacy is established for treatment of major depression, the main risk factor for suicide.

The medication clozapine is approved by the FDA for suicide prevention in people with schizophrenia (Meltzer et al., 2003). Other promising medications and psychosocial treatments for suicidal ideation are being tested. In older adults, polypharmacy is a prevalent problem, and medication reconciliation should include a review of medications that may potentiate suicide risk. In an examination of the FDA Adverse Event Reporting System data from 2004 to 2008, it was found that drugs that were associated with increased suicidal ideation were also associated with increased suicidal attempts or completions, emphasizing the importance of medication reconciliation and screening in the at-risk population (Robertson & Allison, 2009).

Veterans Health Administration

Currently, important program components for suicide prevention within the VHA include the Veterans Crisis Line, Veterans Online Chat service, and VA Suicide Prevention Coordinators at Community Based Outpatient Clinics and VA medical centers (Brenner & Barnes, 2012). The VHA has designated Suicide Prevention Teams (SPTs) at every VA hospital and is committed to the development and implementation of strategies designed to better identify and treat Veterans who are in emotional crisis and to the promotion of change in the environment, the culture, and health care through clinical and public health programs that directly target suicide prevention. The SPTs work collaboratively with mental health, medical, and social work services to identify Veterans at risk and provide enhanced services through case management, clinical service liaisons, crisis recovery groups, monthly newsletters, and collaboration with the National Suicide Hotline and Live Chat line. Eligible Veterans include those who have recently made a suicide attempt, have a history of previous attempt, or have been recently hospitalized for suicidal ideation. High-risk patients may also be flagged if they are identified proactively by their health care providers. Last, through quality improvement and patient safety initiatives, the VHA is committed to conducting root cause analyses on all deaths by suicide and has concurrently established a violent death reporting system in all of its facilities. During the past eight years, suicide rates in the VHA have been stable and are likely attributable to enhancements in detection and treatment of Veterans at risk of suicidal behavior. Most recent VHA efforts have been directed in the area of predictive modeling to identify high-risk patients who were not detected in clinical settings. The results of this initiative could enhance clinical care and guide delivery of preventive interventions.

CONCLUSION

It is encouraging that the number of national initiatives that address suicide risk, intervention, and treatment in community-dwelling older adults has grown during the past decade. A sustained public health approach is required for the development of effective programs in suicide prevention. Importantly, liaisons between national programs servicing Veterans cared for by the VHA, community partners, and mental health and substance abuse providers are essential to ensure that a coordinated range of interventions results in a comprehensive and effective suicide prevention plan.

FUTURE DIRECTIONS

1. Detection of suicide risk of older Veterans who disclose little information and reveal few behavioral clues of intent remains a challenge.
2. Better methods for removing and securing guns and ammunition in at-risk elders' homes need to be developed.
3. Family members, faith-based organizations, and community agencies must be educated about the roles they can play in reducing social isolation and risk of death by suicide among older Veterans.

REFERENCES

American Association for Suicidology. (2008). http://www.suicidology.org/ (accessed 2012).

American Bar Association Commission on Law & Aging, & American Psychological Association. (2008). *Assessment of older adults with diminished capacity: A handbook for psychologists.* Washington, DC: Authors.

Arias, E., Anderson, R. N., Kung, H. C., Murphy, S. L., & Kichanek, K. D. (2003). Deaths: Final data for 2001. *National Vital Statistics Reports,* 52(3), 1–115.

Ayalon, L., Mackin, S., Arean, P. A., Chen, H., & McDonel Herr, E. C. (2007). The role of cognitive functioning and distress in suicidal ideation in older adults. *Journal of the American Geriatrics Society,* 55(7), 1090–1094.

Baglioni, C., Battagliese, G., Feige, B., Spiegelhalder, K., Nissen, C., Voderholzer, U., ... Riemann, D. (2011). Insomnia as a predictor of depression: A meta-analytic evaluation of longitudinal epidemiological studies. *Journal of Affective Disorders,* 135, 10–19.

Barraclough, B. (1971). Suicide in the elderly: Recent developments in psychogeriatrics. *The British Journal of Psychiatry,* (Suppl. 6), 87–97.

Bernert, R., Joiner T. Jr., Cukrowicz, K., Schmidt, N., & Krakow, B. (2005). Suicidal ideation and behavior and sleep disturbances. *Sleep,* 28(9), 1135–1141.

Blixen, C., McDougall, G., & Suen, L. (1997). Dual diagnosis in elders discharged from a psychiatric hospital. *International Journal of Geriatric Psychiatry, 12,* 307–313.

Blow, F. C. (1998). Substance abuse among older adults: Treatment improvement protocol (TIP) Series No. 26. Rockville, MD: US Department of Health and Human Services, Public Health Service, Substance Abuse and Mental Health Services Administration, Center for Substance Abuse Treatment.

Blow, F. C., Brockmann, L. M., & Lawton Barry, K. (2004). Role of alcohol in late-life suicide. *Alcoholism: Clinical and Experimental Research, 28*(5), 48S–56S.

Boehmer, T. K., Flanders, W. D., McGeehin, M. A., Boyle, C., & Barrett, D. H. (2004). Post-service mortality in Vietnam veterans: 30-year follow-up. *Archives of Internal Medicine, 164,* 1908–1916.

Brenner, L. A., & Barnes, S. M. (2012). Facilitating treatment engagement during high risk transition periods: A potential suicide prevention strategy. *American Journal of Public Health, 102*(Suppl. 1), S12–S14.

Britton, P. C., Ilgen, M. A., Valenstein, M., Knox, K., Claassen, C. A., & Conner, K. R. (2012). Differences between veteran suicides with and without psychiatric symptoms. *American Journal of Public Health,102*(Suppl. 1), S125–S130.

Brooks, M., Laditka, S., & Laditka, J. (2008). Long-term effects of military service on mental health among veterans of het Vietnam war era. *Military Medicine, 173,* 570–575.

Brown, G., Ten Have, T., Henriques, G., Xie, S., Hollander, J., & Beck, A. (2005). Cognitive therapy for the prevention of suicide attempts: A randomized controlled trial. *JAMA: Journal of the American Medical Association, 294*(5), 563–570.

Brown, L. M., Bongar, B., & Cleary, K. (2004). Assessing risk for completed suicide in elderly patients: Psychologists' views of critical risk factors. *Professional Psychology: Research and Practice, 35,* 90–96.

Bullman, T. A., & Kang, H. K. (1996). The risk of suicide among wounded Vietnam veterans. *American Journal of Public Health, 86*(5), 662–667.

Center for Suicide Risk Assessment at Columbia University Medical Center. (n.d.). Columbia-suicide severity rating scale (C-SSRS). Retrieved from http://www.cssrs.columbia.edu/

Centers for Disease Control and Prevention. (2003). Behavioral Risk Factor Surveillance System Survey data. Atlanta: US Department of Health and Human Services.

Centers for Disease Control and Prevention. (2012, September 18). Suicide: Risk and protective factors. Retrieved from http://www.cdc.gov/violenceprevention/suicide/riskprotectivefactors.html

Centers for Disease Control and Prevention. (2013). Deaths: Final data 2013. Retrieved from http://www.cdc.gov/nchs/data/nvsr/nvsr64/nvsr64_02.pdf

Centers for Disease Control and Prevention and National Association of Chronic Disease Directors. The State of Mental Health and Aging in America Issue Brief 1: What Do the Data Tell Us? Atlanta, GA: National Association of Chronic Disease Directors; 2008.

Chemtob, C. M., Hamada, R. S., Bauer, G. B., Kinney, B., & Torigoe, R. Y. (1988). Patient suicide: Frequency and impact on psychiatrists. *The American Journal of Psychiatry, 145,* 224–228.

Cherpitel, C. J., Borges, G. L. G., & Wilcox, H. C. (2004). Acute alcohol use and suicidal behavior: A review of the literature. *Alcoholism: Clinical and Experimental Research, 28*(5), 18S–28S.

Conner, K. R., Duberstein, P. R., Conwell, Y., Seidlitz, L., & Caine, E. D. (2001). Psychological vulnerability to completed suicide: A review of empirical studies. *Suicide and Life-Threatening Behavior, 31*(4), 367–385.

Conwell, Y. (2001). Suicide in later life: A review and recommendations for prevention. *Suicide and Life-Threatening Behavior, 31,* 32–47.

Conwell, Y., & Brent, D. (1995). Suicide and aging I: Patterns of psychiatric diagnosis. *International Psychogeriatrics, 7*(2), 149–164.

Conwell, Y., Duberstein, P. R., & Caine, E. D. (2002). Risk factors for suicide in later life. *Biological Psychiatry, 52*(3), 193–204.

Conwell, Y., Duberstein, P. R., Cox, C., Herrmann, J. H., Forbes, N., & Caine, E. D. (1996). Relationships of age and Axis I diagnoses in victims of completed suicide: A psychological autopsy study. *The American Journal of Psychiatry, 153*(8), 1001–1008.

Conwell, Y., Rotenberg, M., & Caine, E. (1990). Completed suicide at age 50 and over. *Journal of the American Geriatrics Society, 38,* 640–644.

Daigle, M. S. (2005). Suicide prevention through means restriction: Assessing the risk of substitution. A critical review and synthesis. *Accident Analysis and Prevention, 37,* 625–632.

Department of Veterans Affairs. (2008). Report of the Blue Ribbon Work Group on Suicide Prevention in the Veteran Population. Washington, DC: Author.

Desai, M. M., Rosenheck, R. A., & Desai, R. A. (2008). Time trends and predictors of suicide among mental health outpatients in the Department of Veterans Affairs. *Journal of Behavioral Health Services and Research, 35*(1), 115–124.

Duberstein, P. R., & Conwell, Y. (2000). Suicide. In S. K. Whitbourne (Ed.), *Psychopathology in later adulthood* (pp. 245–276). New York: Wiley.

Duberstein, P. R., Conwell, Y., Conner, K. R., Eberly, S., & Caine, E. D. (2004). Suicide at 50 years of age and older: Perceived physical illness, family discord and financial strain. *Psychological Medicine, 34*(1), 137–146.

Duberstein, P. R., Conwell, Y., Seidlitz, L., Lyness, J. M., Cox, C., & Caine, E. D. (1999). Age and suicidal ideation in older depressed inpatients. *American Journal of Geriatric Psychiatry, 7*(4), 289–296.

Frierson, R. (1991). Suicide attempts by the old and the very old. *Archives of Internal Medicine, 151*, 141–145.

Garand, L., Mitchell, A. M., Deitrick, A.,Hijjawi, S. P., Pan, D. (2006). Suicide in older adults: Nursing assessment of suicide risk. *Issues in Mental Health Nursing, 27*, 355–370.

Greaney, S. (1995). *Psychologists behaviors and attitudes when working with the nonhospitalized suicidal patient* (Unpublished doctoral dissertation). Pacific Graduate School of Psychology, Palo Alto, CA.

Henriksson, S., Boethius, G., & Isacsson, G. (2001). Suicides are seldom prescribed antidepressants: Findings from a prospective prescription database in Jamtland county, Sweden, 1985–95. *Acta Psychiatrica Scandinavica, 103*, 301–305.

Hoyert, D. L., Kochanek, K. D., & Murphy, S. L. (1999). Deaths: Final data for 1997. *National Vital Statistics Reports, 47*(19), 1–108.

Huh, J., Weaver, C., Martin, J., Caskey, N., O'Riley, A., & Kramer, B. (2012). Effects of late-life suicide risk—assessment training on multidisciplinary healthcare providers. *Journal of the American Geriatrics Society, 60*(4), 775–780.

Ilgen, M., Conner, K., Roeder, K., Blow, F., Austin, K., & Valenstein, M. (2012). Patterns of treatment utilization before suicide among male veterans with substance abuse disorders. *American Journal of Public Health, 102*(Suppl. 1), S88–S92.

Jordan, J., & McIntosh, J. L. (2011). *Grief after suicide: Understanding the consequences and caring for the survivors.* New York: Rutledge.

Kang, H. K., & Bullman, T. A. (2001). Mortality among US veterans of the Persian Gulf War: 7-year follow-up. *American Journal of Epidemiology, 154*(5), 399–405.

Kang, H. K., & Bullman, T. A. (2008). Risk of suicide among US veterans after returning from the Iraq or Afghanistan war zones. *JAMA: Journal of the American Medical Association, 300*(6), 652–653.

Kaplan, M. S., Huguet, N., McFarland, B. H., & Newsom, J. T. (2007). Suicide among male veterans: A prospective population-based study. *Journal of Epidemiology and Community Health, 61*(7), 619–624.

Kaplan, M. S., McFarland, B. H., Huguet, N., & Valenstein, M. (2012). Suicide risk and precipitating circumstances among young, middle-aged, and older male veterans. *American Journal of Public Health, 102*(Suppl. 1), S131–S137.

Kellermann, A., Rivara, F., Somes, G., Reay, D. T., Francisco, J., Gillentine Banton, J., ... Hackman, B. B. (1992). Suicide in the home in relation to gun ownership. *The New England Journal of Medicine, 327*, 467–472.

Ketchum, N. S., & Michalek, J. E. (2005). Postservice mortality of Air Force veterans occupationally exposed to herbicides during the Vietnam war: 20-year follow-up results. *Military Medicine, 170*(5), 406–413.

Khadduri, J., Culhane, D., & Cortes, A. (2010). *Veteran Home-Homeless Assessment.* Report to Congress. Washington, DC.

Knox, K. L., & Bossarte, R. M. (2012). Suicide prevention for veterans and active duty personnel. *American Journal of Public Health, 102*(Suppl. 1), S8–S9.

Knox, K., Stanley, B., Currier, G., Brenner, L., Ghahramanlou-Holloway, M., & Brown, G. (2012). An emergency department-based brief intervention for veterans at risk for suicide (SAFE VET). *American Journal of Public Health, 102*(Suppl. 1), S33–S37.

Lambert, M. T., & Fowler, D. R. (1997). Suicide risk factors among veterans: Risk management in the changing culture of the Department of Veterans Affairs. *Journal of Mental Health Administration, 24*(3), 350–358.

Lim, W., Rubin, E., Coats, M., & Morris, J. (2005). Early-stage Alzheimer disease represents increased suicidal risk in relation to later stages. *Alzheimer Disease & Associated Disorders, 19*, 214–219.

Linehan, M. M., Comtois, K. A., Murray, A. M., Brown, M. Z., Gallop, R. J., Heard, H. L., ... Lindenboim, N. (2006). Two-year randomized controlled trial and follow-up of dialectical behavior therapy vs. therapy by experts for suicidal behaviors and borderline personality disorder. *Archives of General Psychiatry, 63*(7), 757–766.

Lish, J. D., Zimmerman, M., Farber, N. J., Lush, D. T., Kuzma, M. A., & Plescia, G. (1996). Suicide screening in a primary care setting at a Veterans Affairs Medical Center. *Psychosomatics, 37*(5), 413–424.

Lönnqvist, J., Henriksson, M., Isometsä, E., Marttunen, M., Heikkinen, M., Aro, H., & Kuoppasalmi,

K. (1995). Mental disorders and suicide preventions. *Psychiatry and Clinical Neurosciences, 49*(Suppl. 1), S111–S116.

Luoma, J. B, Martin, C. E., & Pearson, J. L. (2002). Contact with mental health and primary care prior to suicide: A review of the evidence. *The American Journal of Psychiatry, 159*(6), 909–916.

Mann, J. J., Apter, A., Bertolote, J., Beautrais, A., Currier, D., Haas, A., . . . Hendin, H. (2005). Suicide prevention strategies: A systematic review. *JAMA: Journal of the American Medical Association, 294*(16), 2064–2074.

McCarthy, J. F., Blow, F. C., Ignacio, R. V., Ilgen, M. A., Austin, K. L., & Valenstein, M. (2012). Suicide among patients in the Veterans Affairs health system: Rural-urban differences in rates, risks, and methods. *American Journal of Public Health, 102*(Suppl. 1), S111–S116.

McIntosh, J. L., Santos, J. F., Hubbard, R. W., & Overholser, J. C. (1994). *Elder suicide: Research, theory, and treatment.*. Washington, DC: American Psychological Association.

McMenamy, J. M., Jordan, J. R., & Mitchell, A. M. (2008). What do suicide survivors tell us they need? Results of a pilot study. *Suicide and Life-Threatening Behavior, 38*(4), 375–389.

Meltzer, H. Y., Alphs, L., Green, A. I., Altamura, A. C., Anand, R., Bertoldi, A., . . . InterSePT Study Group (2003). Clozapine treatment for suicidality in schizophrenia: International suicide prevention trial (InterSePT). *Archives of General Psychiatry, 60*(1), 82–91.

Michalek, J. E., Ketchem, N. S., & Akhtar, F. Z. (1998). Postservice mortality of US Air Force veterans occupationally exposed to herbicides in Vietnam: 15-year follow-up. *American Journal of Epidemiology, 148*(8), 786–792.

Miller, M., Barber, C., Azrael, D., Calle, E. E., Lawler, E., & Mukamal, K. J. (2009). Suicide among US veterans: A prospective study of 500,000 middle-aged and elderly men. *American Journal of Epidemiology, 170*(4), 494–500.

Motto, J. A., & Bostrom, A. G. (2001). A randomized controlled trial of postcrisis suicide prevention. *Psychiatric Services, 52*(6), 828–833.

National Center for Injury Prevention and Control, & Centers for Disease Control and Prevention. (2007). *CDC Web-based Injury Statistics Query and Reporting System (WISQARS)*. Retrieved from www.cdc.gov/injury/wisqars/index.html

Peruzzi, N., & Bongar, B. (1999). Assessing risk for completed suicide in patients with major depression: Psychologists' view of critical factors. *Professional Psychology: Research and Practice, 30*(6), 576–580.

Pigeon, W. R., Britton, P. C., Ilgen, M. A., Chapman, B., & Conner, K. R. (2012). Sleep disturbance preceding suicide among veterans. *American Journal of Public Health, 102*(Suppl. 1), S93–S97.

Robertson, H. T., & Allison, D. B. (2009). Drugs associated with more suicidal ideations are also associated with more suicide attempts. *PLOS One, 4*(10), e7312.

Rurvey, C. L., Conwell, Y., Jonew, M.P., Phillips, C. Simonsick, E. Pearson, J. L., Wallace, R. (2002). Risk factors for late-life suicide: A prospective community-based study. *American Journal of Geriatric Psychiatry, 10*(4), 398–406.

Schinka, J. A., Schinka, K. C., Casey, R. J., Kasprow, W., & Bossarte, R. (2012). Suicidal behavior in a national sample of older homeless veterans. *American Journal of Public Health, 102*(Suppl. 1), S147–S153.

Shein, H. M. (1976). Suicide care: Obstacles in the education of psychiatric residents. *Omega: Journal of Death and Dying, 7*, 75–81.

Simon, G. E., & VonKorff, M. (1995). Recognition, management, and outcomes of depression in primary care. *Archives of Family Medicine, 4*, 99–105.

Simon, R. (2004). *Assessing and managing suicide risk: Guidelines for clinically based risk management*. Arlington, VA: American Psychiatric Publishing.

Skoog, I., Aevarsson, O., Beskow, J., Larsson, L., Palsson, S., Waern, M., Landahl, S., & Ostling, S. (1996). Suicidal feelings in a population sample of non-demented 85-year-olds. *The American Journal of Psychiatry, 153*(8), 1015–1020.

Substance Abuse and Mental Health Services Administration, & Administration on Aging. (2012). *Older Americans behavioral health issue brief 4: Preventing suicide in older Adults*. Retrieved from http://www.ncoa.org/improve-health/center-for-healthy-aging/content-library/Older-Americans-Issue-Brief-4_Preventing-Suicide_508.pdf

Thompson, R., Kane, V. R., Sayers, S. L., Brown, G. K., Coyne, J. C., & Katz, I. R. (2002). An assessment of suicide in an urban VA Medical Center. *Psychiatry, 65*(4), 327–337.

USUSUS Department of Health and Human Services, Office of the Surgeon General, & National Action Alliance for Suicide Prevention. (2012). *National strategy for suicide prevention: Goals and objectives for action*. Washington, DC: US Department of Health and Human Services.

US Department of Health and Human Services, US Food and Drug Administration, & Center for Drug Evaluation and Research. (2012). *Guidance for industry, suicidal ideation and behavior: Prospective assessment of occurrence in clinical trials*. Retrieved from http://www.fda.gov/Drugs/GuidanceComplianceRegulatoryInformation/Guidances/ucm315156.htm

US Department of Veterans Affairs. (2008). *Report of the Blue Ribbon Work Group on Suicide Prevention in the Veteran Population*. Washington, DC: Author.

Valenstein, M., Kim, H. M., Ganoczy, D., McCarthy, J. F., Zivin, K., Austin, K. L., ... Olfson, M. (2009). Higher-risk periods for suicide among VA patients receiving depression treatment: Prioritizing suicide prevention efforts. *Journal of Affective Disorders, 112*(1–3), 50–58.

18

Person-Centered Suicide Prevention in Primary Care Settings

Paul R. Duberstein

Marsha Wittink

Wilfred R. Pigeon

INTRODUCTION

Unlike other physicians who primarily attend to episodes of disease, primary care providers (PCPs)[1] are expected to consider the "whole patient" as well as the broader psychosocial, economic, legal, and cultural contexts of their patients' lives (Hart, 1971; McDaniel, Campbell, Hepworth, & Lorenz, 2005; Starfield, Shi, & Macinko, 2005; Watt, 2002). Patients are seen both when they are experiencing acute symptoms as well as in times of relative quiescence. PCPs are thus granted a privileged perspective on their patients' lives that is rarely available to other physicians or specialty mental health providers. Given that people who die by suicide are more likely to be seen by a PCP than by a specialty mental health provider in the weeks prior to death (Ahmedani et al., 2014; Luoma, Martin, & Pearson, 2002; Pearson et al., 2009), this privilege is accompanied by responsibility for suicide prevention.

Several suicide prevention initiatives have been mounted in primary care. Educating PCPs about suicide risk has been shown to lead to a decrease in suicide risk among women but not men (Rutz, 2001). Moreover, educational initiatives are not sustainable in the longer term. Integrated mental health care is a more desirable option (Working Party Group on Integrated Behavioral Healthcare, 2014). When mental health services are co-located and integrated into the primary care setting, the uptake of mental health treatments is improved (Bartels et al., 2004), which presumably should confer a host of benefits. Indeed, integrated care has been shown to lead to a decrease in suicide ideation (Bruce et al., 2004; Unützer et al., 2006) and all-cause mortality (Raue, Morales, et al., 2010). Data on suicide mortality are unavailable, however.

Integrated mental health care is now commonplace in publicly funded Veterans Health Administration (VHA) clinics (Pomerantz et al., 2010; Tew, Klaus, & Oslin, 2010) and is also available in several private systems, including Kaiser Permanente and Sutter Health. With more than 800 clinics and 8 million enrollees, the VHA's integrated approach to the provision of mental health care in primary care settings plays a key role in suicide risk surveillance and mitigation in Veterans (York, Lamis, Pope, & Egede, 2012). In 2008, the VHA rolled out the Primary Care-Mental Health Initiative to integrate mental health services (e.g., assessment, brief treatment) in all Department of Veterans Affairs (VA) primary care settings. Each VA medical center is mandated to blend these services with a care management program, such as the Behavioral Health Laboratory (Mavandadi, Klaus, & Oslin, 2012; Oslin et al., 2006) and TIDES—Translating Initiatives for Depression into Effective Solutions (VA Health Services Research and Development Service, 2008). The VHA Primary Care Program Office recently developed two new initiatives, Patient-Aligned Care Teams and the Office of Patient Centered Care & Cultural Transformation, to improve the provision of team-based patient-centered care in primary care clinics. The Department of Defense's Tricare Program plays a similar role for active military personnel and their families.

Although the integrated care movement is a positive development for suicide prevention, challenges remain. These are best understood by considering the multiple levels (Figure 18.1) of our health care system (Epping-Jordan, Pruitt, Bengoa, & Wagner, 2004). The outer ring, the macro level, pertains to activities by federal policymakers and regulators. An example of macro-level intervention is the passage of the Patient Protection and Affordable Care Act (PPACA), which incentivizes activities that could indirectly affect the suicide rate. The integrated care movement is primarily a meso level, or "middle ring," innovation. This is where states, insurers, or health systems (VA, Kaiser Permanente) can influence primary care practice. Somewhat ignored amidst the expansion of integrated care and the passage of the PPACA is the inner ring, the clinical microsystem (Nelson et al., 2002; Pomerantz et al., 2010). This is where patients, PCPs, nurses, medical assistants, and office staff interact on a daily basis. The microsystem frames all of the PCP's work activities, including his or her interactions with patients and patients' caregivers, which are depicted in the nucleus of the microsystem (Figure 18.1).

It is tempting to assume that meso-level initiatives, particularly in combination with macro-level initiatives such as the PPACA, will eliminate the need for microsystem innovations. There are two problems with this assumption. First, integrated care is still not widely available. Most PCPs in the United States are self-employed and work with five or fewer partners (Boukus, Cassil, & O'Malley, 2009), not in large systems like the VA or Kaiser Permanente. Moreover, even when integrated care is available, suicide prevention hinges on the capacity of patients and PCPs to engage willingly in difficult conversations about affect-laden topics, such as, financial stressors, domestic violence, cancer prognosis, caregiving for demented family members, and, of course, suicide.

Microsystem innovations are needed to improve communication and decision-making, particularly about marginalized, stigmatized topics. Many PCPs are reluctant or unprepared to have these discussions (Vannoy, Tai-Seale, Duberstein, Eaton, & Cook, 2011), perhaps because they understand that patients require more services (e.g., behavioral, legal, social) than are available in their work setting. Moreover, few patients readily disclose their darkest thoughts. In one study, 13 of 18 veterans who died by suicide explicitly denied having thoughts of suicide in their last contact with a health care provider (Denneson

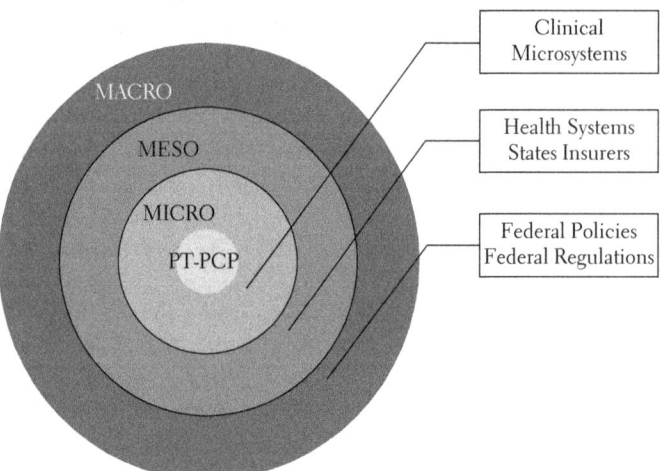

FIGURE 18.1 The layers of the health system. Macro-level activities influence meso-level activities, which in turn affect day-to-day activities in the clinical microsystem. The microsystem, in turn, frames all of the PCP's work activities, including their interactions with patients, patients' family caregivers, and other clinicians. In theory, all of these influences flow in the other direction as well: Microsystems could influence meso-level activities, and meso-level activities could influence macro-influences.

et al., 2010). Other studies show that the content of discussions about suicide in primary care is highly variable (Feldman et al., 2007; Vannoy et al., 2010; Vannoy et al., 2011). Some are adversarial, others end with no plan for follow-up, and others are surprisingly superficial, with meaningful risk assessment buried amidst small talk (Vannoy et al., 2011). When PCPs talk with patients about suicide, they should do so with minimal anxiety and without engaging the patient in a debate about the virtues of living, threatening hospitalization, or automatically prescribing an antidepressant. To achieve better outcomes, PCPs must have access to resources that could help patients resolve their concerns, meaningfully engage patients in discussions about the stressors that are known to confer suicide risk, and develop a plan for managing those stressors.

In this chapter, we offer two suggestions for improving suicide prevention in primary care settings. First, macro-level initiatives that increase the number and quality of PCPs in America are needed. An increase in the number of PCPs could enhance preventive care, disease management, and emotional support. PCPs save lives not just because they prevent and manage disease but because they represent anchors of emotional support for many patients (Epstein et al., 2005; Epstein & Street, 2007; Street & Epstein, 2008). Geographic regions with higher quality primary care have lower all-cause mortality (Bailey & Goodman-Bacon, 2015; Hart et al., 1991; Macinko, Starfield, & Shi, 2007; Starfield et al., 2005), and there is reason to believe that suicide mortality might be lower as well. In one of the more dramatic suicide prevention studies ever conducted (Motto & Bostrom, 2001), the suicide rate decreased after patients in a mental health center were mailed postcards with statements like "If you wish to drop us a note we would be glad to hear from you." The note from the clinic sent a message to the patients: *We truly care about you*. Further studies have shown that the effects of a low-cost postcard intervention can be surprisingly sustainable and cost-effective (e.g., Carter, Clover, Whyte, Dawson, & D'Este, 2013).

Simply increasing the number of PCPs is insufficient. The PCP's work environment must be improved as well. Behavior change interventions are needed to improve the quality of patient–PCP communication, point-of-care decision-making, and the care and management of at-risk patients. Our second recommendation for improving suicide prevention in primary care settings involves reducing the power asymmetries (Van Ryn, 2002; Van Ryn & Fu, 2003) that have historically characterized health care encounters (Fox, 1980; Mishler et al., 1981), equipping patients with tools that enhance their capacity to express their informed priorities, and enabling PCPs to respond effectively to those priorities (Working Party Group on Integrated Behavioral Healthcare, 2014).

This chapter describes a vision of person-centered[2] suicide prevention for primary care settings (Box 18.1). The time is ripe for behavior change in primary care, particularly given the PPACA, early experiences with the patient-centered medical home (Nutting et al., 2011), and growing

---

BOX 18.1 Core Propositions of Person-Centered Suicide Prevention in Primary Care

- The success of any intervention depends on the extent to which fundamental human needs for autonomy, competence, and relatedness have been accommodated.
- Universal prevention efforts will gain traction to the extent that they do not interfere with fundamental human needs.
- Indicated and targeted interventions are presumed to be person-centered when patients receive *the care they need and no less and the care they want and no more*. To provide such care, PCPs must (a) elicit patients' needs and wants in a manner that respects their fundamental needs for autonomy, competence, and relatedness; (b) offer patients opportunities to provide input into and participate in their care; and (c) attempt to enhance the quality of the relationship with the patient.
- In order for PCPs to provide person-centered care, they must work in an environment that accommodates both *their* own needs for autonomy, competence, and relatedness and the fundamental needs of other vested parties (e.g., health care and administrative personnel) and stakeholders.

> **BOX 18.2 When Less (Treatment) Is More (Quality)**
>
> Assuming that a PCP needs more than three hours of sleep a night and has a life outside of work, there are simply not enough hours in the day to provide all the care recommended to meet their patients' need for preventive, acute, and chronic care. Of the 21 hours a day required (Yarnall et al., 2009), some time would be spent completing paperwork and coordinating with physicians in other practices. The average American PCP coordinated with more than 220 physicians annually in more than 110 different practices (Pham, O'Malley, Bach, Saiontz-Martinez, & Schrag, 2009), and that figure excludes care provided for patients not covered by Medicare. Even if a PCP provides just half the care recommended in guidelines, the likelihood of overtreating patients is high. Unnecessary interventions account for 10% to 30% of health care spending in America, or $250 billion to $800 billion annually (Berwick & Hackbarth, 2012). One simulation found that applying individual disease guidelines to a patient with five chronic conditions would result in the prescription of 19 doses of 12 different drugs, taken five times daily and carrying the risk of 10 adverse events or drug interactions (Boyd et al., 2005). Although chronic conditions have different symptomatic manifestations, many share a common underlying systemic disturbance. For example, conditions as seemingly diverse as diabetes, depression, and arthritis are all characterized by neural-immune inflammation. Given this overlap, it is not surprising that the effectiveness of any one treatment often diminishes with the number of treatments received for other conditions (Mold, Hamm, & McCarthy, 2010). Interestingly, when physicians replace guideline concordant care with care that is more attuned to patients' priorities, patients' outcomes are improved. Kurt Stange, MD, PhD, professor of family medicine and editor-in-chief of the *Annals of Family Medicine*, observed that, "[e]vidence-based guidelines are not helpful; in fact, they are potentially harmful for a large proportion of patients seen in primary care" (2009, p. 390). Prioritization, in contrast, can protect patients from overtreatment. One study showed that patient involvement in decision-making about depression, not the provision of guideline concordant care, led to better depression outcomes (Clever et al., 2006). The authors speculated that "physicians' willingness to involve patients in decision-making might have a direct therapeutic effect, because it may signal to patients that their opinions are valuable, thereby improving self-esteem" (p. 403).

frustration with current practice (Box 18.2). In the first section, we describe the two theoretical pillars of our approach to behavior change: systems theory (Engel, 1980; Kizer, 2012a; Litaker, Tomolo, Liberatore, Stange, & Aron, 2006; Plsek & Greenhalgh, 2001; Stange, 2009) and self-determination theory (SDT; Ryan & Deci, 2000). Next, we explain why a person-centered approach to suicide prevention is timely. Following that, we consider the extent to which traditional approaches to suicide prevention in primary care can be viewed as person-centered. To ground this chapter in the clinical reality of primary care practice, we next describe the real-world challenges to the provision of person-centered care by contrasting the activities of PCPSs and specialty providers. Before concluding, we present a hypothetical case to illustrate a person-centered approach to suicide prevention in the primary care setting and how it might differ from the prevailing disease-oriented approach.

THE NEED FOR THEORIES OF BEHAVIOR CHANGE IN PUBLIC HEALTH AND HEALTH POLICY

No matter what their ultimate aim, all public health initiatives must be informed by a theory of human motivation. The reason is simple. Most public health initiatives involve behavior change. Any attempt to modify human behavior does so in the face of inborn resistance (Graham & Martin, 2012). All persons, whether they are intervention targets (patients, at-risk segments of the population) or implementers (e.g., physicians, therapists, administrative and clerical personnel), are active agents who may resist *any* efforts to engender or sustain behavior change. SDT (Ryan & Deci, 2000) suggests that behavior change is more readily accomplished when fundamental human needs for competence, autonomy, and relatedness are accommodated and supported. Too often it has been presumed that it is sufficient to attend only to the motivational needs of the intervention target (e.g., an individual patient). Many interventions have proven to be unsustainable because trialists have not considered the motivational needs of implementers and other stakeholders with vested interests. Systems theory views any intervention target (e.g., a patient, a patient–PCP dyad) as nested in a hierarchy of interdependent systems (e.g., clinical microsystem, meso-level health system) that could influence the effectiveness or sustainability of the targeted intervention.

Systems Theory

As depicted in our adaptation of Engel's (1980) hierarchy of natural systems (Figure 18.2), humans are simultaneously *wholes made up of parts*, and a *part of larger wholes*. This is the primordial duality of our existence. Human behavior is influenced and can be modified by forces lower and higher in the hierarchy of the natural system. On the one hand, an individual's behavior is a product of lower order biochemical and biophysical processes; on the other hand, individual behavior is influenced by the hierarchies (e.g., families, communities, subcultures) in which the individual is nested.

Lower level systems influence higher level systems (bottom-up). Suicide is known to be influenced by the activities of molecular systems (Courtet, Gottesman, Jollant, & Gould, 2011). Moreover, higher level systems influence lower level systems (top-down). Suicide is also influenced by social systems (Fassberg et al., 2012). Bottom-up and top-down influences are evident throughout the hierarchy. Human activity influences the biosphere. Macro-level policy initiatives (e.g., PPACA) influence meso-level and microsystem

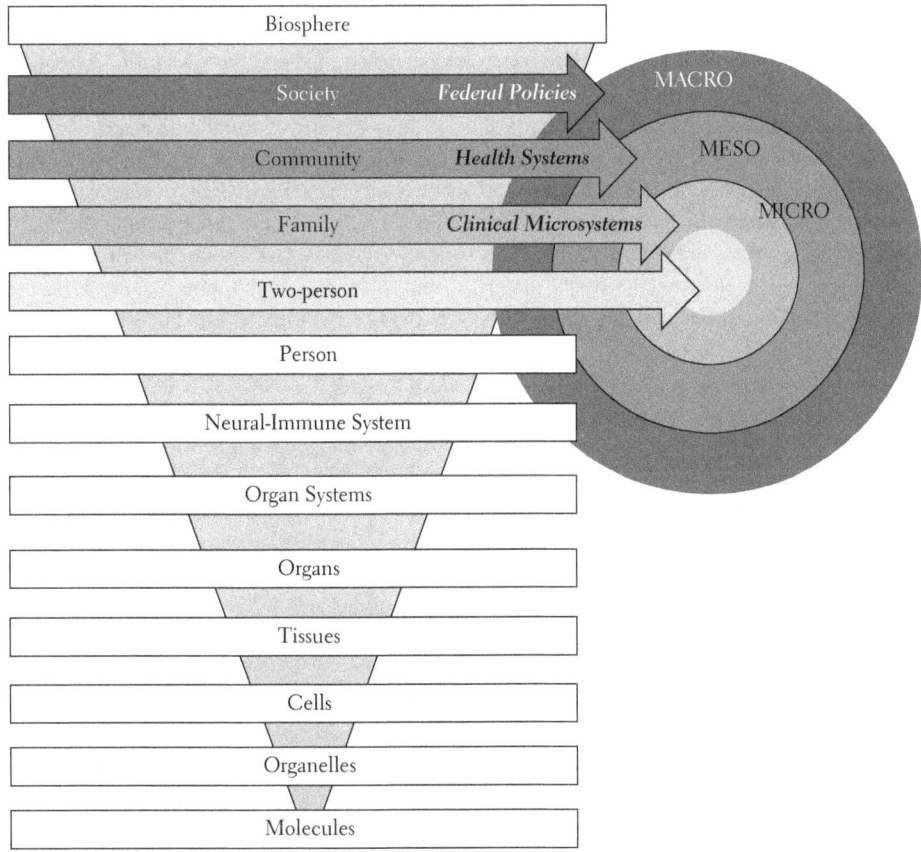

FIGURE 18.2 The layers of the health system (right) overlaid on an adaptation of Engel's (1980) hierarchy of nested natural systems. The health system behaves like any other natural system, and is characterized by bi-directionality and unintended consequences.

activities (Kizer, 2012b). Social systems influence molecular function (Cole et al., 2010).

Bottom-up and top-down responsiveness to input throughout the hierarchy often exacts unintended costs. For example, the PPACA has the potential to have an adverse influence on the provision of integrated care in the VA settings by fragmenting care (Kizer, 2012b). Patients with "behavioral health diagnoses (are) especially vulnerable to untoward consequences resulting from fragmented care" (Kizer, 2012b, p. 789).

From a systems perspective, a PCP's response to a suicidal patient will be influenced by systems lower in the hierarchy (e.g., the PCP's neural systems subserving empathy, the patient's neural systems subserving awareness of painful emotions) and by higher level social systems. If the PCP is nested in a work organization that supports his or her exploration of suicide risk with her patients, he or she is more likely to broach the topic in a calm, empathic, confident manner. In turn, this will have implications for patient outcomes. Consider this comment from a PCP working in an integrated VA clinic: "I am more comfortable asking about suicide and other things I know I can't handle" because mental health specialists "are watching my back" (Pomerantz et al., 2010, p. 122). On the other hand, if the PCP is nested in an organization that threatens the PCP's fundamental needs for competence, autonomy, and relatedness, he or she is less likely to explore suicide risk or difficult life circumstances with patients. "Some conditions of general practice are bad enough to change a good doctor into a bad doctor in a very short period of time" (Collings, as cited in Hart, 1971, p. 406). Far from an academic exercise, therefore, the application of systems theory to suicide prevention initiatives in primary care can inform organizational and workplace interventions designed to enhance the provision of person-centered care that has maximal impact on suicide and other adverse outcomes.

Self-Determination Theory

The part–whole duality is not merely a feature of natural systems. It is also reflected in our inner lives and basic needs. We humans have *a need to be a part* of a social group, and, at the same time, we have a *need to be apart* from others. SDT (Ryan & Deci, 2000), like other major Western and Eastern theories of human motivation, is premised on the idea that humans have fundamental needs for dependence and independence (Angyal, 1982; Bakan, 1966; Blatt & Shichman, 1983; Buber, 1970; Ryan & Deci, 2000; Stephens, Markus, & Fryberg, 2012). These theories have long informed mental health interventions. Only recently have they been applied to the conceptualization of public health or social policy interventions (Moller, Ryan, & Deci, 2006; Stephens et al., 2012).

Decades of basic behavioral research support SDT's core proposition, namely, that outcomes in any life domain (health, education, relationships) are enhanced when fundamental needs for *competence and autonomy* (independence) and *relatedness* (dependence) are accommodated (Ryan & Deci, 2000). Drawing from SDT, we suggest that suicide prevention initiatives will succeed and be sustainable to the extent that they are perceived as posing little threat to the self-determination of all vested parties (Duberstein & Heisel, 2014). Outcomes of any intervention—including initiatives aimed at modifying how patients and PCPs interact with each other—depend on the extent to which fundamental human needs have been accommodated or thwarted.

It is not just the patient's needs for autonomy, competence, and relatedness that are important but the clinician's as well (Bodenheimer & Sinsky, 2014). The latter point bears emphasizing. So many interventions treat clinicians as though they are merely inert substances, without insights and expertise borne of experience, motives and goals that need to be accommodated, and feelings, such as empathy (or the lack thereof) that potentially have implications for patient outcomes. Yet there is growing frustration with this view (Emanuel & Pearson, 2012) and recognition that it must be corrected in order to improve patient outcomes (Bodenheimer & Sinsky, 2014).

Psychologists have argued that manualized treatments have threatened their perceived competence and autonomy by "reducing the clinician to a research assistant who can run subjects in a relatively uniform ... way" (Westen, Novotny, & Thompson-Brenner, 2004, p. 639). Psychiatrists have argued that the research "literature ... often seems to be lacking something. What is missing is what we initially sought in our professional lives—our patients' voices" (Gabbard & Freedman, 2006, p. 184). In other words, relatedness has suffered, undermining job satisfaction. Skepticism among PCPs about the utility of mental health assessments (Gilbody, Sheldon, & Wessely, 2006) can be ascribed in part to their belief that that the screens represent a threat to their autonomy (Leydon et al., 2011), particularly given the ever-growing list of third-party intruders (Baik et al., 2010; Hunt, Kreiner, & Brody, 2012), like mandates (Box 18.2) and direct-to-consumer advertising. Moreover, otherwise well-intentioned integrated care initiatives have often marginalized the PCP by ensuring that care managers, not PCPs, engage patients in discussions about important treatment decisions.

In sum, systems theory encourages interventionists to consider offering interventions at multiple levels of the hierarchy of systems and to anticipate that interventions offered at one level will affect other levels, for better or worse. SDT suggests that initiatives will be sustainable only if the motivational needs of all vested parties have been accommodated.

PERSON-CENTERED PREVENTION: WHY NOW?

A person-centered approach to suicide prevention is timely, as it coincides with the gradual waning of medical paternalism and the rise of patient-centered medical home and patient-centered care (Alston et al., 2012; Institute of Medicine Committee on Quality of Health Care in America, 2001; Nutting et al., 2011), including mental health care (Drake, Cimpean, & Torrey, 2009; Raue, Schulber, et al., 2010). Patient-centered care aims to motivate patients to engage in care. Care is presumed to be patient-centered if patients *receive the care they need and no less and the care they want and no more* (Box 18.1). Another phrase that captures the spirit of patient-centeredness is *no decision about me without me*, coined by philanthropist and advocate Harvey Picker.

Clinicians, clinical microsystems, health care systems, policymakers, and regulators behave in a person-centered manner when, according to one definition (Epstein et al., 2005), they (a) elicit and consider patients' needs and wants, (b) offer patients opportunities to provide input into and participate in their care, and (c) take steps to enhance the quality of patient–clinician relationships. Biomedical concepts (symptoms, risk factors, or lab values) are not mentioned or implied in this definition, but motivational concepts are central. For example, when PCPs offer patients opportunities to provide input into and participate in their care, they are potentially enhancing their patients' beliefs about their own autonomy and competence (Clever et al., 2006).

A rich tradition of humanistic scholarship exists in primary care medicine, but it has been largely marginalized (Sullivan, 2003). Patient-centeredness is a product of this peripheral pedigree, so why is it slowly gaining traction now? Carolyn Clancy (2011), the former director of the Agency for Health Care Research and Quality (AHRQ), identified four broad forces, reducible to a simple mnemonic: *ABCD*.

A is for Airwaves: People have access to more and better information about health and health care. Important developments include increased use of direct to consumer advertising by the pharmaceutical, medical device, and genetic testing industries, the availability of health-related information via the Internet, the Food and Drug Administration's efforts to publicize adverse events, and the ubiquitous presence in the media of alternative treatments, exercise regimes, and diets.

B is for Boomers: Baby boomers are more likely than prior cohorts to desire an active role in their health care. They are more highly educated than prior birth cohorts, and there is a robust association between education and health behavior/self-care (Cutler & Lleras-Muney, 2010).

C is for Chronic Disease: Multimorbidity is on the rise (Tinetti, Bogardus, & Agostini, 2004). More than 10% of Americans, and 40% of those older than 60, are prescribed five or more medications (Gu, Dillon, & Burt, 2010). Yet "treatment associated with one condition has the potential to worsen another" (Fried, McGraw, Agostini, & Tinetti, 2008, p. 1839). Patients are thus triply burdened: burdened by disease, burdened by treatments for diseases, and increasingly burdened by the cognitive and emotional demands of treatment decision-making. Consequently, one is now compelled to ask the "question of whether what is good for the disease is always best for the patient" (Tinetti et al., 2004, p. 2871).

D is for Data: Exponential increases in funding for biomedical research since the 1970s, rapid advances in computing, open-access publishing, and legislation requiring researchers to offer free, publicly accessible reprints of studies reporting the findings from federally funded research have yielded more data than ever before. Given the highly variable applicability of many trials to real-world care (Stange, 2009; Tinetti et al., 2004; Watt, 2002), a major challenge is to transform these data into a form that would be useful for patients and clinicians. A national library, developed with input from diverse constituencies, including patients, has been proposed to meet that that challenge (Lerner, Fox, Ruzek, & Shearer, 2010).

We add a fifth force to Clancy's four: E for Expense. Patient-centeredness may save money and time, a nontrivial concern given escalating health care costs and rates of physician burnout. Patient-centered care can lead to better outcomes (Epstein & Street, 2007; Stewart, 1995), despite fewer specialist referrals (Bertakis & Azari, 2011). Some randomized trials have shown that engaged patients use fewer health services (Stacey et al., 2011; Wennberg, Marr, Lang, O'Malley, & Bennett, 2010), but more data are needed.

ARE TRADITIONAL APPROACHES TO SUICIDE PREVENTION IN PRIMARY CARE PERSON-CENTERED?

Principles of person-centeredness have not influenced, let alone inspired, the design of contemporary suicide prevention initiatives. Still, it is useful to consider whether current approaches meet the criteria outlined in Box 18.1 and provide patients with the care they need and no less and the care they want and no more.

Prevention programs lie along a continuum, ranging from policies that apply to all (universal) to treatments that are offered only to a select few (indicated). *Universal prevention programs* aim to modify health behaviors and health decision-making in broad swaths of the population (Kaplan, 2000). Relevant behaviors and decisions concern alcohol use, tobacco consumption, drug use, eating, exercise, vaccines, and safety with regard to sex, food, medication, transportation, fire prevention, and firearm access and safety (e.g., Walters et al., 2012).

Some universal initiatives are less threatening (e.g., medication safety) to large segments of the population than others (e.g., firearm sales). To minimize resistance, a person-centered approach encourages the devotion of limited resources to initiatives that deliver the care people want *and no more* while accommodating their needs for self-determination. In the United Kingdom, legislation concerning access to over-the-counter medications has proven effective in suicide prevention (Hawton, 2002; Hawton et al., 2009). To prevent people from swallowing an entire bottle of pills, lawmakers insisted that drug manufacturers use blister packs for certain products. Few people are emotionally invested in decisions about the packaging of pills, and most do not care whether their pills are packaged in blister packs or bottles. In contrast, firearms legislation in America and bridge barriers initiatives worldwide are more controversial, because, according to SDT, they represent a greater threat to autonomy, competence, or relatedness.

Beyond means restriction, other policy innovations are less threatening and might be just as effective, if not more so. Now that there is good evidence that the roll-out of community health centers in the United States led to a reduction in all-cause mortality in high-risk demographic groups (Bailey & Goodman-Bacon, 2015), efforts to increase the number and quality of PCPs might lead to reductions in suicide, perhaps especially among those at increased risk. In America, geographic regions with poorer quality primary care have elevated rates of all-cause mortality (Macinko et al., 2007; Starfield et al., 2005). By one estimate, more than 125,000 deaths could be prevented annually simply by adding one PCP for every 10,000 residents (Macinko et al., 2007). Given that depression rates are higher in regions with poorer quality primary care (Shi, Starfield, Politzer, & Regan, 2002), many of these prevented deaths would be suicides.

On the other side of the prevention spectrum, indicated interventions are offered to primary care patients with one or more known clinical risk factor for suicide. One type of indicated intervention, integrated care, has been shown to reduce the severity of suicide ideation (Bruce et al., 2004; Unützer et al., 2006) and decrease risk for all-cause mortality (Raue, Morales, et al., 2010). In order to be enrolled in an indicated intervention, eligible patients must first be identified. Patients can self-refer, and clinicians can make referrals. Screening instruments are also used. In VA clinics, screens for suicide ideation, depression, posttraumatic stress disorder, and traumatic brain injury are offered annually. When patients screen positive they are referred for intervention, often some form of integrated care. It is likely that this surveillance and treatment program has saved lives.

In the middle of the prevention spectrum are targeted interventions, which are aimed at mitigating risk markers for suicide prior to the development of the clinical risk factors that compel indicated interventions. PCPs play a role in targeted prevention by counseling patients about health behaviors (e.g., smoking, exercise, alcohol consumption) that are known to be associated with suicide risk. Screening instruments are commonly used to identify at-risk patients. For example, primary care practices in the VA conduct annual screens for alcohol misuse. When a patient screens positive, PCPs are obligated to provide brief counseling and consider other interventions, such as referrals for specialized treatments.

A targeted intervention called stepped care aims to hinder the progression of mild symptoms into more severe symptoms and to ensure that subsyndromal symptoms do not blossom into full-blown syndromes. Two well-established VA programs, the Behavioral Health Lab (Oslin et al., 2006; Tew et al., 2010) and the White River Model (Pomerantz et al., 2010) can be considered stepped care programs. In stepped care for depression (Franx, Oud, de Lange, Wensing, & Grol, 2012), the PCP and patient first have the option to engage in watchful waiting. As a next step, the PCP can prescribe low-intensity treatments, such as bibliotherapy or exercise. At a still higher step, the PCP can prescribe a higher intensity treatment (e.g., psychotherapy). If there is still no improvement, or if symptoms worsen, more intensive treatment options (e.g., antidepressants, electroconvulsive therapy) are offered.

There is much to like about stepped care. Whereas patients suffering from mental health conditions are rarely offered more than three options (medications, psychotherapy, or a combination), stepped care recognizes the need to offer patients greater variety. Moreover, many of these interventions are low intensity. For example, in VA settings behavioral medicine specialists are now available to offer counsel about nontraditional, low-intensity mental health interventions such as exercise and healthy sleeping habits. Consequently, targeted interventions like stepped care have greater potential to ensure patients receive the care they want *and no more*. Titration is one of its greatest strengths.

Notwithstanding the promise of both indicated and targeted interventions, there is room for improvement in two areas. First, indicated and targeted interventions have rarely accommodated implementers' (staff, therapists, physicians, etc.) needs for competence and autonomy, nor have interventions explicitly attended to implementers' needs to interact effectively with patients and each other. PCPs, psychologists, and psychiatrists have all expressed concerns about threats to their fundamental motivational needs (Baik et al., 2010; Dowrick et al., 2009; Gabbard & Freedman, 2006; Leydon et al., 2011; Westen et al., 2004). Reservations have also been expressed about whether clinicians can work together to co-manage patient problems without feeling that their autonomy or competence is threatened (Tew et al., 2010). Concerns about power, territory, and communication among health care providers are important, as some evidence suggests that suicide risk is heightened during care transitions, perhaps due to miscommunications (Gunnell et al., 2012; Kales, Kim, Austin, & Valenstein, 2010; Valenstein et al., 2009; Valenstein, Kim et al., 2009). Workplace innovations designed to build more effective primary care teams are needed, especially given the expectation that health care personnel will work more closely together than ever before (Nutting et al., 2011; Working Party Group on Integrated Behavioral Healthcare, 2014). One promising intervention is Team Strategies & Tools to Enhance Performance & Patient Safety (Team STEPPS), developed by the Department of Defense in collaboration with AHRQ (Clancy & Tornberg, 2007). Team-building initiatives have been applied to surgery and other medical specialties but have lagged behind in primary care where the issues are arguably more complex and the resources are less abundant.

Second, targeted and indicated interventions typically focus on symptoms and disease, not necessarily patients' wants and needs. In America, insurers pay PCPs mainly to manage symptoms, treat disease, and, in rare instances, prevent disease. Professional role socialization (Fox, 1980) and the disease-centered biomedical model (Mishler et al., 1981) reinforce this focus, with implications for the types of interventions PCPs deem acceptable (Duberstein & Wittink, 2015; Wittink, Givens, Knott, Coyne, & Barg, 2011). Members of primary care teams that were part of an implementation evaluation of a depression treatment study felt that a low-intensity intervention—prescription of a self-help manual—"is not general medicine" (Franx et al., 2012, p. 8). Yet other PCPs disagreed, arguing that labeling and treating symptoms of depression as a disease "could have the negative effect of adopting too narrow of an approach to the patient's problems, offering medical solutions without considering the patient's story and contextual factors" (p. 5).

PCPs' beliefs about the scope of their job could undermine suicide prevention efforts in primary care, as many potent clinical risk markers for suicide are not diseases in the usual sense of the term. It is increasingly difficult to ignore the data challenging the biomedical, disease-centered vision of what general medicine is or ought to be. Lack of social connectedness, life event stressors, and health-damaging personality traits are not diseases, but they can confer risk for premature mortality. Associations between suicide mortality and social disconnectedness, as indexed by marital status, social network size, loneliness, and social support, have been documented (Duberstein et al., 1998; Duberstein, Conwell, Conner, Eberly, Evinger, et al., 2004; Fassberg et al., 2012). Suicide risk is elevated following interpersonal conflicts, relationship disruptions, family discord, financial difficulties, job problems, and a host of other life event stressors (Conner, Duberstein, & Conwell, 2000; Conner, Cox, et al., 2001; Duberstein, Conwell, Conner, Eberly, & Caine, 2004). More than 100 studies have shown that personality is associated with suicide mortality (Brezo, Paris, & Turecki, 2006; Conner, Duberstein, Conwell, Seidlitz, & Caine, 2001; Duberstein & Conwell, 1997; Duberstein & Witte, 2009). Self-consciousness, aggression, impulsivity, and hopelessness all confer risk, as do antisocial, avoidant, borderline, and schizoid personality disorders.

Lack of social connectedness (Holt-Lunstad, Smith, & Layton, 2010), life event stressors, and health-damaging personality traits (Chapman, Roberts, & Duberstein, 2011) are all strongly associated with all-cause mortality as well as suicide. More important, the magnitudes of some of these associations are comparable to those attributed to chronic diseases. Yet social disconnectedness, life event exposures, and health-damaging personality traits do not appear on any lists of leading causes of death. PCPs are expected to offer whole-person care, but most American insurers and employers of physicians do not pay PCPs to counsel patients about personality traits or consider how patients' responses to their interventions might be influenced by personality. PCPs are expected to play a vital role in suicide prevention, but they are not paid to help patients feel more socially connected or to weather life's stressors (although many PCPs do this anyway). Even mental health clinics often fail to provide these services. In one qualitative study, patients expressed "concern with the reluctance of the clinics to engage with distressing and seemingly intractable problems in the patient's life" (Morgan, 1999, p. 446). It is difficult to envision how care can be person centered when both the topic of suicide and the difficult circumstances of people's lives that lead them to contemplate suicide are marginalized, ignored, or minimized in the primary care encounter (Ganzini et al., 2013; Vannoy et al., 2011).

In summary, suicide prevention could be more person-centered by being more mindful of the needs and wants of all stakeholders across the prevention continuum. Three recommendations follow from these observations. First, the precious time and effort devoted to universal initiatives might be used more wisely by anticipating affect-laden counterarguments and identifying less controversial initiatives (Duberstein & Heisel, 2014). If compromise is impossible (Gutmann & Thompson, 2012), advocates of controversial initiatives must use proven tactics of persuasion to marshal public support and disarm critics (Oliver, 2006). In this regard, we recognize that our call for more PCPs is at odds with market forces, which currently favor specialty care (Schwartz, 2012). Second, workplace innovations are needed to accommodate implementers' needs for autonomy, competence, and relatedness. The successful sustainability of any preventive intervention hinges on the day-to-day engagement of all personnel (Pomerantz et al., 2010).

Third, and perhaps more significantly, we hypothesize that improvements in the way PCPs and patients communicate and make decisions at the point of care will prevent suicides (Duberstein & Wittink, 2015) while improving other outcomes (Kelley, Kraft-Todd, Schapira, Kossowsky, & Riess, 2014). Current institutional arrangements—the biomedical model and its economic cousin, fee-for-service payment systems—have conferred benefits to many stakeholders. Unfortunately, they have also made it more difficult for PCPs to prevent suicides by responding to patients' needs and wants. Some may argue that this problem will not go away without changing how PCPs are paid and what they are paid to do (Relman, 1994). Certainly, those changes would be helpful, but there are some incremental fixes that may be viable given developments in technology and five forces (ABCDE) driving interest in person-centeredness. To enhance the provision of person-centered care, the prevailing challenges must be exposed.

PERSON-CENTERED CARE: A CHALLENGE FOR PCPS AND PATIENTS

When specialty care physicians (e.g., oncologist, cardiologist, gastrointestinal surgeon), see patients, they can be relatively certain why the patient has sought care or counsel. The patient's story has already been "filtered" (Marino, Gallo, Ford, & Anthony, 1995). The frame of the encounter permits the specialist a sort of cognitive economy that PCPs might envy. Many uncertainties have been extracted, cast aside, or resolved. Specialists are expected to engage patients in a discussion about whether intervention for a particular disease (or risk marker) is warranted and, if so, which intervention is preferred. In comparison to PCPs, specialists are more concerned with disease-specific outcomes and adherence to guidelines (Starfield et al., 2005).

Comprehensivist PCPs see patients with unfiltered stories and correspondingly numerous uncertainties. They can engage patients in discussions about multiple diseases, risk factors for multiple diseases, psychosocial circumstances, and general well-being. They also have more options for when they can talk about a particular issue: this visit, the next, or a subsequent visit. Moreover, patients often present with multiple, vague, uncharacterized symptoms. Whereas the top six diagnostic clusters account for up to 90% of patient visits to specialty providers, the top 20 diagnostic clusters account for roughly half the patient visits to PCPs (Stange et al., 1998).

Deprived of the specialist's filter, gradients of uncertainty loom in virtually every aspect of primary care practice, including suicide risk assessment. A clinician can be relatively less uncertain about his or her options when a patient (a) readily discloses thoughts of suicide, (b) has only one condition that confers risk for suicide (e.g., major depression, single episode, moderate), (c) experiences minimal life stress (e.g., no marital or employment problems), and (d) is motivated to adhere to specialty mental health treatments. Affordable, safe, effective treatments are available for the few patients in these circumstances.

Unfortunately, many primary care patients at risk for suicide suffer from multiple chronic conditions, and psychological, social, or economic turmoil are common. PCPs have understandable reservations about the capacity of controlled trials to yield actionable findings that will enable them to provide *the care patients need and no less* because clinical trials and even observational studies often exclude the types of patients they see every day (Watt, 2002). Moreover, few of the treatments studied are readily available, accessible, or affordable. As a result, PCPs work in a state of chronic equipoise: many treatment options appear to be available, none of which is demonstrably superior to others, readily accessible, or customized to the unique needs of the individual patient.

Physicians are urged to provide *the care patients want and no more*, but this ideal also remains elusive. Health care encounters have historically been characterized by power asymmetries (Mishler et al., 1981). Patients have rarely been empowered to exercise their autonomy and competently express their wants. The physician is a powerful authority who starts the encounter, frames the options for discussion, and ends the encounter (Coulter & Collins, 2011; Wirtz, Cribb, & Barber, 2006). Power asymmetries could stoke adversarial mistrust and enable PCPs' undesirable personal qualities (e.g., arrogance; Duberstein & Wittink, 2015) to affect patient care. PCP personality traits and personal experiences with depression are known to influence their approach to the assessment and treatment of depression (Duberstein, Chapman, Epstein, McCollumn, & Kravitz, 2008; Kravitz et al., 2006; Lampe et al., 2013) as well as their patients' experience of care (Duberstein, Meldrum, Fiscella, Shields, & Epstein, 2007). Similar effects have been documented in the psychotherapy literature (Ackerman & Hilsenroth, 2003; Flückiger, Del Re, Wampold, Symonds, & Horvath, 2012; Hilsenroth, Cromer, & Ackerman, 2012) and in research on end-of-life care (Wilkinson & Truog, 2013). Despite the waning of paternalism, the deferential behavior of intimidated patients underscores their strong desire to avert their physicians' displeasure (Frosch, May, Rendle, Tietbohl, & Elwyn, 2012).

In disease-centered care, powerful (supply-side) PCPs and dependent (demand-side) patients (Relman, 1994) adopt ascribed roles. The PCPs counsel patients about their options. Consequently, and particularly in fee-for-service settings, supply-side forces could unduly influence patient care and outcomes. The Dartmouth Atlas (Wennberg, Fisher, Goodman, & Skinner, 2008) has documented the unsettling influence of the mere availability of a particular test, treatment, or service on health care decisions, as well as health care utilization, cost, and quality.

It is the rare patient who, risking his PCP's disapproval, expresses his genuine wants to a PCP. He might express his wants when interacting with a barber or when surfing the Internet for a deal on a home item. He might demand refunds from a store manager when a purchase fails to meet his expectations. In contrast, he rarely expresses his health-related wants and expectations to his PCP. Believing that medical science is esoteric and beyond his ken, he is insufficiently motivated to advocate for himself. He may associate self-assertion within the medical context with shame or embarrassment.

Patients may not possess specialized medical knowledge, but they are experts in their own circumstances (Coulter & Collins, 2011). Evidence is mounting that these circumstances often bode poorly for patient outcomes. Financial stressors, family discord, domestic violence, social isolation, legal imbroglios, and other adversities have been linked to suicide (Conner et al., 2012; Duberstein, Conwell, Conner, Eberly, Caine, 2004). In a truly person-centered primary care treatment setting, patients would be informed that adverse circumstances and historically stigmatized personal concerns (e.g., suicide, sexuality, transgressions in battle) represent legitimate conversational fare and that relevant services are available, accessible, and affordable.

The clinical micro system must cater to all of the patient's health-related needs and wants, not merely those that are currently on the "primary care menu." Patients need more options for care and more combinations of options. We use the term "menu" to underscore that patients (consumers) should be empowered to behave as they might in other markets. A restaurant or food superstore metaphor is thus preferable (Wittink, Duberstein, & Lyness, 2013) to the home and neighborhood metaphors that are in vogue (Nutting et al., 2011). A downside of person-centered health care or choice in any consumer setting is uncertainty (Iyengar & Lepper, 2000) as the following hypothetical case illustrates.

A CASE STUDY: CONTRASTING DISEASE- AND PERSON-CENTERED OPTIONS

In any care setting, no single practitioner (PCP, psychotherapist) can be more effective than the available resources allow. For the sake of illustration, we assume a better-case scenario: Mr. A. receives his primary care in a VHA clinic, and Dr. B has been his PCP for 15 years. Most primary care practices in America do not formally assess depression, anxiety, or suicide ideation, but these data are gathered annually in VHA clinics, and behavioral interventions are available.

Mr. A. is a 67-year-old divorced, retired auto mechanic and Vietnam-era veteran who was diagnosed with Stage II non-small cell lung cancer two months ago. He is currently receiving adjuvant chemotherapy and is prescribed a β-blocker (for hypertension) as well as a nonsteroidal anti-inflammatory drug and a prn muscle relaxer (recurrent back pain from injuries sustained in battle). His body mass index is 28 (overweight). Mr. A. has never screened positive for suicide risk or depression. The last screen was conducted seven months ago. He presents for routine follow-up of his hypertension. On review of systems his only complaint is fatigue and mild sleep disturbance.

Psychosocial

Mr. A. no longer sees or speaks to his ex-wife, but both his daughters live within a 30-minute drive of his rural home where he resides with his eight-year-old dog. Although Dr. B. has never met any of Mr. A.'s family members or friends, Mr. A. has repeatedly informed Dr. B. that he has a "good" relationship with his daughters, and he frequently refers to his dog and his "buddies." Favorite activities include seeing his three grandchildren, daily walks with his dog, weight-lifting at the gym, and a weekly breakfast meeting with friends, most of whom are veterans. Mr. A. is not the type to discuss his emotions. Still, in a rare moment of vulnerability at the time of his divorce 15 years ago, he mentioned that his wife complained about his "mood swings," which Dr. B. documented.

Health Behavior

Dr. B. has repeatedly advised the patient to stop smoking, lose weight, and cut down on his drinking and recreational drug use (cannabis), but Mr. A. has been unmoved by this counsel. Taking pride in his ability to care for himself, Mr. A. has claimed that his habits have not caused any difficulties, so there is no need to change anything. On several occasions, he claimed that Dr. B.'s proposed interventions would deprive him of some of life's few pleasures, and he once joked, "I would kill myself before getting rid of these," gesturing to the cigarettes in his shirt pocket.

From Dr. B.'s Perspective, What Is Certain About Mr. A's Suicide Risk?

Lung cancer, pain, smoking, excessive alcohol consumption (Ilgen et al., 2010), recreational drug use, and sleep disturbance (Pigeon, Britton, Ilgen, Chapman, & Conner, 2012; Pigeon, Pinquart, & Conner, 2012) all increase suicide risk. Suicide risk in cancer is highest shortly after diagnosis. Suicide risk is higher among divorced people, rural dwellers, and individuals who live alone. Given that many risk factors are definitely present, Mr. A.'s suicide risk is nontrivial.

From Dr. B.'s Perspective, What Is Uncertain About Mr. A's Suicide Risk?

It is not known whether Mr. A. is experiencing significant symptoms of depression, anxiety, or thoughts of suicide. If these symptoms are present, it is not certain whether Mr. A. is interested in receiving specialty mental health treatment. Dr. B. does not know what Mr. A. knows of his cancer prognosis, nor is Dr. B. able to forecast how Mr. A. would react to bad news about his cancer in the event that the cancer progressed. Dr. B. is also uncertain whether Mr. A. feels socially isolated, the extent to which he derives emotional support from his daughters and friends, whether Mr. A. has made plans to take his own life, or whether he has access to lethal means (firearms, stockpiled medications).

Generating Hypotheses About Suicide Risk Based on Limited Data

If no more data are gathered, the available data could lead Dr. B. to generate hypotheses that are consistent with elevated suicide risk, beyond the certainties listed previously. First, Mr. A.'s wife had once registered concern about his moodiness. Second, Mr. A. has not always heeded Dr. B.'s counsel. A pattern of nonadherence might signal increased suicide risk. Third, Mr. A. once threatened suicide, albeit jokingly. Fourth, older adults who have been diagnosed with a grave illness are frequently accompanied by a friend or relative when visiting the PCP (Wolff & Roter, 2011), yet Mr. A. has always appeared at the office solo. Dr. B. might justifiably wonder whether Mr. A. fears becoming a burden to his daughters (Joiner, 2005; Van Orden et al., 2010; Wilson, Curran, & McPherson, 2005). Mr. A. has worked in professions (the military, auto repair) that place more of a premium on autonomous, team-oriented, proactive problem-solving than on relationships and dependency. He might view suicide as an instrumental, preemptive solution to a prospective problem.

Despite the known risk factors, the same data could lead Dr. B. to generate hypotheses that are consistent with minimal risk. Mr. A. remains physically active, has good relationships with his daughters, and enjoys spending time with his grandchildren, friends, and dog. He has not spontaneously expressed any active thoughts of suicide and does not appear visibly distressed or anxious. There is no known (documented) family history of suicide, and Mr. A. has never screened positive for suicide risk. Suicide is a rare event, and death by suicide is much less likely than death by natural causes, even among people who are believed to have multiple risk factors for suicide.

Disease-Centered (Biomedical) Approach

The disease-centered approach exploits the power asymmetries in the patient–PCP relationship by presuming the scientific validity of screening data and the utility of expert clinical judgment. Patients who score above prespecified thresholds are referred for interventions. Patients who score below thresholds but are nonetheless judged by a PCP to be at-risk can also be referred. In this case, Dr. B. uses his clinical judgment and refers Mr. A to the Behavioral Health Laboratory, and Mr. A. consents to seek treatment.

Advantages of the disease-centered (biomedical) approach. Uncertainty is a universally aversive experience. The use of screening instruments can minimize (if not fully resolve) many uncertainties. When uncertainties linger, PCPs could use their judgment and make a referral in the absence of a positive screen (as in this case). When a PCP makes a referral, few will second-guess the decision. Patients who are presumed to be in need of treatment will be referred for treatments that are reasonably effective. By reducing uncertainty, the disease-centered approach confers a potent short-term psychological advantage to patients and physicians alike (Duberstein & Wittink, 2015).

Disadvantages of the disease-centered (biomedical) approach. PCPs have expressed skepticism about the capacity of controlled trials to inform their decision-making because trialists often exclude the types of patients they see every day. Screening instruments are imperfect, yielding false positives, false negatives, and unintended consequences, and there is a dearth of clinical trials to support the net benefits of screening for depression (Thombs et al., 2012) or suicide. The U.S. Preventive Services Task Force currently gives an "I" recommendation (insufficient evidence to recommend for or against) for suicide risk screening (LeFevre & U.S. Preventive Services Task Force, 2014). False positives burden the health care system and drive up costs (Franks, Clancy, & Nutting, 1992; Palmer & Coyne, 2003; Woolf & Harris, 2012). The Blue Ribbon Work Group on Suicide Prevention in the Veteran Population (2008) highlighted the problem of false positives when it recommended that the VA re-evaluate its policy regarding mandated suicide screening. False negatives may be more harmful to patients. For example, it would not be surprising if Mr. A. experienced thoughts of suicide but, unsure of how Dr. B. would respond, refrained from endorsing these thoughts on a screening instrument. Mr. A. might fear being judged or being referred for hospitalization. Unproductive, adversarial exchanges between patients and PCPs occur in part because PCPs feel compelled to act, debate, or prescribe rather than listen and slow down (Back, Bauer-Wu, Rushton, & Halifax, 2009; Duberstein & Wittink, 2015; Vannoy et al., 2011).

Mr. A. agreed to initiate mental health treatment, but he may not follow up on the referral (Oslin et al., 2006; van Geffen et al., 2009; Wittink et al., 2005). Even if he does show up for the first appointment, high rates of treatment nonadherence beg for an explanation. Expressed desires for treatment are fallible and vulnerable to bias (Swindell, McGuire, & Halpern, 2010). Mr. A. may agree to initiate treatment mainly to please the PCP (acquiescence bias), happen to be in a particularly bad or good mood at the moment the decision is made (projection bias, Loewenstein, 2005), or only consider short-term gains but not longer term consequences and costs. When deciding about a treatment for a particular condition, few patients thoroughly consider the opportunity costs or the decision's impact on their other treatments and other activities. For example, Mr. A. may claim he desires psychotherapy but not realize that adding 2.5 hours in his schedule every week (with travel time) means that he needs to sacrifice 2.5 hours somewhere else (perhaps walking his dog). There are many other biases (Swindell et al., 2010), such as the availability heuristic, impact bias, and focalism, all of which may be exacerbated in distressed, at-risk individuals (Clark et al., 2011; Denburg et al., 2009; Hoerger, Chapman, Epstein, & Duberstein, 2012).

Person-Centered Approach

First and foremost, a person-centered approach to indicated or targeted suicide preventions requires actively engaging patients in their own care. This must start with an assessment of patients' prioritized needs and wants. Mr. A. may desire symptom improvement (reduction of fatigue), but he may also want assistance with everyday activities, family issues, or navigating the health care delivery system. All of these issues have significant implications for Mr. A's quality of life and outcomes. In this case, Mr. A. desires assistance with transportation to his cancer treatment, pet-sitting while he is receiving the treatments, pointers on how to talk to his daughters about his disease, and advice about how to talk with his oncologist about his prognosis. By hewing to a standard biomedical model, a PCP is unlikely to learn about all of these concerns.

In an effort to offer Mr. A *the care he wants and no more*, Dr. B. can ask Mr. A, "What do you want? What are your priorities?" Intuitively, one might think an assessment of patients' priorities ought to be straightforward. Although it is vital for PCPs to elicit patients' wants in a manner that enables them to prioritize one outcome over another (Fried, Tinetti, & Iannone, 2011; Fried et al., 2011; Reuben & Tinetti, 2012), it is often quite difficult for people, particularly distressed individuals, to prioritize their goals when trade-offs are involved. Research has shown that patients can prioritize their goals (Fried, Tinetti, Agostini, Iannone, & Towle, 2011) and can be trained to ask simple questions of their doctor (Clayton et al., 2007; Shepherd et al., 2011), but the cognitive demands required to make decisions involving multiple trade-offs typically exceed the capacity of the human brain (Gigerenzer, 2007) and may be particularly onerous for individuals experiencing symptoms of depression or anxiety. Although people are well aware that most treatments have "side effects," few anticipate the magnitude of the trade-offs involved. For example, Mr. A. might not anticipate that opting for psychotherapy would cut into the quality time he spends with his dog.

The decisions are not computationally intractable, however. A development in marketing research called adaptive conjoint analysis, a method that forces respondents to make trade-offs in options, has been used in online retail (e.g., on Edmunds.com) and is currently being tested to see if it improves decision-making in specialty oncology settings (Saigal et al., 2012) and communication about mental health needs in primary care (M.N. Wittink, PI, R34MH101236; clinical trials.gov# NCT02100982). Research has documented improvements in patients' experience of care following more modest communication interventions (Carter et al., 2013; Flückiger et al., 2012; Motto & Bostrom, 2001; Young, Bell, Epstein, Feldman, & Kravitz, 2008).

An assessment of Mr. A's priorities could be done online prior to his appointment with Dr. B., or, in the primary care waiting room, using computer kiosks, laptop computers, or smartphones. No matter where the assessment is completed, Mr. A. could receive output listing his priorities along with examples of specific questions he can ask Dr. B as well as resources he can access. By preparing in advance for his meeting with Dr. B., both parties will be able to use the brief time allotted in a manner that meets their mutual goals. One recent study, designed to study the effect of an interactive motivational computer program (IMCP) deployed in primary care waiting rooms reported surprisingly promising findings regarding suicide (Shah et al., 2014). The IMCP was designed to motivate patients with at least mild depression symptoms to discuss symptoms of depression with their PCP and encourage patients to be receptive to treatment offers (Kravitz et al., 2013). The IMCP provided text, audio, and video messages tailored to patient presentation (symptom level, visit agenda); causal explanations of depression; and views about mental health treatment. Despite little suicide-specific content, the IMCP led to increased clinician inquiry about suicidal thoughts without disturbing workflow. It required only five minutes (median) to complete (Shah et al., 2014).

Earlier we noted that a central feature of person-centered primary care might be a "superstore" that offers a variety of services, including pharmacy, psychotherapy, financial counseling as well as social, behavioral health, recreational, legal, and employment services. Given his concerns, Mr. A. might benefit from referral to a behavioral health specialist with expertise in health communication.

Advantages of a person-centered approach. The empowerment of patients minimizes the adverse effects of supply-side forces and power asymmetries. It can potentially improve outcomes at a fraction of the cost of disease-centered care (Bertakis & Azari, 2011; Berwick & Hackbarth, 2012; Stacey et al., 2011; Wennberg et al., 2010). A person-centered approach can also minimize the influence of undesirable PCP traits and elicit more desirable behavior from the PCP. This will enhance the quality of the patient–PCP relationship while increasing patient autonomy, competence, and access to services (e.g., psychosocial, behavioral, legal, financial, etc.), all of which could confer positive outcomes, including decreased suicide risk.

Disadvantages of a person-centered approach. The person-centered approach is largely unfamiliar to patients, PCPs, and other stakeholders. Moreover, it encourages more conversational options and a greater array of interventions: the more options, the greater the uncertainty. Given that uncertainty is often experienced as aversive, it is not surprising that physicians are trained to root it out (Fox, 1980), and patients often expect nothing less.

Many patients may not want to take a more active role in their care. This may be especially true of patients who prefer to cede decision-making authority to the PCP in order to reduce their own uncertainty. Patients who might otherwise benefit from standard specialty mental health treatments might not be referred for it. The optimal method for eliciting patients' priorities is unknown; this is a topic of active investigation.

Because a person-centered approach is quite new, it faces numerous pragmatic and policymaking obstacles. For better and worse, the disease-centered approach has been codified in laws, regulations, medical education, quality assurance evaluations, and credentialing practices (e.g., mandated screening for depression or suicide ideation in some practice settings). Consequently, a PCP who elicits patients' goals and provides person-centered care may perform more poorly on standard quality metrics when the patients' goals are not aligned with those of the guidelines (Reuben & Tinetti, 2012).

A person-centered approach recognizes that primary care must expand its reach beyond standard biomedical interventions to mitigate the health-damaging effects of adverse living circumstances. Unfortunately, the infrastructure (workforce, institutions, payment systems, links to community, social, financial, and legal services) required to accomplish this goal is not yet established. The advent of the patient-centered medical home is encouraging, however (Nutting et al., 2011), as it suggests that stakeholders are receptive to making significant changes in health services and everyday practice.

CONCLUSION

By calling for a person-centered approach, we do not mean to imply that disease-centered approaches ought to be abandoned (Duberstein & Wittink, 2015). PCPs see patients who are afflicted with many diseases that confer suicide risk, including schizophrenia, cancer, heart disease, lung disease, multiple sclerosis, and Huntington's Disease—to name just a few. Incurable diseases should be managed. Curable diseases in an otherwise healthy person should be treated. Disease treatment and management requires clinical microsystems that are equipped to assess and cater to the prioritized needs and wants of all patients.

Discerning who needs and wants intervention for suicide risk, and what type, is more difficult than televised ads for antidepressants might lead one to presume. When greeting the PCP, few patients will announce, "Doc, I'm suicidal; I have major depression, and am ready to begin that antidepressant I saw advertised on TV. I would also like to start psychotherapy." Ethical considerations compel approaches that meet the needs of all, not just those who happen to appear in primary care offices requesting an antidepressant, readily acknowledging his or her thoughts of suicide. Primary care practices must be responsive to the needs and wants of these patients—and to the needs and wants of everyone else (Hart, 1971; Watt, 2002). Yet these needs and wants are rarely elicited. Power asymmetries, supply-side economic factors, and the disease-centered approach—all bedfellows— mitigate some of the uncertainties of patienthood and doctoring (Fox, 1980). Unfortunately, they have compromised patient outcomes by undermining patient engagement, particularly with respect to communication and decision-making. Patients at risk for suicide rarely disclose their intent. When they do, the ensuing discussions are not as helpful as they could be.

Privileging the patients' needs and wants requires greater attentiveness to communication and decision-making processes. PCPs must think not only about lab values, diagnostic tests, and symptoms but also about the psychosocial, legal, and economic circumstances that are known to influence suicide and other patient outcomes. Patients must think about their priorities, but this is easier said than done. Numerous biases are now known to undermine prioritization and health decision-making more broadly. Technology has been used to deploy screening instruments in primary care, and person-centered care can similarly take advantage of exciting technological innovations to elicit primary care patients' prioritized needs and wants. No less an authority than William Osler said, "It is more important to know what type of person has the disease than to know that type of disease the person has." What better way to learn about "what type of person has the disease" than to know what his or her priorities in life are?

Imploring PCPs to elicit patients' priorities is insufficient, however. Theoretically informed innovations are needed to overcome sociohistorically conditioned power asymmetries, market forces, and psychological reinforcers that lead PCPs (and patients) to prioritize one biomedical topic over another and to privilege biomedical over psychosocial topics. Without theory, misalignments and ad hoc decision-making ensue.

Drawing from systems theory and self-determination theory, we offer the following recommendations to increase the capacity of PCPs to engage patients in discussions about suicide and the often stigmatized adverse circumstances that confer risk.: (a) patients must be empowered to express all of their health-related priorities, not just those deemed "biomedical"; (b) PCPs must be equipped to respond effectively to these priorities; and (c) workplace and workforce innovations are needed to ensure responsiveness to patients' expressed priorities. These innovations should accommodate the needs of all stakeholders (patients, PCPs, other health care personnel) for autonomy, competence, and relatedness.

We began this chapter by noting that PCPs bear some responsibility for suicide prevention. That has been the received wisdom, but it is insufficiently nuanced. Comprehensivist PCPs help patients manage a triple burden: the burden of disease, the burden of treatment decision-making, and the burden of the treatments themselves. As such, they must frequently work with patients to trade-off one outcome against others. Universal initiatives designed to increase the number of PCPs must therefore be accompanied by meso-level and microsystem improvements in the way

PCPs and patients communicate and make decisions about indicated and targeted interventions. These improvements are expected to mitigate suicide risk and improve other patient outcomes as well as PCP job satisfaction.

FUTURE DIRECTIONS

1. *Unintended consequences.* Systems theory suggests that any perturbation of a system could have both intended and unintended consequences. Unintended consequences cannot be predicted with certainty, but they can be anticipated. Unintended consequences of routine or mandated screening for depression and/or suicide risk include undermining clinician autonomy and alienating some patients. Having ancillary personnel, rather than PCPs, administer screening instruments could deprive clinicians of an opportunity to leverage the therapeutic relationship and provide care that is responsive to patients' circumstances.
2. *Therapeutic alliance.* A meta-analysis of the psychotherapy literature showed that the patient–clinician alliance is associated with outcomes, independent of specific intervention techniques (Flückiger et al., 2012). Outside the mental health literature, the importance of the patient–clinician alliance is increasingly recognized. For example, a meta-analysis of intervention studies designed to modify patient–clinician interventions (Kelley et al., 2014) showed that patient–clinicians interactions have implications both for objective outcomes (e.g., blood pressure) as well as subjective outcomes (e.g., pain, quality of life). More research on the patient–clinician alliance in primary and specialty medical (nonpsychiatric) settings is needed. A report in the oncology literature suggests that the patient–clinician bond might mitigate suicide risk (Trevino et al., 2014), and it would be interesting to see if similar effects are observed in primary care.
3. *Patients' needs and wants.* A mantra of the patient empowerment movement is that patients should get the "care they need and no less, the care they want and no more." Most biomedically oriented suicide prevention initiatives presume that at-risk patients need to receive treatment for mental disorders, but is that their most pressing need? For example, patients with financial strains are likely to prioritize a job search over a counseling appointment, and those caring for children, grandchildren, or an ill relative might prioritize caregiving activities over their own self-care. And what do at-risk patients want? By focusing more on suffering persons and less on diagnosing and treating disease, this chapter has attempted to call attention to a *phenomenological imperative*: the need for suicide prevention initiatives to be aligned more with the public's native understanding of suicide (Owens et al., 2009; Owens et al., 2011) and with the phenomenological experiences of our patients (Epstein et al., 2010). Pioneering suicidologist Edwin Shneidman (1992) put it this way: "A focus on mental illness is often misleading. Physicians and other health professionals need the courage and wisdom to work on a person's suffering at the phenomenological level" (p. 890). In this regard, we need to be more mindful of the sources of suffering and stress in people's lives. Researchers have identified numerous life circumstances that confer suicide risk, including family discord, financial difficulties, job loss, legal imbroglios, social isolation, concerns about sexual identity, and many others. Research is needed to identify the life circumstances individuals want help managing and to explore whether offering such help can reduce suicide risk while minimizing unintended consequences.
4. *Attempted versus completed suicide.* Most clinicians have had far more experience with patients who have attempted suicide than those who have gone on to take their lives, and there is more clinical research on attempted suicide that research on suicide mortality. Two bodies of research cast doubt on hypotheses generated about approaches to suicide prevention based on clinical experiences and research focused on individuals who engage in nonfatal self-harm. First, epidemiologic and sociological studies have long suggested that suicide attempters and people who die by suicide are drawn from two distinct, albeit overlapping, populations of at-risk individuals. For example, rates of attempted suicide are highest in young women; in most

countries worldwide suicide rates are highest in older men. People who have never made a prior suicide attempt account for the majority of suicides in many demographic groups. Men are nearly 70% more likely than women to have their first attempt be their last. Second, more recently, clinical research has shown that suicide attempters and people who die by suicide share some clinical features but are also quite different (Giner et al., 2013; Hirvikoski & Jokinen, 2012; Innamoratie et al., 2008; Overholser, Stockmeier, Dilley, & Freiheit, 2002; Tsoh et al., 2005; Useda et al., 2007; Younes et al., 2015). Patients who ultimately die by suicide may interact with primary care offices less frequently than those who engage in nonfatal self-harm. Moreover, when they do present, they may appear somewhat healthier psychologically than patients who engage in nonfatal self-harm (Useda et al., 2007; Younes et al., 2015). We are aware of no prior reviews of clinical research comparing people who engage in nonfatal self-harm and those who die by suicide. Such a review is long overdue, and more research is needed on the differences between people who die by suicide and those who survive suicide attempts. The implications of these differences for clinical and public health interventions warrant consideration.

5. *Gender differences*. Prior biomedically oriented primary care–based suicide prevention initiatives have been shown to decrease suicide mortality in women but not men (e.g., Rutz, 2001), perhaps because these interventions have not adequately recognized that men and women may benefit from different approaches to intervention. Gender-based identities and cultural scripts (Canetto & Leter, 1998) of suicidal behavior and help-seeking can influence how men and women interact with PCPs, regulate and express emotions, and respond to primary care–based interventions. There is a pressing need to develop primary care–based interventions that are effective for both women and men. Ideally, these interventions would be informed by research on the phenomenological experience of suicide desire in men and women as well as gender differences in the reporting of suicide ideation and intent in primary care settings (Shand et al., 2015; Vannoy & Robins, 2011).

6. *Workforce considerations*. Discussions about suicide prevention in primary care have largely ignored the size and configuration of the primary care workforce. We have suggested that policy initiatives designed to increase the number of PCPs could affect the suicide rate. There is cause for optimism, as other policy changes (e.g., detoxification of domestic gas; modification of over-the-counter medication packaging) have been associated with reductions in suicide rates. Beyond increasing the number of PCPs, policies and regulations designed to increase the number of psychologists working in primary care ought to be considered, along with initiatives to increase their scope of practice (e.g., medication prescribing).

ACKNOWLEDGMENTS

Work on this chapter was supported by the University of Rochester Department of Psychiatry's Hendershott Research Fund (Drs. Duberstein and Wittink) and by the VA Center of Excellence for Suicide Prevention (Drs. Pigeon and Wittink). The contents of this chapter do not represent the views of the Department of Veterans Affairs or the U.S. government.

NOTES

1. For our purposes, we define primary care providers as physicians who have been certified and credentialed in family medicine and general internal medicine. Others have defined primary care in terms of the amount of time devoted to cognitive activities (evaluation and management) as opposed to procedures. For example, the PPACA mandates bonuses for primary care providers, defined as those physicians who derive 60% or more of their revenue from cognitive activities. One analyst concluded that neurology and psychiatry would both meet this criterion (Sigsbee, 2011). It must also be recognized that primary care providers include nurse practitioners as well as physicians who have specialized in pediatrics/adolescent medicine. Moreover, many women receive primary care from their obstetrician/gynecologist, and patients with heart disease or cancer may view their cardiologist or oncologist as their primary care provider.

2. Despite the cachet associated with the term "patient-centered," we use the term "person-centered" for three reasons. First, in order to be effective, suicide

prevention initiatives must reach well beyond those who self-define as "a patient" and show up in medical clinics or facilities for intervention. Second, many patients are accompanied by caregivers when visiting the PCP, and PCPs interactions with these third parties can influence patient outcomes. Third, both systems theory and self-determination theory suggest that interventions targeting patients in health care settings will not reach their full potential unless they account for the needs of nonpatients, namely, health care providers and other personnel in the health care system. In other words, patients will receive person-centered care only insofar as the needs of the persons involved in care provision (e.g., PCPs, administrative personnel) are accommodated.

REFERENCES

Ackerman, S. J., & Hilsenroth, M. J. (2003). A review of therapist characteristics and techniques positively impacting the therapeutic alliance. *Clinical Psychology Review, 23*(1), 1–33.

Ahmedani, B. K., Simon, G. E., Stewart, C., Beck, A., Waitzfelder, B. E., Rossom, R., ... Solberg, L. I. (2014). Health care contacts in the year before suicide death. *Journal of General Internal Medicine, 29*(6), 870–877. doi:10.1007/s11606-014-2767-3

Alston, C., Paget, L., Halvorson, G., Novelli, B., Guest, J., McCabe, P., ... Von Kohorn, I. (2012). *Communicating with patients on health care evidence*. (Discussion Paper). Washington, DC.: Institute of Medicine.

Angyal, A. (1982). *Neurosis and treatment: A holistic theory*. Cambridge, MA: Da Capo Press.

Back, A. L., Bauer-Wu, S. M., Rushton, C. H., & Halifax, J. (2009). Compassionate silence in the patient-clinician encounter: A contemplative approach. *Journal of Palliative Medicine, 12*(12), 1113–1117. doi:10.1089/jpm.2009.0175

Baik, S. Y., Gonzales, J. J., Bowers, B. J., Anthony, J. S., Tidjani, B., & Susman, J. L. (2010). Reinvention of depression instruments by primary care clinicians. *Annals of Family Medicine, 8*(3), 224–230. doi:10.1370/afm.1113

Bailey, M. J., & Goodman-Bacon, A. (2015). The War on Poverty's experiment in public medicine: Community health centers and the mortality of older Americans. *The American Economic Review, 105*(3), 1067–1104. doi:10.1257/aer.20120070

Bakan, D. (1966). *The duality of human existence: Isolation and communion in Western man*. Boston: Beacon Press.

Bartels, S. J., Coakley, E. H., Zubritsky, C., Ware, J. H., Miles, K. M., Areán, P. A., ... Costantino, G. (2004). Improving access to geriatric mental health services: A randomized trial comparing treatment engagement with integrated versus enhanced referral care for depression, anxiety, and at-risk alcohol use. *The American Journal of Psychiatry, 161*(8), 1455–1462.

Bertakis, K. D., & Azari, R. (2011). Patient-centered care is associated with decreased health care utilization. *Journal of the American Board of Family Medicine, 24*(3), 229–239. doi:10.3122/jabfm.2011.03.100170

Berwick, D. M., & Hackbarth, A. D. (2012). Eliminating waste in US health care. *JAMA: Journal of the American Medical Association, 307*(14), 1513–1516. doi:10.1001/jama.2012.362

Blatt, S. J., & Shichman, S. (1983). Two primary configurations of psychopathology. *Psychoanalysis & Contemporary Thought, 6*(2), 187–254.

Blue Ribbon Work Group on Suicide Prevention in the Veteran Population. (2008). *Report of the Blue Ribbon Work Group on Suicide Prevention in the Veteran Population*. Retrieved from http://www.mentalhealth.va.gov/suicide_prevention/Blue_Ribbon_Report-FINAL_June-30-08.pdf

Bodenheimer, T., & Sinsky, C. (2014). From triple to quadruple aim: Care of the patient requires care of the provider. *Annals of Family Medicine, 12*(6), 573–576. doi:10.1370/afm.1713

Boukus, E. R., Cassil, A. C., & O'Malley, A. S. (2009). *2008 Health Tracking Physician Survey*. (No. 35). Washington, DC: Center for Studying Health System Change.

Boyd, C. M., Darer, J., Boult, C., Fried, L. P., Boult, L., & Wu, A. W. (2005). Clinical practice guidelines and quality of care for older patients with multiple comorbid diseases: Implications for pay for performance. *JAMA: Journal of the American Medical Association, 294*(6), 716–724. doi:10.1001/jama.294.6.716

Brezo, J., Paris, J., & Turecki, G. (2006). Personality traits as correlates of suicidal ideation, suicide attempts, and suicide completions: A systematic review. *Acta Psychiatrica Scandinavica, 113*(3), 180–206. doi:10.1111/j.1600-0447.2005.00702.x

Bruce, M. L., Ten Have, T. R., Reynolds, C. F. I. I. I., Katz, I. I., Schulberg, H. C., Mulsant, B. H., ... Alexopoulos, G. S. (2004). Reducing suicidal ideation and depressive symptoms in depressed older primary care patients: A randomized controlled trial. *JAMA: Journal of the American Medical Association, 291*(9), 1081–1091. doi:http://dx.doi.org/10.1001/jama.291.9.1081

Buber, M. (1970). *I and thou*. Philadelphia: Scribner's.

Canetto, S. S., & Lester, D. (1998). Gender, culture, and suicidal behavior. *Transcultural Psychiatry, 35*(2), 163–190.

Carter, G. L., Clover, K., Whyte, I. M., Dawson, A. H., & D'Este, C. (2013). Postcards from the EDge: 5-year outcomes of a randomised controlled trial for hospital-treated self-poisoning. *The British Journal of Psychiatry*, 202(5), 372–380. doi:10.1192/bjp.bp.112.112664

Chapman, B. P., Roberts, B., & Duberstein, P. (2011). Personality and longevity: Knowns, unknowns, and implications for public health and personalized medicine. *Journal of Aging Research*. Retrieved from http://www.hindawi.com/journals/jar/2011/759170. doi:10.4061/2011/759170

Clancy, C. M. (2011). Researching and regulating for patient-centeredness. Retrieved from https://www.ecri.org/Conferences/Pages/Annual_Conference_2011_Agenda.aspx

Clancy, C. M., & Tornberg, D. N. (2007). TeamSTEPPS: Assuring optimal teamwork in clinical settings. *American Journal of Medical Quality*, 22(3), 214–217. doi:10.1177/1062860607300616

Clark, L., Dombrovski, A. Y., Siegle, G. J., Butters, M. A., Shollenberger, C. L., Sahakian, B. J., & Szanto, K. (2011). Impairment in risk-sensitive decision-making in older suicide attempters with depression. *Psychology and Aging*, 26(2), 321–330. doi:10.1037/a0021646

Clayton, J. M., Butow, P. N., Tattersall, M. H. N., Devine, R. J., Simpson, J. M., Aggarwal, G., . . . Noel, M. A. (2007). Randomized controlled trial of a prompt list to help advanced cancer patients and their caregivers to ask questions about prognosis and end-of-life care. *Journal of Clinical Oncology*, 25(6), 715–723. doi: 10.1200/JCO.2006.06.7827

Clever, S., Ford, D., Rubenstein, L., Rost, K., Meredith, L., Sherbourne, C., . . . Cooper, L. (2006). Primary care patients' involvement in decision-making is associated with improvement in depression. *Medical Care*, 44(5), 398–405. doi:10.1097/01.mlr.0000208117.15531.da

Cole, S. W., Arevalo, J. M., Takahashi, R., Sloan, E. K., Lutgendorf, S. K., Sood, A. K., . . . Seeman, T. E. (2010). Computational identification of gene-social environment interaction at the human IL6 locus. *Proceedings of the National Academy of Sciences of the United States of America*, 107(12), 5681–5686. doi:10.1073/pnas.0911515107

Conner, K. R., Cox, C., Duberstein, P. R., Tian, L., Nisbet, P. A., & Conwell, Y. (2001). Violence, alcohol, and completed suicide: A case-control study. *The American Journal of Psychiatry*, 158(10), 1701–1705. doi:http://dx.doi.org/10.1176/appi.ajp.158.10.1701

Conner, K. R., Duberstein, P. R., & Conwell, Y. (2000). Domestic violence, separation and suicide in young men with early onset alcoholism: Reanalyses of Murphy's data. *Suicide and Life-Threatening Behavior*, 30(4), 354–359.

Conner, K. R., Duberstein, P. R., Conwell, Y., Seidlitz, L., & Caine, E. D. (2001). Psychological vulnerability to completed suicide: A review of empirical studies. *Suicide and Life-Threatening Behavior*, 31(4), 367–385. doi:http://dx.doi.org/10.1521/suli.31.4.367.22048

Conner, K. R., Houston, R., Swogger, M., Conwell, Y., You, S., Gamble, S., . . . Duberstein, P. R. (2012). Stressful life events and suicidal behavior: Role of event type, severity, and timing. *Drug and Alcohol Dependence*, 120, 155–161.

Coulter, A., & Collins, A. (2011). *Making shared decision-making a reality: No decision about me, without me.* London: The King's Fund.

Courtet, P., Gottesman, I. I., Jollant, F., & Gould, T. D. (2011). The neuroscience of suicidal behaviors: What can we expect from endophenotype strategies? *Translational Psychiatry*, 1, 1–7. doi:10.1038/tp.2011.6

Cutler, D. M., & Lleras-Muney, A. (2010). Understanding differences in health behaviors by education. *Journal of Health Economics*, 29(1), 1–28. doi:10.1016/j.jhealeco.2009.10.003

Denburg, N., Weller, J., Yamada, T., Shivapour, D., Kaup, A., LaLoggia, A., . . . Bechara, A. (2009). Poor decision making among older adults is related to elevated levels of neuroticism. *Annals of Behavioral Medicine*, 37(2), 164–172.

Denneson, L. M., Basham, C., Dickinson, K. C., Crutchfield, M. C., Millet, L., Shen, X., & Dobscha, S. K. (2010). Suicide risk assessment and content of VA health care contacts before suicide completion by veterans in Oregon. *Psychiatric Services*, 61(12), 1192–1197. doi:10.1176/appi.ps.61.12.1192

Dowrick, C., Leydon, G. M., McBride, A., Howe, A., Burgess, H., Clarke, P., . . . Kendrick, T. (2009). Patients' and doctors' views on depression severity questionnaires incentivised in UK quality and outcomes framework: Qualitative study. *BMJ*, 338, 1–6. doi: 10.1136/bmj.b663

Drake, R. E., Cimpean, D., & Torrey, W. C. (2009). Shared decision making in mental health: Prospects for personalized medicine. *Dialogues in Clinical Neuroscience*, 11(4), 455–463.

Duberstein, P. R., Chapman, B. P., Epstein, R. M., McCollumn, K. R., & Kravitz, R. L. (2008). Physician personality characteristics and inquiry about mood symptoms in primary care. *Journal of General Internal Medicine*, 23(11), 1791–1795. doi:http://dx.doi.org/10.1007/s11606-008-0780-0

Duberstein, P. R., Conwell, Y., Conner, K. R., Eberly, S., & Caine, E. D. (2004). Suicide at 50 years of age and older: Perceived physical illness, family discord and financial strain. *Psychological Medicine, 34*(1), 137–146. doi:http://dx.doi.org/10.1017/S0033291703008584

Duberstein, P. R., Conwell, Y., Conner, K. R., Eberly, S., Evinger, J. S., & Caine, E. D. (2004). Poor social integration and suicide: Fact or artifact? A case-control study. *Psychological Medicine, 34*(7), 1331–1337. doi:http://dx.doi.org/10.1017/S0033291704002600

Duberstein, P. R., & Conwell, Y. (1997). Personality disorders and completed suicide: A methodological and conceptual review. *Clinical Psychology: Science and Practice, 4*(4), 359–376.

Duberstein, P.R., Conwell, Y., & Cox, C. (1998). Suicide in widowed persons: A psychological autopsy comparison of recently and remotely bereaved older subjects. *American Journal of Geriatric Psychiatry, 6,* 328–334.

Duberstein, P. R., & Heisel, M. J. (2014). Person-centered prevention of suicide among older adults. In M. Nock (Ed.), *Oxford handbook of suicide and self-injury* (pp. 113–132). New York: Oxford University Press.

Duberstein, P., Meldrum, S., Fiscella, K., Shields, C. G., & Epstein, R. M. (2007). Influences on patients' ratings of physicians: Physicians' demographics and personality. *Patient Education and Counseling, 65*(2), 270–274. doi:http://dx.doi.org/10.1016/j.pec.2006.09.007

Duberstein, P., & Witte, T. K. (2009). Suicide risk in personality disorders: An argument for a public health perspective. In P. M. Kleespies (Ed.), *Behavioral emergencies: An evidence-based resource for evaluating and managing risk of suicide, violence, and victimization* (pp. 257–286). Washington, DC: American Psychological Association. doi:http://dx.doi.org/10.1037/11865-012

Duberstein, P. R., & Wittink, M. (2015). Person-centered suicide prevention for older adults seen in primary care settings. In B. Bensadon (Ed.), *Psychology and geriatrics: Integrated care for an aging nation* (pp. 153–181). San Diego: Elsevier.

Emanuel, E. J., & Pearson, S. D. (2012). Physician autonomy and health care reform. *JAMA: Journal of the American Medical Association, 307*(4), 367–368. doi:10.1001/jama.2012.19; 10.1001/jama.2012.19

Engel, G. L. (1980). The clinical application of the biopsychosocial model. *The American Journal of Psychiatry, 137*(5), 535–544.

Epping-Jordan, J., Pruitt, S., Bengoa, R., & Wagner, E. (2004). Improving the quality of health care for chronic conditions. *Quality and Safety in Health Care, 13*(4), 299–305.

Epstein, R. M., Duberstein, P. R., Feldman, M. D., Rochlen, A. B., Bell, R. A., Kravitz, R. L., ... Paterniti, D. A. (2010). "I didn't know what was wrong:" How people with undiagnosed depression recognize, name and explain their distress. *Journal of General Internal Medicine, 25*(9), 954–961. doi:10.1007/s11606-010-1367-0

Epstein, R. M., Franks, P., Fiscella, K., Shields, G. C., Meldrum, S. C., Kravitz, R. L., & Duberstein, P. R. (2005). Measuring patient-centered communication in patient-physician consultations: Theoretical and practical issues. *Social Science & Medicine, 61*(7), 1516–1528. doi:http://dx.doi.org/10.1016/j.socscimed.2005.02.001

Epstein, R. M., & Street, R. L. Jr. (2007). *Patient-centered communication in cancer care: Promoting healing and reducing suffering.* Bethesda, MD: National Cancer Institute.

Fassberg, M. M., van Orden, K. A., Duberstein, P., Erlangsen, A., Lapierre, S., Bodner, E., ... Waern, M. (2012). A systematic review of social factors and suicidal behavior in older adulthood. *International Journal of Environmental Research and Public Health, 9*(3), 722–745. doi:10.3390/ijerph9030722.

Feldman, M. D., Franks, P., Duberstein, P. R., Vannoy, S., Epstein, R., & Kravitz, R. L. (2007). Let's not talk about it: Suicide inquiry in primary care. *Annals of Family Medicine, 5*(5), 412–418. doi:http://dx.doi.org/10.1370/afm.719

Flückiger, C., Del Re, A. C., Wampold, B. E., Symonds, D., & Horvath, A. O. (2012). How central is the alliance in psychotherapy? A multilevel longitudinal meta-analysis. *Journal of Counseling Psychology, 59*(1), 10–17. doi:10.1037/a0025749

Flückiger, C., Del Re, A. C., Wampold, B. E., Znoj, H., Caspar, F., & Joerg, U. (2012). Valuing clients' perspective and the effects on the therapeutic alliance: A randomized controlled study of an adjunctive instruction. *Journal of Counseling Psychology, 59*(1), 18–26. doi:10.1037/a0023648

Fox, R. (1980). The evolution of medical uncertainty. *Milbank Memorial Fund Quarterly-Health and Society, 58*(1), 1–49. doi:10.2307/3349705

Franks, P., Clancy, C. M., & Nutting, P. A. (1992). Gatekeeping revisited—protecting patients from overtreatment. *The New England Journal of Medicine, 327*(6), 424–429. doi:10.1056/NEJM199208063270613

Franx, G., Oud, M., de Lange, J., Wensing, M., & Grol, R. (2012). Implementing a stepped-care approach in primary care: Results of a qualitative study. *Implementation Science, 7,* 1–13. doi:10.1186/1748-5908-7-8

Fried, T. R., Tinetti, M. E., & Iannone, L. (2011). Primary care clinicians' experiences with treatment decision making for older persons with multiple conditions. *Archives of Internal Medicine, 171*(1), 75–80. doi:10.1001/archinternmed.2010.318

Fried, T. R., McGraw, S., Agostini, J. V., & Tinetti, M. E. (2008). Views of older persons with multiple morbidities on competing outcomes and clinical decision-making. *Journal of the American Geriatrics Society*, 56(10), 1791–1990. doi:10.1111/j.1532-5415.2008.01923.x

Fried, T. R., Tinetti, M., Agostini, J., Iannone, L., & Towle, V. (2011). Health outcome prioritization to elicit preferences of older persons with multiple health conditions. *Patient Education and Counseling*, 83(2), 278–282. doi:10.1016/j.pec.2010.04.032

Frosch, D. L., May, S. G., Rendle, K. A. S., Tietbohl, C., & Elwyn, G. (2012). Authoritarian physicians and patients' fear of being labeled "difficult" among key obstacles to shared decision making. *Health Affairs*, 31(5), 1030–1038. doi:10.1377/hlthaff.2011.0576

Gabbard, G. O., & Freedman, R. (2006). Psychotherapy in the journal: What's missing? *The American Journal of Psychiatry*, 163(2), 182–184. doi:10.1176/appi.ajp.163.2.182

Ganzini, L., Denneson, L. M., Press, N., Bair, M. J., Helmer, D. A., Poat, J., & Dobscha, S. K. (2013). Trust is the basis for effective suicide risk screening and assessment in veterans. *Journal of General Internal Medicine*, 28(9), 1215–1221. doi:10.1007/s11606-013-2412-6; 10.1007/s11606-013-2412-6

Gigerenzer, G. (2007). *Gut feelings: The intelligence of the unconscious*. New York: Penguin.

Gilbody, S., Sheldon, T., & Wessely, S. (2006). Should we screen for depression? *BMJ*, 332(7548), 1027–1030. doi:10.1136/bmj.332.7548.1027

Giner, L., Blasco-Fontecilla, H., Perez-Rodriguez, M. M., Garcia-Nieto, R., Giner, J., Guija, J. A., ... Baca-Garcia, E. (2013). Personality disorders and health problems distinguish suicide attempters from completers in a direct comparison. *Journal of Affective Disorders*, 151(2), 474–483.

Graham, R. G., & Martin, G. I. (2012). Health behavior: A Darwinian reconceptualization. *American Journal of Preventive Medicine*, 43(4), 451–455. doi: 10.1016/j.amepre.2012.06.016

Gu, Q., Dillon, C. F., & Burt, V. L. (2010). *Prescription drug use continues to increase: U.S. prescription drug data for 2007–2008*. (No. 42). Hyattsville, MD: Centers for Disease Control and Prevention.

Gunnell, D., Metcalfe, C., While, D., Hawton, K., Ho, D., Appleby, L., & Kapur, N. (2012). Impact of national policy initiatives on fatal and non-fatal self-harm after psychiatric hospital discharge: Time series analysis. *The British Journal of Psychiatry*, 201(3), 233–238. doi:10.1192/bjp.bp.111.104422

Gutmann, A., & Thompson, D. (2012). *The spirit of compromise: Why governing demands it and campaigning undermines it*. Princeton, NJ: Princeton University Press.

Hart, J. T. (1971). Inverse care law. *Lancet*, 1(7696), 405–412.

Hart, J. T., Thomas, C., Gibbons, B., Edwards, C., Hart, M., Jones, J., ... Walton, P. (1991). 25 years of case finding and audit in a socially deprived community. *BMJ*, 302(6791), 1509–1513.

Hawton, K. (2002). United Kingdom legislation on pack sizes of analgesics: Background, rationale, and effects on suicide and deliberate self-harm. *Suicide and Life-Threatening Behavior*, 32(3), 223–229.

Hawton, K., Bergen, H., Simkin, S., Brock, A., Griffiths, C., Romeri, E., ... Gunnell, D. (2009). Effect of withdrawal of co-proxamol on prescribing and deaths from drug poisoning in England and Wales: Time series analysis. *BMJ*, 338, b2270. doi:10.1136/bmj.b2270

Hilsenroth, M. J., Cromer, T. D., & Ackerman, S. J. (2012). How to make practical use of therapeutic alliance research in your clinical work. In R. A. Levy (Ed.), *Psychodynamic psychotherapy research: Evidence-based practice and practice-based evidence* (pp. 361–380). London: Springer Science+Business Media.

Hirvikoski, T., & Jokinen, J. (2012). Personality traits in attempted and completed suicide. *European Psychiatry*, 27(7), 536–541.

Hoerger, M., Chapman, B. P., Epstein, R. M., & Duberstein, P. R. (2012). Emotional intelligence: A theoretical framework for individual differences in affective forecasting. *Emotion*, 12, 716–725.

Holt-Lunstad, J., Smith, T. B., & Layton, J. B. (2010). Social relationships and mortality risk: A meta-analytic review. *PLoS Medicine*, 7(7), 1–20. doi:10.1371/journal.pmed.1000316

Hunt, L. M., Kreiner, M., & Brody, H. (2012). The changing face of chronic illness management in primary care: A qualitative study of underlying influences and unintended outcomes. *Annals of Family Medicine*, 10(5), 452–460. doi:10.1370/afm.1380

Ilgen, M. A., Bohnert, A. S. B., Ignacio, R. V., McCarthy, J. F., Valenstein, M. M., Kim, H. M., & Blow, F. C. (2010). Psychiatric diagnoses and risk of suicide in veterans. *Archives of General Psychiatry*, 67(11), 1152–1158.

Innamorati, M., Pompili, M., Masotti, V., Persone, F., Lester, D., Tatarelli, R., ... Amore, M. (2008). Completed versus attempted suicide in psychiatric patients: A psychological autopsy study. *Journal of Psychiatric Practice*, 14(4), 216–224. doi:10.1097/01.pra.0000327311.04153.01 [doi]

Institute of Medicine Committee on Quality of Health Care in America. (2001). *Crossing the quality chasm: A new health system for the 21st century.* Washington, DC: National Academy Press.

Iyengar, S., & Lepper, M. (2000). When choice is demotivating: Can one desire too much of a good thing? *Journal of Personality and Social Psychology, 79*(6), 995–1006. doi:10.1037//0022-3514.79.6.995

Joiner, T. (2005). *Why people die by suicide.* Cambridge, MA: Harvard University Press.

Kales, H. C., Kim, H. M., Austin, K. L., & Valenstein, M. (2010). Who receives outpatient monitoring during high-risk depression treatment periods? *Journal of the American Geriatrics Society, 58*(5), 908–913. doi:10.1111/j.1532-5415.2010.02810.x

Kaplan, R. M. (2000). Two pathways to prevention. *The American Psychologist, 55*(4), 382–396.

Kelley, J. M., Kraft-Todd, G., Schapira, L., Kossowsky, J., & Riess, H. (2014). The influence of the patient-clinician relationship on healthcare outcomes: A systematic review and meta-analysis of randomized controlled trials. *PloS One, 9*(4), e94207. doi:10.1371/journal.pone.0094207 [doi]

Kizer, K. W. (2012a). Observations about achieving "Systemness" in healthcare. Retrieved from https://www.ecri.org/Conferences/Pages/Annual_Conference_2012_Speakers.aspx

Kizer, K. W. (2012b). Veterans and the Affordable Care Act. *JAMA: The Journal of the American Medical Association, 307*(8), 789–790.

Kravitz, R. L., Franks, P., Feldman, M., Meredith, L. S., Hinton, L., Franz, C., ... Epstein, R. M. (2006). What drives referral from primary care physicians to mental health specialists? A randomized trial using actors portraying depressive symptoms. *Journal of General Internal Medicine, 21*(6), 584–589. doi:http://dx.doi.org/10.1111/j.1525-1497.2006.00411.x

Kravitz, R. L., Franks, P., Feldman, M. D., Tancredi, D. J., Slee, C. A., Epstein, R. M., ... Jerant, A. (2013). Patient engagement programs for recognition and initial treatment of depression in primary care: A randomized trial. *JAMA: Journal of the American Medical Association, 310*(17), 1818–1828. doi:10.1001/jama.2013.280038; 10.1001/jama.2013.280038

Lampe, L., Fritz, K., Boyce, P., Starcevic, V., Brakoulias, V., Walter, G., ... Malhi, G. (2013). Psychiatrists and GPs: Diagnostic decision making, personality profiles and attitudes toward depression and anxiety. *Australasian Psychiatry, 21*(3), 231–237.

LeFevre, M. L., & US Preventive Services Task Force. (2014). Screening for suicide risk in adolescents, adults, and older adults in primary care: U.S. Preventive Services Task Force recommendation statement. *Annals of Internal Medicine, 160*(10), 719–726.

Lerner, J. C., Fox, D. M., Ruzek, S. B., & Shearer, G. E. (2010). The case for a National Patient Library. *Health Affairs, 29*(10), 1914–1919. doi:10.1377/hlthaff.2010.0631; 10.1377/hlthaff.2010.0631

Leydon, G. M., Dowrick, C. F., McBride, A. S., Burgess, H. J., Howe, A. C., Clarke, P. D., ... QOF Depression Study Team. (2011). Questionnaire severity measures for depression: A threat to the doctor-patient relationship? *The British Journal of General Practice: The Journal of the Royal College of General Practitioners, 61*(583), 117–123. doi: 10.3399/bjgp11X556236

Litaker, D., Tomolo, A., Liberatore, V., Stange, K. C., & Aron, D. (2006). Using complexity theory to build interventions that improve health care delivery in primary care. *Journal of General Internal Medicine, 21*(Suppl. 2), S30–S34. doi:10.1111/j.1525-1497.2006.00360.x

Loewenstein, G. (2005). Projection bias in medical decision making. *Medical Decision Making, 25*(1), 96–105. doi:10.1177/0272989X04273799

Luoma, J. B., Martin, C. E., & Pearson, J. L. (2002). Contact with mental health and primary care providers before suicide: A review of the evidence. *The American Journal of Psychiatry, 159*(6), 909–916.

Macinko, J., Starfield, B., & Shi, L. (2007). Quantifying the health benefits of primary care physician supply in the United States. *International Journal of Health Services, 37*(1), 111–126. doi:10.2190/3431-G6T7-37M8-P224

Marino, S., Gallo, J. J., Ford, D., & Anthony, J. C. (1995). Filters on the pathway to mental-health-care, 1: Incident mental-disorders. *Psychological Medicine, 25*(6), 1135–1148.

Mavandadi, S., Klaus, J. R., & Oslin, D. W. (2012). Age group differences among veterans enrolled in a clinical service for behavioral health issues in primary care. *American Journal of Geriatric Psychiatry, 20*(3), 205–214. doi:10.1097/JGP.0b013e3181ec828a

McDaniel, S. H., Campbell, T. L., Hepworth, J., & Lorenz, A. (2005). *Family-oriented primary care* (2nd ed.). New York: Springer.

Mishler, E. G., AmaraSingham, L. R., Hauser, S. T., Liem, R., Osherson, S. D., & Waxler, N. E. (1981). *Social contexts of health, illness, and patient care.* New York: Cambridge University Press.

Mold, J. W., Hamm, R. M., & McCarthy, L. H. (2010). The law of diminishing returns in clinical medicine: How much risk reduction is enough? *Journal of the American Board of Family Medicine, 23*(3), 371–375. doi:10.3122/jabfm.2010.03.090178

Moller, A. C., Ryan, R. M., & Deci, E. L. (2006). Self-determination theory and public policy: Improving the quality of consumer decisions without using coercion. *Journal of Public Policy & Marketing*, 25(1), 104–116. doi:http://dx.doi.org/10.1509/jppm.25.1.104

Morgan, D. (1999). "Please see and advise": A qualitative study of patients' experiences of psychiatric outpatient care. *Social Psychiatry and Psychiatric Epidemiology*, 34(8), 442–450.

Motto, J. A., & Bostrom, A. G. (2001). A randomized controlled trial of postcrisis suicide prevention. *Psychiatric Services*, 52(6), 828–833.

Nelson, E. C., Batalden, P. B., Huber, T. P., Mohr, J. J., Godfrey, M. M., Headrick, L. A., & Wasson, J. H. (2002). Microsystems in health care: Part 1. Learning from high-performing front-line clinical units. *The Joint Commission Journal on Quality Improvement*, 28(9), 472–493.

Nutting, P. A., Crabtree, B. F., Miller, W. L., Stange, K. C., Stewart, E., & Jaen, C. (2011). Transforming physician practices to patient-centered medical homes: Lessons from the national demonstration project. *Health Affairs*, 30(3), 439–445. doi:10.1377/hlthaff.2010.0159

Oliver, T. (2006). The politics of public health policy. *Annual Review of Public Health*, 27, 195–233. doi:10.1146/annurev.publhealth.25.101802.123126

Oslin, D. W., Ross, J., Sayers, S., Murphy, J., Kane, V., & Katz, I. R. (2006). Screening, assessment, and management of depression in VA primary care clinics: The Behavioral Health Laboratory. *Journal of General Internal Medicine*, 21(1), 46–50. doi: 10.1111/j.1525-1497.2005.00267.x

Overholser, J. C., Stockmeier, C., Dilley, G., & Freiheit, S. (2002). Personality disorders in suicide attempters and completers: Preliminary findings. *Archives of Suicide Research*, 6(2), 123–133.

Owens, C., Owen, G., Belam, J., Lloyd, K., Rapport, F., Donovan, J., & Lambert, H. (2011). Recognising and responding to suicidal crisis within family and social networks: Qualitative study. *BMJ*, 343, d5801. doi:10.1136/bmj.d5801

Owens, C., Owen, G., Lambert, H., Donovan, J., Belam, J., Rapport, F., & Lloyd, K. (2009). Public involvement in suicide prevention: Understanding and strengthening lay responses to distress. *BMC Public Health*, 9, 308. doi:10.1186/1471-2458-9-308

Palmer, S. C., & Coyne, J. C. (2003). Screening for depression in medical care: Pitfalls, alternatives, and revised priorities. *Journal of Psychosomatic Research*, 54(4), 279–287.

Pearson, A., Saini, P., Da Cruz, D., Miles, C., While, D., Swinson, N., ... Kapur, N. (2009). Primary care contact prior to suicide in individuals with mental illness. *British Journal of General Practice*, 59(568), 825–832. doi:10.3399/bjgp09X472881

Pham, H. H., O'Malley, A. S., Bach, P. B., Saiontz-Martinez, C., & Schrag, D. (2009). Primary care physicians' links to other physicians through Medicare patients: The scope of care coordination. *Annals of Internal Medicine*, 150(4), 236–242.

Pigeon, W. R., Britton, P. C., Ilgen, M. A., Chapman, B., & Conner, K. R. (2012). Sleep disturbance preceding suicide among veterans. *American Journal of Public Health*, 102, S93–S97. doi:10.2105/AJPH.2011.300470

Pigeon, W. R., Pinquart, M., & Conner, K. R. (2012). Meta-analysis of sleep disturbance and suicidal thoughts and behaviors. *Journal of Clinical Psychiatry*, 73(9),1160–1167.

Plsek, P., & Greenhalgh, T. (2001). Complexity science: The challenge of complexity in health care. *BMJ*, 323(7313), 625–628.

Pomerantz, A. S., Shiner, B., Watts, B. V., Detzer, M. J., Kutter, C., Street, B., & Scott, D. (2010). The White River Model of colocated collaborative care: A platform for mental and behavioral health care in the medical home. *Families Systems & Health*, 28(2), 114–129. doi:10.1037/a0020261

Raue, P. J., Morales, K. H., Post, E. P., Bogner, H. R., Have, T. T., & Bruce, M. L. (2010). The wish to die and 5-year mortality in elderly primary care patients. *The American Journal of Geriatric Psychiatry*, 18(4), 341–350. doi:10.1097/JGP.0b013e3181c37cfe; 10.1097/JGP.0b013e3181c37cfe

Raue, P. J., Schulberg, H. C., Lewis-Fernandez, R., Boutin-Foster, C., Hoffman, A. S., & Bruce, M. L. (2010). Shared decision-making in the primary care treatment of late-life major depression: A needed new intervention? *International Journal of Geriatric Psychiatry*, 25(11), 1101–1111. doi:10.1002/gps.2444

Relman, A. S. (1994). The impact of market forces on the physician-patient relationship. *Journal of the Royal Society of Medicine*, 87(Suppl. 22), 22–24; discussion 24–25.

Reuben, D. B., & Tinetti, M. E. (2012). Goal-oriented patient care—An alternative health outcomes paradigm. *The New England Journal of Medicine*, 366(9), 777–779.

Rutz, W. (2001). Preventing suicide and premature death by education and treatment. *Journal of Affective Disorders*, 62(1–2), 123–129.

Ryan, R. M., & Deci, E. L. (2000). Self-determination theory and the facilitation of intrinsic motivation, social development, and well-being. *The American Psychologist*, 55(1), 68–78. doi:http://dx.doi.org/10.1037/0003-066X.55.1.68

Saigal, C., Kaplan, R., Crespi, C., Garcia, E., Lambrechts, S., & Dahan, E. (2012). Impact of a novel method of patient preference elicitation on decision quality in men with prostate cancer: Pilot data. *Journal of Urology*, 187(4), E266–E267.

Schwartz, M. D. (2012). Health care reform and the primary care workforce bottleneck. *Journal of General Internal Medicine*, 27(4), 469–472. doi:10.1007/s11606-011-1921-4

Shah, R., Franks, P., Jerant, A., Feldman, M., Duberstein, P., Fernandez y Garcia, E., ... Kravitz, R. L. (2014). The effect of targeted and tailored patient depression engagement interventions on clinician inquiry about suicidal thoughts: A randomized control trial. *Journal of General Internal Medicine*, 29, 1148–1154.

Shand, F. L., Proudfoot, J., Player, M. J., Fogarty, A., Whittle, E., Wilhelm, K., ... Christensen, H. (2015). What might interrupt men's suicide? Results from an online survey of men. *BMJ Open*, 5(10), e008172-2015-008172. doi:10.1136/bmjopen-2015-008172

Shepherd, H. L., Barratt, A., Trevena, L. J., McGeechan, K., Carey, K., Epstein, R. M., ... Tattersall, M. H. N. (2011). Three questions that patients can ask to improve the quality of information physicians give about treatment options: A cross-over trial. *Patient Education and Counseling*, 84(3), 379–385. doi:10.1016/j.pec.2011.07.022

Shi, L., Starfield, B., Politzer, R., & Regan, J. (2002). Primary care, self-rated health, and reductions in social disparities in health. *Health Services Research*, 37(3), 529–550. doi: 10.1111/1475-6773.t01-1-00036

Shneidman, E. S. (1992). Rational suicide and psychiatric disorders. *The New England Journal of Medicine*, 326(13), 889–890; author reply 890–891. doi:10.1056/NEJM199203263261311

Sigsbee, B. (2011). The income gap: Specialties vs. primary care or procedural vs. nonprocedural specialties? *Neurology*, 76(10), 923–926. doi:10.1212/WNL.0b013e31820f2dfd

Stacey, D., Bennett, C. L., Barry, M., Col, N. F., Eden, K. B., Holmes-Rovner, M., ... Thomson, R. (2011). Decision aids for people facing health treatment or screening decisions. *Cochrane Database of Systematic Reviews*, 10(CD001431) doi:10.1002/14651858.CD001431.pub3

Stange, K., Zyzanski, S., Jaen, C., Callahan, E., Kelly, R., Gillanders, W., ... Goodwin, M. (1998). Illuminating the "black box": A description of 4454 patient visits to 138 family physicians. *Journal of Family Practice*, 46(5), 377–389.

Stange, K. C. (2009). A science of connectedness. *Annals of Family Medicine*, 7(5), 387–395. doi:10.1370/afm.990

Starfield, B., Shi, L. Y., & Macinko, J. (2005). Contribution of primary care to health systems and health. *Milbank Quarterly*, 83(3), 457–502. doi:10.1111/j.1468-0009.2005.00409.x

Stephens, N. M., Markus, H. R., & Fryberg, S. A. (2012). Social class disparities in health and education: Reducing inequality by applying a sociocultural self model of behavior. *Psychological Review*, 119(4), 723–744.

Stewart, M. A. (1995). Effective physician-patient communication and health outcomes: A review. *CMAJ: Canadian Medical Association Journal = Journal De l'Association Medicale Canadienne*, 152(9), 1423–1433.

Street, R. L. J., & Epstein, R. M. (2008). Key interpersonal functions and health outcomes: Lessons from theory and research on clinician-patient communication. In K. Glanz, B. K. Rimer, & K. Viswanath (Eds.), *Health behavior and health education: Theory, research, and practice* (4th ed., pp. 237–269). San Francisco, CA: Jossey-Bass.

Sullivan, M. (2003). The new subjective medicine: Taking the patient's point of view on health care and health. *Social Science & Medicine*, 56(7), 1595–1604.

Swindell, J., McGuire, A. L., & Halpern, S. D. (2010). Beneficent persuasion: Techniques and ethical guidelines to improve patients' decisions. *The Annals of Family Medicine*, 8(3), 260–264.

Tew, J., Klaus, J., & Oslin, D. W. (2010). The Behavioral Health Laboratory: Building a stronger foundation for the patient-centered medical home. *Families Systems & Health*, 28(2), 130–145. doi:10.1037/a0020249

Thombs, B. D., Coyne, J. C., Cuijpers, P., de Jonge, P., Gilbody, S., Ioannidis, J. P. A., ... Ziegelstein, R. C. (2012). Rethinking recommendations for screening for depression in primary care. *Canadian Medical Association Journal*, 184(4), 413–418.

Tinetti, M. E., Bogardus, S. T., & Agostini, J. V. (2004). Potential pitfalls of disease-specific guidelines for patients with multiple conditions. *The New England Journal of Medicine*, 351(27), 2870–2874. doi:10.1056/NEJMsb042458

Trevino, K. M., Abbott, C. H., Fisch, M. J., Friedlander, R. J., Duberstein, P., & Prigerson, H. G. (2014) Patient-oncologist alliance as protection against suicidal ideation in young adults with advanced cancer. *Cancer*, 120, 2272–2281.

Tsoh, J. M. Y., Chiu, H. F. K., Duberstein, P. R., Chan, S. S. M., Chi, I., Yip, P., & Conwell, Y. (2005). Attempted suicide in elderly Chinese persons: A multi-group controlled study. *American Journal of Geriatric Psychiatry*, 13, 562–571.

Unützer, J., Tang, L., Oishi, S., Katon, W., Williams, J. W. Jr., Hunkeler, E., ... IMPACT Investigators. (2006). Reducing suicidal ideation in depressed older primary care patients. *Journal of the American Geriatrics Society*, 54(10), 1550–1556. doi:10.1111/j.1532-5415.2006.00882.x

Useda, J. D., Duberstein, P. R., Conner, K. R., Beckman, A., Franus, N., Tu, X., & Conwell, Y. (2007). Personality differences in attempted suicide versus suicide in adults 50 years of age or older. *Journal of Consulting and Clinical Psychology*, 75(1), 126–133. doi:http://dx.doi.org/10.1037/0022-006X.75.1.126

VA Health Services Research and Development Service. (2008). *Collaborative care for depression in the primary care setting. A primer on VA's Translating Initiatives for Depression into Effective Solutions (TIDES) project*. Washington, DC: Department of Veterans Affairs.

Valenstein, M., Eisenberg, D., McCarthy, J. F., Austin, K. L., Ganoczy, D., Kim, H. M., ... Blow, F. C. (2009). Service implications of providing intensive monitoring during high-risk periods for suicide among VA patients with depression. *Psychiatric Services*, 60(4), 439–444.

Valenstein, M., Kim, H. M., Ganoczy, D., McCarthy, J. F., Zivin, K., Austin, K. L., ... Olfson, M. (2009). Higher-risk periods for suicide among VA patients receiving depression treatment: Prioritizing suicide prevention efforts. *Journal of Affective Disorders*, 112(1–3), 50–58. doi: 10.1016/j.jad.2008.08.020

van Geffen, E. C., Gardarsdottir, H., van Hulten, R., van Dijk, L., Egberts, A. C., & Heerdink, E. R. (2009). Initiation of antidepressant therapy: Do patients follow the GP's prescription? *The British Journal of General Practice: The Journal of the Royal College of General Practitioners*, 59(559), 81–87. doi: 10.3399/bjgp09X395067; 10.3399/bjgp09X395067

Van Orden, K. A., Witte, T. K., Cukrowicz, K. C., Braithwaite, S. R., Selby, E. A., & Joiner, T. E., Jr. (2010). The interpersonal theory of suicide. *Psychological Review*, 117(2), 575–600. doi:10.1037/a0018697; 10.1037/a0018697

Van Ryn, M. (2002). Research on the provider contribution to race/ethnicity disparities in medical care. *Medical Care*, 40(1), 140–151.

Van Ryn, M., & Fu, S. S. (2003). Paved with good intentions: Do public health and human service providers contribute to racial/ethnic disparities in health? *American Journal of Public Health*, 93(2), 248–255.

Vannoy, S. D., Fancher, T., Meltvedt, C., Unützer, J., Duberstein, P., & Kravitz, R. L. (2010). Suicide inquiry in primary care: Creating context, inquiring, and following up. *Annals of Family Medicine*, 8(1), 33–39. doi:10.1370/afm.1036

Vannoy, S. D., & Robins, L. S. (2011). Suicide-related discussions with depressed primary care patients in the USA: Gender and quality gaps. A mixed methods analysis. *BMJ Open*, 1(2), e000198-2011-000198. doi:10.1136/bmjopen-2011-000198 [doi]

Vannoy, S. D., Tai-Seale, M., Duberstein, P., Eaton, L. J., & Cook, M. A. (2011). Now what should I do? Primary care physicians' responses to older adults expressing thoughts of suicide. *Journal of General Internal Medicine*, 26(9), 1005–1011. doi:10.1007/s11606-011-1726-5

Walters, H., Kulkarni, M., Forman, J., Roeder, K., Travis, J., & Valenstein, M. (2012). Feasibility and acceptability of interventions to delay gun access in VA mental health settings. *General Hospital Psychiatry*, 34(6), 692–698. doi:10.1016/j.genhosppsych.2012.07.012

Watt, G. (2002). The inverse care law today. *Lancet*, 360(9328), 252–254. doi:10.1016/S0140-6736(02)09466-7

Wennberg, J. E., Fisher, E. S., Goodman, D. C., & Skinner, J. D. (2008). *Tracking the care of patients with severe chronic illness: The 2008 Dartmouth Atlas of Health Care* (L. Bronner, Ed.), Hanover, NH: The Dartmouth Institute for Health Care Policy & Clinical Practice.

Wennberg, D. E., Marr, A., Lang, L., O'Malley, S., & Bennett, G. (2010). A randomized trial of a telephone care-management strategy. *The New England Journal of Medicine*, 363(13), 1245–1255. doi:10.1056/NEJMsa0902321; 10.1056/NEJMsa0902321

Westen, D., Novotny, C. M., & Thompson-Brenner, H. (2004). The empirical status of empirically supported psychotherapies: Assumptions, findings, and reporting in controlled clinical trials. *Psychological Bulletin*, 130(4), 631–663. doi:10.1037/0033-2909.130.4.631

Wilkinson, D. J., & Truog, R. D. (2013). The luck of the draw: Physician-related variability in end-of-life decision-making in intensive care. *Intensive Care Medicine*, 39(6), 1128–1132.

Wilson, K. G., Curran, D., & McPherson, C. J. (2005). A burden to others: A common source of distress for the terminally ill. *Cognitive Behaviour Therapy*, 34(2), 115–123.

Wirtz, V., Cribb, A., & Barber, N. (2006). Patient-doctor decision-making about treatment within the consultation—a critical analysis of models. *Social Science & Medicine*, 62(1), 116–124.

Wittink, M. N., Duberstein, P., & Lyness J. (2013). Late life depression in the primary care setting: Building the case for customized care. In H. Lavretsky, M. Sajatovic & C. Reynolds (Eds.), *Late life mood disorders* (pp. 500–515). New York: Oxford University Press.

Wittink, M. N., Givens, J. L., Knott, K. A., Coyne, J. C., & Barg, F. K. (2011). Negotiating depression treatment with older adults: Primary care providers' perspectives. *Journal of Mental Health, 20*(5), 429–437. doi:10.3109/09638237.2011.556164

Wittink, M. N., Oslin, D., Knott, K. A., Coyne, J. C., Gallo, J. J., & Zubritsky, C. (2005). Personal characteristics and depression-related attitudes of older adults and participation in stages of implementation of a multi-site effectiveness trial (PRISM-E). *International Journal of Geriatric Psychiatry, 20*(10), 927–937. doi:10.1002/gps.1386

Wolff, J. L., & Roter, D. L. (2011). Family presence in routine medical visits: A meta-analytical review. *Social Science & Medicine, 72*(6), 823–831. doi:10.1016/j.socscimed.2011.01.015

Woolf, S. H., & Harris, R. (2012). The harms of screening: New attention to an old concern. *JAMA: Journal of the American Medical Association, 307*(6), 565–566.

Working Party Group on Integrated Behavioral Healthcare. (2014). Joint principles: integrating behavioral health care into the patient-centered medical home. *Annuals of Family Medicine, 12*(2), 183–185. doi:10.1370/afm.1633.

Yarnall, K. S. H., Ostbye, T., Krause, K. M., Pollak, K. I., Gradison, M., & Michener, J. L. (2009). Family physicians as team leaders: "Time" to share the care. *Preventing Chronic Disease, 6*, A59. Retrieved from http://www.cdc.gov/pcd/issues/2009/apr/pdf/08_0023.pdf

York, J. A., Lamis, D. A., Pope, C. A., & Egede, L. E. (2012). Veteran-specific suicide prevention. *Psychiatry Quartlerly, 84*(2), 219–238. doi: 10.1007/s11126-012-9241-3

Younes, N. Melchior, M., Turbelin, C., Blanchon, T., Hanslik, T., Chee, C. Chan (2015) Attempted and completed suicide in primary care: Not what we expected? *Journal of Affective Disorders, 170*, 150–154.

Young, H. N., Bell, R. A., Epstein, R. M., Feldman, M. D., & Kravitz, R. L. (2008). Physicians' shared decision-making behaviors in depression care. *Archives of Internal Medicine, 168*(13), 1404–1408. doi:10.1001/archinte.168.13.1404

19

Caring Letters for Military Suicide Prevention

David D. Luxton

"The forces that bind us willingly to life are mostly those exerted by our relationships with other people, whether they be intimately involved in our lives or influence us by other psychological processes." (Jerome A. Motto, MD)

INTRODUCTION

Patients who are discharged from psychiatric treatment are a high-risk group for suicide and repeat suicide attempts (Geddes, Juszczak, O'Brien, & Kendrick, 1997; Goldacre, Seagroatt, & Hawton, 1993; Ho, 2003; Luxton, June, & Comtois, 2013). Several studies have shown that the greatest risk for posthospitalization suicide occurs soon after discharge (Appleby, Shaw, et al., 1999; Gunnell et al., 2008; Hunt et al., 2008; Meehan et al., 2006). For example, Appleby, Shaw, and colleagues assessed individuals (*n* = 2,370) who had contact with mental health professionals within 12 months prior to death by suicide and found that 519 (24%) of patients discharged from inpatient treatment had died by suicide within three months after discharge. One hundred eighty-six (41%) of the suicide deaths occurred before the first scheduled aftercare appointment. Furthermore, Meehan and colleagues examined risk of suicide during the three months following hospital discharge and found that out of 1,100 patients who died by suicide within this period, 337 (32%) patients died by suicide within the first two weeks and 32 (3%) died on the first day after discharge. Three hundred and ninety-seven deaths (40%) occurred before the patients' first postdischarge mental health follow-up appointment in the community. Similar data have been reported by Valenstein and colleagues (2009) who examined suicide rates among depressed U.S. veterans who were discharged from psychiatric hospitalization. Suicide rates were 568 per 100,000 person-years during the 12 weeks following psychiatric hospitalization—a rate that was approximately five times the base rate for non-hospitalized depressed veterans and 34 times that of the general U.S. population's suicide rate. Luxton, Trofimovich, and Clark (2013) examined suicide rates among active duty U.S. service members who were discharged from psychiatric-related hospitalizations between 2001 and 2011 and found the rate to be 66.4 per 100,000 person-years. The suicide rate for the overall active component from 1998 to 2011 was 13.7 per 100,000 (Armed Forces Health Surveillance Center (AFHSC), 2012), which indicates that posthospitalized service members are at approximately five times the risk for suicide than the general active duty population during the same time period. As noted by Luxton and colleagues, these data may underestimate the true rate of suicide in this cohort because the sample did not include service members treated outside of the U.S. military health system or those who were discharged from military service within the surveillance time frame.

There are several issues that may increase risk for suicide following psychiatric treatment. Many psychiatric patients who die by suicide are not found to be at high or immediate risk at last contact with mental health providers (Appleby, Dennehy, Thomas,

Faragher, & Lewis, 1999; Appleby, Shaw, et al., 1999). The perceived loss of support or reduction in the level of clinical supervision after discharge may increase suicide risk (Appleby, Shaw, et al., 1999; Gunnell et al., 2008; Meehan et al., 2006). Brief hospitalizations are also common and may not allow adequate time for effective treatment of suicidality (Knesper, 2010). Some studies have shown that a shorter length of hospital stay is associated with greater suicide risk after discharge (Desai, Dausey, & Rosenheck, 2005). A significant number of high-risk patients are also discharged from emergency departments, and suicide risk assessments may not be possible if patients leave without staff evaluation (Bennewith, Gunnel, Peters, Hawton, & House, 2004; Bennewith, Peters, Hawton, House, & Gunnell, 2005; Hickey, Hawton, Fagg, & Weitzel, 2001). Noncompliance with after-care treatment, including medication regimens, is also associated with increased risk for suicide and suicide behaviors (Maris, Berman, & Silverman, 2000). Limited access to care due to geographical distance from treatment services, transportation issues, and financial costs of treatment can also increase risk for suicide among high-risk individuals (Knesper, 2010). Furthermore, the transition back into the community can be especially difficult for postdischarged psychiatric patients due to return to the same environment and stressors that may have contributed to the initial admission into psychiatric treatment (Gunnell et al., 2008). Social factors, such as living alone (Roy, 1982) and low levels of social support and social integration, are also factors linked to suicide risk that can be a significant problem for patients (Linehan, 1981; Neeleman & Wessely, 1999; Trout, 1980).

Stigma associated with seeking psychiatric care is another factor that may increase suicide risk, especially among military personnel. Many service members who experience psychological problems may not seek help from a mental health professional because of the perceived stigma associated with the disclosure of problems and the seeking of treatment (Hoge et al., 2004). Some data have shown that U.S. soldiers typically report more discomfort with discussion of psychological problems than medical problems and are less likely to follow through with a psychological referral (Greene-Shortridge et al., 2007). Stigma is thus a serious problem that requires interventions and prevention programs that address the issue.

Given the heightened level of suicide risk among postdischarged patients and the numerous factors that contribute to this risk, it makes sense to test and implement suicide prevention interventions that specifically target this high-risk group. While some psychotherapy and pharmacotherapy interventions have been shown to help reduce the rate of suicide attempts among posthospitalized patients (Comtois & Linehan, 2006; Goldsmith et al., 2002; Mann et al., 2005), only a few posttreatment approaches have been shown to prevent death by suicide (Comtois & Linehan, 2006; Goldsmith et al., 2002; Linehan, 2008). One promising suicide intervention with growing empirical support for its effectiveness is the "caring letters" or "caring contacts" concept (Luxton, June, et al., 2013; Motto, 1976).

THE CARING LETTERS CONCEPT

The caring letters concept is a suicide prevention intervention that involves the routine sending of brief expressions of care to patients following hospitalization. The caring letters concept was first examined more than four decades ago by psychiatrist Jerome Motto (1976), and additional studies since then have tested similar contact interventions. Luxton, June, et al. (2013) reviewed and evaluated published empirical studies of caring contact interventions with self-directed violence (suicide, attempts, and ideation) as outcomes. Study populations included inpatient psychiatric or emergency department patients discharged without further hospitalization. A total of eight original studies, two follow-up studies, and one secondary analysis study were identified. Contact modalities included phone, postal letter, postcards, in-person, email, and texting. Two of these studies were shown to prevent deaths by suicide: Motto's caring letters (1976) and a World Health Organization (WHO) study reported by Fleishmann et al. (2008).

Motto's Caring Letters Study

Motto's 1976 paper described a caring letters study that was underway at nine psychiatric facilities in San Francisco, California. For the study, 3,006 patients who were admitted for treatment between 1968 and 1974 because of a "depressive or suicide state" were identified as high risk for suicide. After hospital discharge, 843 of these patients who declined additional treatment or dropped out of treatment within 30 days were randomly divided into two subgroups: a contact group that received the contact intervention and a no-contact group. The contact group received a series of short letters and sometimes a brief phone call. The letters were sent from the research staff member who interviewed the patient in the hospital, and the letters were individually typed and included responses to comments from the patients if comments were received. The letters consisted of simple expressions of concern regarding personal well-being and invited a response if the patient wished to do so. Motto emphasized the importance of not using the caring letters to gather test data or information but to let patients know that the staff remembered them and had positive feelings about them. Also, Motto theorized that one note would not have much influence but that the cumulative effect of multiple caring contacts would potentially have the greatest influence. Contacts were scheduled monthly for four months, then every two months for eight months, and finally every three months for four years (i.e., 24 contacts during five years).

In 2001, Motto and Bostrom (2001) reported the outcome results of the caring letters study. The number of suicides in the no-contact group was found to be more than twice that of the contact group for the first two years. Although the data showed that the reduction of suicide mortality rates was not statistically significant after two years, the data did show a long-term trend in the reduction of the number of suicides that tapered off toward the end of the 15-year follow-up. The significant differences in suicide rates during the first two years occurred when the letters were most frequent, indicating that the frequency of the contacts may be important (Luxton, June, et al., 2013).

WHO SUPRE-MISS Study

The only other study to have shown a reduction is suicide mortality rates was the international WHO SUPRE-MISS study (Fleischmann et al., 2008). The study was a large five-country (Campinas, Brazil; Chennai, India; Colombo, Sri Lanka; Karaj, Islamic Republic of Iran; and Yuncheng, China) randomized controlled trial that compared a group of previous suicide attempters who received a series of personalized follow-up contacts to a noncontact treatment as usual (TAU) control group. The suicide attempters were identified by medical staff in emergency care settings.

A total of nine follow-up contacts (telephone calls or in-person visits) were conducted by a person with clinical experience (e.g. nurse, doctor, or psychologist). The contacts were made according to a specific timeline for up to 18 months (at 1, 2, 4, 7, and 11 weeks, and at 4, 6, 12, and 18 months). The study results showed that at 18-months follow-up, more patients had died from suicide in the TAU group than in the contacts group.

The WHO study has several limitations that influence the interpretation of results (Luxton, June, et al. 2013). The cause of death information was derived from reports made by informants (e.g., relatives of patients) and was not based on standardized data such as death certificates. There was also significant variance in the sample sizes and proportion of losses at follow-up across each study location. Furthermore, there was a difference in suicide and all-cause mortality rates in favor of the intervention but also greater nonsuicide mortality in the intervention group although the study authors did not provide any additional analysis or explanation for this finding. Bertolote et al. (2010) conducted a secondary analysis of the WHO SUPRE-MISS data that examined repeated suicide attempts. The results indicated that rates were lower in the follow-up contact group compared to the TAU group in two countries (Sri Lanka and India). Data from Brazil and the Islamic Republic of Iran, however, showed higher rates in the intervention group compared to the TAU group whereas data from China showed reduction in repeat suicide attempts in both groups.

Other Caring Contact Studies

Three studies reported in the literature used postcards for posthospitalization caring contacts (Beautrais et al., 2010; Carter et al., 2005; Hassanian-Moghaddam et al., 2011). For example, Carter et al. tested a postcards intervention (postcards from the EDge project) with patients who were discharged from hospitalization for deliberate self-poisoning in Australia. One group was randomized to receive caring postcards ($n = 378$) and was compared to a TAU control group ($n = 394$). Eight word-processed postcards were mailed in envelopes monthly for four months and then once every two months for up to one year. The study sample was not large enough to evaluate suicide mortality rates; however, results showed that at 12 months the postcards group had lower rates of repeat self-poisoning episodes and hospital days than compared to the control group. In a follow-up study, Carter and colleagues (2007) evaluated the outcomes of the postcards intervention 24 months later and did not find a difference between the control group and intervention group in the proportion of repeat self-harm episodes at two-years. Patients in the postcard group, however, had significantly fewer readmissions of deliberate self-poisoning and used less than half of the hospitalization days as compared to the control group.

Other studies have evaluated interventions that specifically used telephone calls for posttreatment follow-ups (Bertolote et al., 2010; Cedereke, Monti, & Ojenhagen, 2002; Vaiva et al., 2006). For example, Vaiva et al. tested the effectiveness of telephone contacts with patients who were discharged from emergency departments) in France following suicide attempts by self-poisoning. Six hundred five patients were randomized to either a group that received telephone contact at one month, a group that received telephone follow-up at three months, or a no-contact group (TAU). The phone calls, made by psychiatrists who had not met with the patients, involved review of recommended treatments and empathetic psychological support. Intent to treat analyses results indicated that there were not any differences between conditions in the proportion of adverse outcomes. However, the number of patients contacted at one month who reattempted suicide was significantly lower than that of patients in the no-contact control group. This difference was observed over the six months after the telephone contact was made. The number who attempted suicide did not significantly differ compared to controls for participants reached at three months. Although the results of this particular study are limited, they suggest that patient contact at one month after hospitalization for deliberate self-harm may help to reduce subsequent suicide attempts.

The results of the Luxton and colleagues (2012) review showed that repeated follow-up contacts may exert a preventative effect on suicidal behaviors. The aforementioned studies, Motto's caring letters (1976) and the WHO study reported by Fleishmann et al. (2008), were the only studies shown to prevent deaths by suicide. Three (i.e., Motto & Bostrom, 2001; Fleishmann et al., 2008; Vaiva et al., 2006) of the 11 studies showed a statistically significant reduction in repeat suicide attempts, and four studies (Bertolote et al., 2010; Carter et al., 2005; Hassanian-Moghaddam et al., 2011; Termansen & Bywater, 1975), including one of the follow-up studies (Motto & Bostrom, 2001) and a study reporting secondary analyses (Carter et al., 2007), showed mixed or nonconclusive results but also showed trends toward a preventative effect. Two studies did not show preventative effects for the follow-up interventions (Beautrais et al., 2010; Cedereke et al., 2002).

With the exception of Motto's caring letters (1976) and the WHO (2008) study, the published contact studies that included mortality rates as outcomes are statistically underpowered, and thus meaningful interpretations and conclusions regarding their ability to prevent suicide are limited (Luxton, June, et al., 2013). Studies with very large sample sizes are needed, which may explain why the original Motto caring letters intervention has not been retested in a full replication trial. Also, the majority of the follow-up contact studies in the literature examined repeated suicide attempts as the primary outcome. Although intentional self-harm is a significant risk factor for suicide, it is not directly comparable to completed suicide (Luxton, June, et al., 2013). The best evidence that these types of interventions are effective for suicide prevention are reductions in suicide mortality rates; however, the low base rate of suicide makes the outcome difficult to study in both retrospective and prospective study designs (Goldney, 1998; Motto & Bostrom, 2001). Nonetheless, these studies provide evidence that this type of postdischarge intervention can be effective at reducing suicide deaths, attempts, and ideation.

THEORETICAL BASIS FOR THE CARING LETTERS INTERVENTION

At the core of the caring letters intervention is the interpersonal bond that the contacts may facilitate. Motto (1976) emphasized the importance of the interpersonal connection and hypothesized that a suicidal person may be encouraged to retain an interest in living when another person initiates regular and long-term contact that espouses nondemanding care and concern about the other's well-being. The importance of interpersonal connectedness is consistent with several psychological theories that emphasize the human need to form interpersonal bonds and to maintain social belongingness. Baumeister and Leary (1995), for example, hypothesized that people have an innate drive to form and maintain strong, long-term bonds with others that stems from an evolutionary origin with survival and reproductive benefits. Baumeister and Leary (1995) proposed that the need to belong has two primary features: The first feature is a person needs frequent personal contacts or interactions with another person that are pleasant, positive, and free from conflict and negative feelings. Second, a person needs to perceive that there is an interpersonal bond or relationship that has stability, affective concern, and foreseeable continuation. These features are purported to create a relational context for one's interactions with another person. Baumeister and Leary also suggest that to satisfy the need to belong, the person must believe that the other person cares about his or her welfare and likes (or loves) him or her.

The interpersonal-psychological theory of suicide (Joiner, 2005; Van Orden et al., 2010) also emphasizes the importance of belongingness in suicide risk. The theory states that individuals with an acquired capability for self-harm, perceptions of being a burden, and perceptions of thwarted belongingness are more likely to attempt or complete suicide than are others. Thwarted belongingness refers to a person's belief that he or she has infrequent positive social interactions and the perception that he or she is not cared for by others (Baumeister & Leary, 1995). Empirical evidence supports a relationship between thwarted belongingness and suicides or suicide attempts (Conner, Britton, Sworts, & Joiner, 2007; Joiner, Hollar, & Van Orden, 2006). Individuals who die by suicide often harbor feelings of disconnect from others in such a way that they may believe that there is nobody who truly cares about them or that no one can relate to them and understand their situation—beliefs that lead to isolation. It is possible that caring letters written by behavioral health providers who have seen the participants at their most vulnerable (and yet still espouse care) could provide a sense of belonging and acceptance that would help to mitigate suicidal behavior (Luxton et al., 2011).

The literature on the role of social support and mental health also provides some theoretical basis for the caring letters intervention concept. The association between social support and mental health, such as depression, is well documented (Ingram, Miranda, & Segal, 1998). In general, low levels of social support and social integration (connection to one's community and peers) have been shown to be strongly associated with depression as well as suicide risk. For example, Nisbet (1996) examined data from the Epidemiological Catchment Area study and found that the social support of friends and family was associated with a lower risk of suicide. Darke and colleagues (2005) found an association between social isolation and subsequent suicide attempts following drug treatment among a sample of 495 heroin users, even after controlling for both diagnosis and a history of suicide attempts. Further, Desai and colleagues conducted a prospective mortality study of psychiatric inpatients from 128 U.S. Department of Veterans Affairs (VA) hospitals across the United States and found that states with a higher rating of social capital (defined as the level of social cohesiveness and trust in a community) had fewer suicides. You, Van Orden, and Conner (2011) found a connection between thwarted belongingness, perceived social support, and living alone as significant indicators of suicide attempts among patients in substance-use treatment programs. These studies point to the importance of social connections to prevent suicide and the sense of belonging and acceptance that caring letters may provide.

Another explanation for how consistent contacts from care providers may reduce suicide behaviors is that they may be instrumental in getting patients reconnected with help services (Luxton, June, et al., 2013; Luxton et al., 2012). The contacts may serve as reminders of available treatment and also serve as an avenue for patients to contact a care provider or treatment facility if they feel that they need help. Lists of available treatment resources, such as suicide prevention and counseling hotlines or other support information, can also provide recipients with a route to seek help if they are in crisis (Luxton et al., 2012). Further, follow-up contacts may help patients to feel better about treatment and therefore motivate them to seek or adhere to treatment (Luxton, June, et al., 2013). Follow-up contact has been effective in promoting adherence to a variety of health-related behaviors including cancer screening (Somkin et al., 1997), diabetes treatment (Tran & Billups, 2008), and cardiovascular disease care (Hunt, Siemienczuk, Touchette, & Payne, 2004). Also, caring notes in the form of emails or postal mail can be kept and reread if a patient wishes to do so. The manner in which patients perceives their treatment experience, however, may also influence the effectiveness of the posttreatment contact intervention (Luxton, June, et al., 2013). Follow-up messages from a treatment provider that gave less than optimal care may not be effective or possibly have a negative influence on the patient (Kapur et al., 2010).

In sum, there are several theoretical mechanisms that may underlie the influence that follow-up contacts may have on suicide and help-seeking behavior. It is possible that repeat caring contacts that espouse care and concern for the patient may provide a sense of belonging that provides a buffer against suicide and self-harm behaviors. These contacts may also provide a route for recipients to seek help if needed. Despite the promising data of previous caring contacts interventions, there have not been any full replications of Motto's (1976) original intervention since the initial findings were reported. Given the rise in overall military suicide rates since the beginning of the Global War on Terror (rates that have surpassed that of the general population), it is important to implement empirically supported interventions that have the potential to reduce the occurrence of suicides and suicide behaviors. The Department of Defense (DOD) and the VA have initiated several programs and policies to reduce suicide and suicide behavior among service members and veterans; however, there are few empirically supported interventions to further guide and support these initiatives. The caring letters intervention, one of the only suicide prevention interventions with any initial empirical support, is now being tested in the DOD and VA.

Caring Letters Project Pilot Study

Luxton and colleagues (2012) at the National Center for Telehealth & Technology conducted a pilot caring letters program, called the Caring Letters Project, in order to evaluate the feasibility of the caring letters concept with a military population and to provide initial data to inform expansion to a full-scale randomized controlled trial (RCT). Specific objectives for the pilot study were to (a) evaluate the program to determine how best to tailor the caring letters intervention to the military setting, (b) explore whether the Caring Letters Project intervention showed trends toward reductions in self-injury and/or suicide, (c) compare the use of hand written letters versus email correspondence, and (d) determine the feasibility of translating the intervention to a multisite RCT. The pilot study did not include a matched no-contact control condition.

The decision to test the caring letters intervention with email contacts was based on the popularity of email and preferences of service members. Luxton and colleagues (2012) conducted a volunteer focus group comprised of active duty soldiers. The soldiers indicated that the caring letters concept would be welcomed by service members and that the project would likely be helpful. The soldiers also highly recommended the use of an email version of caring letters because of the ease of use, higher likelihood of reading the messages, and possible delay of regular mail because of military deployments or change of addresses. This recommendation is consistent with survey data indicating high prevalence of personal technology use, such as smartphones, among service members both at home and on combat deployments (Bush, Fullerton, Crumpton, Metzger-Abamukong, & Fantelli, 2012). Because the goal was to design a protocol that would closely replicate the original Motto (1976) intervention, Motto and other subject matter experts were consulted with regarding study procedures (Motto & Bostrom, 2001).

The pilot study participants were recruited at a large military treatment facility psychiatric inpatient unit. Participants provided consent to participate in the program and then completed a semistructured interview with a research staff member. The interview, which took approximately one hour to complete, consisted of measures of depression symptoms, suicidal thinking and history, as well as an interview designed to learn more about the patients' positive assets (e.g., hobbies, family support, social group membership, etc.). The interview also served as a way to create an initial caring connection between the research staff members and the patient. The research assistant also checked in with and said hello to patients during their stay on the inpatient unit. After discharge from the hospital, participants were sent brief, personalized caring letters or emails at regular intervals for two years. Participants were given the choice to receive the letters by either email or mail. The choice provided the research team to assess preferences as well as any potential differences in procedures and outcomes. As with Motto's study, the letters were intended to be simple expressions of good will and not to ask questions or make any demands that would require any kind of reply. When writing the letters, study staff reviewed the participants' personal information, including information that was collected during the interview regarding support networks, coping skills, hobbies, and other activities. Personalization was added to each letter based on information that was discussed during the interview. Personalization was very basic such as "I hope your move went well" or "I hope that you have a nice birthday." A list of several local and national help resources for the military and veteran population including crisis hotlines and other psychological health resources were also included. Although the inclusion of this information was not done in the original Motto study, the rationale was to include them as a reminder of the available help resources that could be kept for retrieval if the patient so wished to use the information. Letters were sent at regular intervals on a schedule similar to that in Motto's study. In Motto's (1976) study, 12 letters were sent over the first two years of the trial. This was modified slightly by adding one additional during the first week following discharge because the risk for suicide is expected to be highest shortly after hospital discharge. Thus letters were sent on the following schedule: one letter within one week of discharge, four more following at one-month intervals, then four more at two-month intervals, and finally four more at three-month intervals for a total of 13 letters or emails over two years.

Any replies to the letters from patients were tracked for analysis. During the first year of the pilot study, more than 436 letters and emails were sent to 111 participants. Although the pilot study did not include a no-contact control group and was underpowered to assess suicide rates, the results showed that 15 (14%) participants were subsequently readmitted after enrollment in the program compared to 20 (8%) nonparticipating inpatients during the first year of the program. Luxton and colleagues (2012) note that it was possible that the intervention may have influenced the readmission rate although the small sample size precluded determination of such an influence. The results also showed that the majority of participants (63%) preferred to receive follow-up contacts via email versus postal mail, and only two emails were returned as undeliverable during the first year. All of the patients enrolled in the pilot study were still living at the end of the two-year intervention period.

In sum, the pilot caring letters study showed that the protocol was feasible to conduct at a military treatment facility. The pilot supported the feasibility of using email as the mode of contacts. The pilot study showed that participants responded well to the caring letters interview and contact process. The protocol also did not cause any extra time burden on treatment staff nor did it cause any deviations from the normal standard of care. Furthermore, all of the participant responses to the contacts were positive, and there were not any adverse events that occurred during the pilot study. Ultimately, the pilot study provided useful data that informed the implementation of a large RCT that will determine the effectiveness of the caring letters intervention, including the effect on readmittance, treatment utilization, self-harm behavior, and completed suicides among U.S. service members and veterans.

A Randomized Controlled Trial

Luxton and colleagues (2014) are conducting a grant-funded five-year multisite trial that is evaluating the effect of a caring letters intervention on suicide mortality as compared to usual care in a group of posthospitalized military personnel and veterans. The trial is grant funded by the Military Operational Medicine Research Program and is a partial replication of Motto's (1976) original work that is updated via the use of email to send the caring contacts. Participants are recruited from inpatient psychiatry units at four military (three Army, one Navy) and two VA medical centers. The target sample size is 4,730 participants (2,365 per study group), which will provide sufficient statistical power to determine whether the caring letters intervention can prevent suicide and suicide behaviors in the military and veteran population. The research team's primary hypothesis is that during the two-year follow-up after the index hospital discharge, the frequency of suicide will be significantly lower in the intervention group compared to those in the usual care group. The research team also hypothesizes that the frequency of medically admitted self-inflicted injuries will be lower in the intervention group compared to the usual care group and that the time to suicidal behavior, if any, will be longer among participants in the intervention group compared to the usual care group.

As with the caring letters pilot study, patient participants in the clinical trial provide consent to study procedures and then complete a semistructured psychosocial interview with a trained research assistant. Participants are then randomized to either a group that receives caring emails or a usual care group that does not. Identical to the pilot study, participants in the intervention group will be emailed letters for two years. Suicide and all-cause mortality rates will be assessed at the end of the two-year intervention period. In addition, the number of suicide attempts, time to suicide behavior, and mental health treatment utilization rates will be compared across conditions. Suicide counts will be based on death certificates recorded in the Center for Disease Control and Prevention's National Death Index Plus (NDI-Plus). The NDI-Plus provides mortality information to include whether persons in studies have died, the names of the states in which those deaths occurred, the dates of death, and the corresponding death certificate numbers. Because the data available from the NDI-Plus is delayed by two years, the Social Security Administrations Death Master File, which is updated more frequently than the NDI-Plus, will be searched. After the two-year intervention period, all participants (both those in the intervention group and the usual care groups) will be sent a link to a brief secure web-based survey or receive a paper-and-pencil version in the mail in order to obtain information regarding treatment utilization, rehospitalizations, and suicidal thinking and behaviors. Participant feedback on whether this program should be used with other service members and veterans will also be assessed. Medical records of participants receiving care in the DOD or VA will also be reviewed to assess the accuracy of the self-reported data.

The randomized controlled trial will be completed in 2017. The results of the study will fill an important gap in the evidence base for the caring letters intervention and will help inform larger program implementation. The study will also help to determine moderators of treatment effect and determine for whom this intervention may be particularly well suited. In particular, subpopulation analyses are planned to compare suicide and self-harm rates among groups that continued in care after discharge to those who did not as well as differences between genders and other demographic variables. Study participants are also asked to complete the Interpersonal Needs Questionnaire and the Acquired Capability for Suicide Scale (Van Orden, Witte, Gordon, Bender, & Joiner, 2008) as well as the Soldier's Perceptions of Unit Cohesion Scale (Wright et al., 2009) in order to examine how thwarted belongingness, perceived burdensomeness, capability for lethal self-injury, and perceived unit cohesion and stigma (active duty military participants only) may influence outcomes. If the results of this RCT are promising, the intervention can be expanded and potentially become a standard of care in the DOD and VA health care systems.

CONSIDERATIONS FOR IMPLEMENTING CARING LETTERS INTERVENTIONS

Safety and Ethical Considerations

There are safety, ethical, and legal issues that should also be considered in caring follow-up contact programs. A safety plan that address procedures for scenarios when a threat of self-harm or harm to others is received in a follow-up contact reply or if the patient is in crisis is recommended. The safety protocol used for both the pilot study and RCT calls for immediate notification of the inpatient treatment team when such a reply is received so that they may contact the patient and assist with getting them additional care. In the caring letters project pilot study, there were just three occasions when a patient replied to an email with information that indicated that the patient was potentially in or may become in a crisis state. In all cases the safety plan was followed and the situation was resolved without incident. Also, unless a participant response includes notification of a potential or current crisis state, the plan specifies that the next letter will be sent according to the predetermined schedule. Similar to Motto's (1976) study, the subsequent letter briefly mentions the previous response for the patient. The inclusion of links to additional support resources can also be included. It is also important to not reinforce excessive responses and interchanges that are not consistent with the intervention intent. Excessive responses were not a problem in the Caring Letters Project pilot study.

Motto and Bostrom (2001) did not report any adverse events related to discontinuing the letters at two years, nor there were any adverse events at the end of the Caring Letters Project pilot study. However, it is important to consider that the discontinuation of the letters may be experienced negatively by some participants. In the Caring Letters Project pilot study and RCT, the letter frequency is decreased over time to reduce any sense of loss that may be experienced by the patient. The fact that a final email will be sent was also disclosed during the study consent process and the last two letters (at 21 and 24 months) mention that the letters will soon discontinue. The 12th letter includes the paragraph, "Just to let you know, we will send out one last note after this one. Please know that it has been a pleasure for us to reach out to you and we hope you are out in the world doing well for yourself." The 13th letter includes the paragraph:

> This is officially our last email to you and we hope that you have enjoyed getting our little quick notes over this time. As we said in our last letter, it was been a true joy for us to drop you a few lines here and there to let you know we truly care about you and hope you are getting along well in your life. If you wish to write us and let us know how you have been doing over the years, we would be glad to hear from you. (Luxton et al., 2014)

Technology Considerations

The updating of the original caring letters intervention from postal mail to electronic format is logical given the low cost and popularity of email. One of the primary advantages of electronic forms of correspondence is that it can reach service members no matter where they are located—even on deployments to combat zones. The use of smartphones and other mobile devices also allows users to access their email anywhere and at any time. The use of text messaging may also be a promising way to follow up with patients (Berrouiquet et al., 2014: Chen, Mishara, & Liu, 2010; Conner & Simons, 2014; Luxton, June, et al., 2013). Texting may be especially beneficial given the popularity, convenience, and low cost of text messaging. The lab of Comtois and colleagues, for example, is presently testing an texting contact intervention at military outpatient clinics (Conner & Simons, 2014).

Unlike written letters that take time to write and be received through the postal mail, electronic forms of communication, such as email and texting, are immediate. Thus it is important to consider the expectations of users of email and texting regarding this. The following statement was included in all emails in the Caring Letters Project pilot study and RCT: "Please know that I make every attempt to read my emails each business day. If for some reason you need immediate assistance, please reach out to the resources listed above." It is also possible that some patients may send and expect an excessive amount of correspondence. This was not an issue in the Caring Letters Pilot study; however, the best practice may be to respond to emails when appropriate and necessary but to maintain adherence to the planned schedule in order to not reinforce the behavior.

Also, not all eligible patients that may benefit from the caring letters intervention have access to email. Although all active duty service members have access to email, not all patients in the VA population do. A portion of the veteran population may also be homeless with very limited if any access to the internet. However, the VA recently implemented a program called "My Healthy Vet" that provides all veterans receiving VA care with an email account. Whether or not veterans will access their email on a regular basis is unknown; however, this will be assessed with the follow-up survey in the RCT. It is also a best practice to ask patients to provide an alternate email or contact address in case the account is closed or if there is a change of address.

Frequency and Timing of Contacts

Luxton, June, et al. (2013) point out that although the studies they reviewed involved posttreatment follow-up contact interventions aimed to prevent suicidal behavior, the interventions varied in regard to the timing, frequency, and total number of the follow-up contacts. For example, two of the studies (Fleishmann et al., 2008; Termansen & Bywater, 1975) initiated contact within the first week postdischarge and, another within two weeks, and the remaining had initial follow-up contacts ranging from one month to four months after hospital discharge. The number of follow-up contacts also varied from one to as many as 24 follow-up contacts with time durations ranging from one month to five years. There are not any specific conclusions from these studies regarding the optimal timing and frequency of follow-up contacts. However, data show that the highest risk for suicide is in first month postdischarge. Thus implementing the intervention immediately after discharge from psychiatric hospitalization is logical.

The optimal time period for sending caring contacts may vary based on population and other characteristics. For example, patients with the highest levels of risk, with a history of suicidal behavior, or who do not continue in care may benefit from extended contact periods. A sensitivity analysis conducted with data that result from the RCT will provide information to help determine the optimal and most efficient time frame to send the caring letters. It may also make the most sense to adjust the frequency of contact on an individual basis and to increase the frequency of contact when there are increases in other risk factors (Luxton, June, et al., 2013; Vaiva et al., 2011). Optimal modalities may also vary based on patient population characteristics. A tiered approach to the contact intervention, such as a hybrid caring followups program that is custom tailored to the type of patient (e.g., level of risk), may prove to be the most optimal. The ALGOS algorithm (Vaiva et al., 2011), for example, describes use of telephone contacts for repeat attempters, crisis cards for first attempters, and postal letters or postcards for patients who refuse or do not continue in treatment. The crisis card concept (Evans, Evans, Morgan, Hayward, & Gunnell, 2005), involves giving discharged patients a small green card that has contact information for suicide prevention organizations and care center hotlines. A RCT of the ALGOS algorithm is currently underway and its results will determine whether providing subgroups with specific interventions is effective at both short and longer time periods.

Level of Personalization

As discussed by Luxton, June, et al. (2013), the degree of personalization of the contacts varies between the interventions published in the literature. The studies that involved in-person and direct phone contacts were likely to be more personal because they involved direct and interactive conversation with patients. Motto's (Motto, 1976; Motto & Bostrom, 2001) intervention was personalized because information that was collected or received from patients who replied to the contacts was included in the letters. The three studies that used postcards (Beautrais et al., 2010; Carter et al., 2005; Hassanian-Moghaddam et al., 2011) sent standardized messages although the messages were written to express care. Also, with exception of one postcard study (Beautrais et al., 2010), all studies involved some form of initial interview or baseline assessment with patients. It may not be necessary, however, to conduct additional interviews or collect additional information from patients to implement caring contacts programs. That is, it may be adequate to send the contacts with the standard information that is learned about the patient during their treatment. Luxton, June, et al. speculate, however, that the degree to which follow-up contact providers get to know patients before making the contacts may be important for the effectiveness of the intervention. Moreover, who is sending the contacts, whether a single person, a treatment team, or an organization, may also influence personalization and perceived connection the patient has with the sender (Luxton, June, et al., 2013). For example, a generic letter sent from an entire treatment unit may be perceived as less personal or genuine than one that is sent and signed by one or two members of the treatment staff that the patient met. Although the optimal level of personalization is not known, the data available to date suggest that some level of personalization is important.

CONCLUSION AND FUTURE DIRECTIONS

This chapter provided an overview of the caring letters suicide prevention intervention—one of the only interventions shown to reduce suicide mortality rates. The multi-site Caring Letters RCT has the potential to reduce suicide and suicidal behaviors among service members and Veterans known to be at high-risk for suicide. Should it prove effective, this simple and inexpensive intervention could be expanded throughout the DOD and VA healthcare systems. As discussed earlier in this chapter, the current and additional research on the caring follow-ups intervention will help to identify the optimal modalities, timeframes, and types of messages (i.e., level of personalization) that are sufficient and effective in preventing suicide and suicide behavior.

One of the primary strengths of the intervention is its simplicity—it entails the sending of brief expressions of concern about the well-being of others. The intervention is also easy to implement and can be used in various treatment settings that encounter high risk for suicide patients. Although psychiatric hospitals are logical settings for implementing caring contact interventions, implementation in outpatient mental health settings, as well as primary care and specialty clinics may also be of value. Data reported by Trofimovich, Skopp, Luxton, and Reger (2012) shows that during 2001 through 2010, 45% of U.S.

active-duty service members who died by suicide and 75% of those who self-harmed had outpatient encounters within 30 days prior to suicide or self-harm. Primary care was the most frequently visited medical specialty prior to both suicide and self-harm. As compared to their counterparts, service members with suicidal behavior had a higher rate of outpatient visit rates within 60 days of their deaths or self-harm injuries. These data suggest that outpatient and primary care patients may benefit from additional suicide risk screening and follow-up contact interventions.

Caring contacts may also reduce costs associated with repeat hospitalizations. In particular, their preventative effect on suicidal behavior may reduce rehospitalization costs and also improve treatment compliance (Cedereke, Monti, & Öjehagen, 2002). Some data has indicated that hospitalized patients who did not comply with at least one outpatient appointment after discharge were two times more likely to be rehospitalized than those who kept at least one appointment after discharge (Nelson, Maruish, & Axler, 2000). By preventing suicide behaviors and encouraging help-seeking behaviors, the intervention may ultimately help reduce costs associated with rehospitalizations and lost productivity over the long run. It is important to note that although the caring letters intervention is simple, it nonetheless takes some resources to implement these types of programs. Implementation requires personnel to manage sending the correspondence as well as monitoring, tracking, and responding to them when necessary. The amount of resources needed to implement this type of program depends on the volume of patients as well as the length in time and number of contacts.

In conclusion, caring contact interventions may help prevent posttreatment suicide and suicide behaviors among our nation's military service members and veterans. The intervention has the potential to be especially useful at targeting service members and veterans who choose to not continue in care or who may not have easy access to care. For example, many service members may not seek care in the VA after discharge from military service or they may reside in medically underserved areas (e.g., National Guard or reservists living in remote areas). Moreover, this non-intrusive intervention may not have the stigma associated with it that other psychological interventions do. At first consideration the simplicity of the caring contacts intervention may cast doubt as to whether it can prevent suicidal behavior; however, data show that this simple, compassionate intervention may be an effective approach to preventing suicide and suicide behaviors. This intervention leaves the door open to communication between patients and providers of care, connects patients in crisis to care services, and facilitates interpersonal connectedness and belongingness that may help to save lives.

REFERENCES

Appleby, L., Dennehy, J. A., Thomas, C. S., Faragher, E. B., & Lewis, G. (1999). Aftercare and clinical characteristics of people with mental illness who commit suicide: A case-control study. *Lancet, 352,* 1397–1400.

Appleby, L., Shaw, J., Amos, T., McDonnell, R., Harris C., McCann, K, ... Parsons, R. (1999). Suicide within 12 months of contact with mental health services: National clinical survey. *BMJ, 318,* 1235–1239.

Armed Forces Health Surveillance Center (AFHSC). (2012). Deaths by suicide while on Active Duty, Active and Reserve Components, U.S. Armed Forces, 1998–2011. *MSMR, 19*(6), 7–10.

Baumeister, R. F., & Leary, M. R. (1995). The need to belong: Desire for interpersonal attachments as a fundamental human motivation. *Psychological Bulletin, 117,* 497–529.

Beautrais, A. L., Gibb, S. J., Faulkner, A., Fergusson, D. M., & Mulder, R. T. (2010). Postcard intervention for repeat self-harm: Randomized controlled trail. *The British Journal of Psychiatry, 197,* 55–60.

Bennewith, O., Gunnel, D., Peters, T. J., Hawton, K., & House, A. (2004). Variations in the hospital management of self-harm in adults: an observational study. *BMJ, 328,* 1108–1109.

Bennewith, O., Peters, T., Hawton, K., House, A., & Gunnell, D. (2005). Factors associated with the non-assessment of self-harm patients attending an accident and emergency department: Results of a national study. *Journal of Affective Disorders, 89,* 91–97.

Berrouiguet, S., Gravey, M., Le Galudec, M., Alavi, Z., & Walter, M. (2014). Post-acute crisis text messaging outreach for suicide prevention: A pilot study. *Psychiatry Research, 217*(3), 154–157.

Bertolote, J. M., Fleishmann, A., De Leo, D., Phillips, M. R., Botega, N.J., Vijayajumar, L, ... Wasserman, D. (2010). Repetition of suicide attempts: Data from emergency care settings in five culturally different low- and middle-income countries participating in the WHO SUPRE-MISS study. *Crisis: The Journal of Crisis Intervention and Suicide Prevention, 31,* 194–201.

Bush, N. E., Fullerton, N., Crumpton, R., Metzger-Abamukong, M., & Fantelli, E. E. (2012). Soldiers' personal technologies on deployment and at home. *Telemedicine and eHealth, 18,* 253–263.

Carter, G., Clover, K., Whyte, I., Dawson, A., & D'Este C. (2005). Postcards from the EDge project: Randomized controlled trail of an intervention using postcards to reduce repetition of hospital treated deliberate self poisoning. *BMJ, 331.* doi:10.1136/bmj.38579.455266.E0.

Carter, G., Clover, K., Whyte, I., Dawson, A., & D'Este, C. (2007). Postcards from the EDge project: 24-month outcomes of a randomized controlled trail for hospital-treated self poisoning. *The British Journal of Psychiatry, 191,* 548–553.

Cedereke, M., Monti, K., & Öjehagen, A. (2002). Telephone contact with patients in the year after a suicide attempt: Does it affect treatment attendance and outcome? A randomized controlled study. *European Psychiatry, 17,* 82–91.

Chen, H., Mishara, B. L., & Liu, X. X. (2010). A pilot study of mobile telephone message interventions with suicide attempters in China. *Crisis: The Journal of Crisis Intervention and Suicide Prevention, 31,* 109–112.

Comtois, K. A., & Linehan, M. M. (2006). Psychosocial treatments of suicidal behaviors: A practice-friendly review. *Journal of Clinical Psychology, 62,* 161–170.

Conner KR, Britton PC, Sworts LM, Joiner TE. (2007). Suicide attempts among individuals with opiate dependence: the critical role of belonging. *Addictive behaviors, 32,* 1395–1404.

Conner, K. R., & Simons, K. (2015). State of innovation in suicide intervention research with military populations. *Suicide and Life Threatening Behavior, 45,* 281–292.

Darke, S., Williamson, A., Ross, J., & Teesson, M. (2005). Attempted suicide among heroin users: 12-month outcomes from the Australian Treatment Outcome Study (ATOS). *Drug and Alcohol Dependence, 78,* 177–186.

Desai, R. A., Dausey, D. J., Rosenheck, R. A. (2005). Mental health service delivery and suicide risk: The role of individual patient and facility factors. *The American Journal of Psychiatry, 162,* 311–318.

Evans, J., Evans, M., Morgan, H. G., Hayward, A., & Gunnell, D. (2005). Crisis card following self-harm: 12-month follow-up of a randomized controlled trial. *The British Journal of Psychiatry, 187,* 186–187.

Fleischmann, A., Bertolote, J. M., Wasserman, D., De Leo, D., Bolhari, J., Botega, N. J, … Thanh, H. T. T. (2008). Effectiveness of brief intervention and contact for suicide attempters: A randomized control trial in five countries. *Bulletin of the World Health Organization, 86,* 703–709.

Geddes, J. R., Juszczak, E., O'Brien, F., & Kendrick, S. (1997). Suicide in the 12 months after discharge from psychiatric inpatient care, Scotland 1968–92. *Journal of Epidemiology and Community Health, 51,* 430–434.

Goldacre, M., Seagroatt, V., & Hawton, K. (1993). Suicide after discharge from psychiatric inpatient care, *Lancet, 342,* 283–286.

Goldney, R. D. (1998). Suicide prevention is possible: A review of recent studies. *Archives of Suicide Research, 4*(4), 329–339.

Goldsmith, S. K., Pellmar, T. C., Kleinman, A. M., & Bunney, W. E. (Eds.). (2002). *Reducing suicide: A national imperative.* Washington, DC: National Academies Press.

Greene-Shortridge, T. M., Britt, T. W., Castro, C. A. (2007). The stigma of mental health problems in the military. *Military Medicine 172,* 157–161.

Gunnell, Hawton, Ho, Evans, O'Connor, Potokar, Donovan, & Kapur, N. (2008). Hospital admissions for self-harm after discharge from psychiatric inpatient care: cohort study. *British Medical Journal Online, 337.* Retrieved from https://www.researchgate.net/publication/23482266_Hospital_admission_for_self_harm_after_discharge_from_inpatient_care_Cohort_study

Hassanian-Moghaddam, H., Saeedeh, S., Kolahi, A., & Carter, G. L. (2011). Postcards in Persia: Randomised controlled trial to reduce suicidal behaviors 12 months after hospital-treated self-poisoning. *The British Journal of Psychiatry, 198*(4), 309–316. doi:10.1192.bjp.bp.109.067199.

Hickey, L., Hawton, K., Fagg, J., & Weitzel, H. (2001). Deliberate self-harm patients who leave the accident and emergency department without a psychiatric assessment: A neglected population at risk of suicide. *Journal of Psychosomatic Research, 50,* 87–93.

Ho, T. P. (2003). The suicide risk of discharged psychiatric patients. *Journal of Clinical Psychiatry, 64*(6), 702–707.

Hoge, C. W., Castro, C. A., Messer, S. C., McGurk, D., Cotting, D. I., & Koffman, R. L. (2004). Combat duty in Iraq and Afghanistan, mental health problems, and barriers to care. *The New England Journal of Medicine, 351*(1), 13–22.

Hunt, I., Kapur, N., Webb, R., Robinson, J., Burns, J., Shaw, J., & Appleby, L. (2008). Suicide in recently discharged psychiatric patients: A case-control study. *Psychological Medicine, 39*, 443–449.

Ingram, R. E., Miranda, J., & Segal, Z. V. (1998). *Cognitive vulnerability to depression.* New York: Guildford Press.

Joiner, T.E. (2005). *Why people die by suicide.* Cambridge, MA: Harvard University Press.

Joiner, T. E. Jr., Hollar, D., & Van Orden, K. A. (2006). On Buckeyes, Gators, Super Bowl Sunday, and the Miracle on Ice: "Pulling together" is associated with lower suicide rates. *Journal of Social and Clinical Psychology, 25*, 180–196.

Kapur, N., Cooper, J., Urara, H., May, C., Appleby, L., House, A. (2004). Emergency department management and outcome for self-poisoning: A cohort study. *Gen Hosp Psychiatry, 26*, 36–41.

Knesper, D. J., American Association of Suicidology, & Suicide Prevention Resource Center. (2010) *Continuity of care for suicide prevention and research: Suicide attempts and suicide deaths subsequent to discharge from the emergency department or psychiatry inpatient unit.* Newton, MA: Education Development Center.

Linehan, M. M. (1981). A social-behavioral analysis of suicide and parasuicidal implications for clinical assessment and treatment. In H. Glazier & J. Clarkin (Eds.), *Depression: Behavioral and directive intervention strategies* (pp. 229–294). New York: Garland.

Linehan, M. M. (2008). Suicide intervention research: A field in desperate need of development. *Suicide and Life-Threatening Behavior, 38*, 483–485.

Luxton, D. D., June, J. D. & Comtois, K. A. (2013). Can post-discharge follow-up contacts prevent suicide and suicide behavior? A review of the evidence. *Crisis: The Journal of Crisis Intervention and Suicide Prevention, 34*, 32–41.

Luxton, D. D., Kinn, J. T., June, J. D., Pierre, L. W., Reger, M. A., & Gahm, G. A., (2011). The Caring Letter Project: A military suicide prevention pilot program. *Crisis: The Journal of Crisis Intervention and Suicide Prevention, 33*, 5–12.

Luxton, D. D., Thomas, E. K., Chipps, J., Relova, R. M., Brown, D., McLay, R., ... Smolenski, D. J., (2014). Caring Letters for suicide prevention: Implementation of a multi-site randomized clinical trial in the US military and Veteran Affairs healthcare systems. *Contemporary Clinical Trials, 37*(2), 252–260. doi:10.1016/j.cct.2014.01.007

Luxton D. D., Trofimovich, L., & Clark, L. L. (2013). Suicide risk among US service members following psychiatric hospitalization, 2001–2011. *Psychiatric Services, 64*, 626–629.

Mann, J. J., Apter, A., Bertolote, J., Beautrais, A., Currier, D., Haas, A, ... Hendin, H. (2005). Suicide prevention strategies: A systematic review. *JAMA: Journal of the American Medical Association, 294*, 2064–2074.

Maris, R. W., Berman, A. L., & Silverman, M. M. (2000). Treatment and prevention of suicide. In R. W. Maris, A. L. Berman, & M. M. Silverman (Eds.), *Comprehensive textbook of suicidology.* (pp. 509–535). New York: Guilford Press.

Meehan, J., Kapur, N., Hunt I. M., Turnbull, P., Robinson, J. Bickley, H., ... Appleby, L. (2006). Suicide in mental health inpatients and within 3 months of discharge: National clinical survey. *The British Journal of Psychiatry, 188*, 129–134.

Motto, J. A. (1976). Suicide prevention for high-risk persons who refuse treatment. *Suicide and Life-Threatening Behavior, 6*, 223–230.

Motto, J. A., & Bostrom, A. G. (2001). A randomized controlled trial of postcrisis suicide prevention. *Psychiatric Services, 52*, 828–833.

Neeleman, J., & Wessely, S. (1999). Ethnic minority suicide: A small area geographical study in south London. *Psychological Medicine, 29*, 429–436.

Nelson, E. A., Maruish, M. E., & Axler, J. L., (2000). Effects of discharge planning and compliance with outpatient appointments on readmission rates. *Psychiatric Services, 51*, 885–889.

Nisbet, P. A. (1996). Protective factors for suicidal black females. *Suicide and Life-Threatening Behavior, 26*, 325–341.

Roy A. (1982). Risk factors for suicide in psychiatric patients. *Archives of General Psychiatry, 39*, 1089–1095.

Somkin, C. P., Hiatt, R. A., Hurley, L. B., Gruskin, E., Ackerson, L., & Larson, P. (1997). The effect of patient and provider reminders on mammography and papanicolaou smear screening in a large healthcare maintenance organization. *Archives of Internal Medicine, 157*, 1658–1664.

Tran, M. T., & Billups, S. J. (2008). Can guideline adherence be improved by letter outreach? *Diabetes Care, 31*(4), e22.

Termansen, P. E., & Bywater, C. (1975). S.A.F.E.R.: A follow-up service for attempted suicide in Vancouver. *The Canadian Psychiatric Association Journal, 20*, 29–34.

Trofimovich, L., Skopp, N. A., Luxton, D. D., & Reger, M. A., (2012). Health care experiences prior to suicide and self-inflicted injury, active component, US Armed Forces, 2001–2010. *Medical Surveillance Monthly Report, 19*, 2–6.

Trout, D. L. (1980). The role of social isolation in suicide. *Suicide and Life-Threatening Behavior, 10,* 10–23.

Vaiva, G., Vaiva, G., Ducrocq, F., Meyer, P., Mathieu, D., Philippe, A. ... Goudemand, M. (2006). Effect of telephone contact on further suicide attempts in patients discharged from an emergency department: Randomised controlled study. *BMJ, 332*(7552), 1241–1245.

Vaiva, G., Walter, M., Al Arab, A.S., Courtet, P., Bellivier, F., Demarty, A.L, ... Libersa C. (2011). ALGOS: The development of a randomized controlled trial testing a case management algorithm designed to reduce suicide risk among suicide attempters. *BMC Psychiatry, 1,* 1–7.

Van Orden, K. A., Witte, T. K., Cukrowicz, K. C., Braithwaite, S., Selby, E. A., & Joiner, T. E. (2010). The Interpersonal Theory of Suicide. *Psychological Review, 117*(2), 575–600. http://doi.org/10.1037/a0018697

Van Orden, K. A., Witte, T. K., Gordon, K. H., Bender, T. W., & Joiner, T. E. Jr. (2008). Suicidal desire and the capability for suicide: Tests of the Interpersonal-Psychological Theory of Suicidal Behavior among adults. *Journal of Consulting and Clinical Psychology, 76,* 72–83.

Valenstein, M., Kim, H. M., Ganoczy, D., McCarthy, J. F., Zivin, K., Austin, K. L., ... Olfson, M. (2009). Higher-risk periods for suicide among VA patients receiving depression treatment: Prioritizing suicide prevention efforts, *Journal of Affective Disorders, 112*(1–3), 50–58.

Wright, K. M., Cabrera, O. A., Bliese, P. D., Adler, A. B., Hoge, C. W., & Castro, C. A. (2009). Stigma and barriers to care in soldiers postcombat. *Psychological Services, 6,* 108–116.

You, S., Van Orden, K., & Conner, K. (2011). Social connections and suicidal thoughts and behavior. *Psychology of Addictive Behaviors, 25,* 180–184. doi:10.1037/a0020936.

Index

Page references for figures are indicated by *f*, for tables by *t*, and for boxes by *b*.

ABCDE forces, 219–220
ABCD model, cognitive restructuring, 111
Abuse, childhood, 46, 56
Access to care
　mental health care in, 2, 207
　for OEF/OIF/OND veterans, 29–30
　for older veterans, 207
ACE education program, 48
Acquiescence bias, 226
Acquired capability for suicide, 54*f*, 55, 64
　combat exposure on, 104
　combat-related killing and, 70–71, 90–91
　in military and veterans, 66–67
　military service on, 55–57
　previous suicide attempts in, 56
　self-injurious behaviors in, 56
Acquired Capability for Suicide Scale (ACSS), 65, 91
Adaptive disclosure, 72
Adler, L. E., 184
Advance directives, 17
Afghanistan conflict veterans, 23–34. *See also* Operation Enduring Freedom (OEF) and Operation Iraqi Freedom (OIF)/Operation New Dawn (OIF/OND) veterans; *specific topics*
　Army National Guard in, 39–49 (*See also* National Guard suicides, of Army OEF/OIF/OND veterans)
　barriers to care in, 2
　combat-related trauma exposure in, 57
　stress exposure in, 53
　suicide rates of, 1
African Americans, 16, 81
Age. *See also* Older veterans, suicide management
　in suicide risk, 41*t*, 42*t*, 43, 45
Air Force, OEF/OIF/OND veteran suicide rates in, 25–26

Alcohol abuse. *See also* Substance abuse and substance use disorders
　killing and, 71, 72
　in Special Operations Force, 191
　on suicide risk in older veterans, 204
ALGOS algorithm, 249
All-volunteer force, 23–24
Alteration of consciousness/mental state (AOC)
　assessment of, 180
　in traumatic brain injury, 178–179, 179*t*
Alterations
　marked, 132
　negative, 132
Altieri, J., 131
Altruism, traumatic stress and, 18
Altruistic suicide, 15
Amitriptyline, for PTSD, 134
Anestis, M. D., 56–57
Anomic suicide, 15
Antiquity, 10–11
Anxiety, 96
Apter, A., 206, 207
Aristotle, 11, 15
Army, suicide rates
　historic, 40–41
　in OEF/OIF/OND veterans, 25
Army National Guard, 39–49. *See also* National Guard suicides, of Army OEF/OIF/OND veterans
　deployment of, OEF/OIF/OND, 39
　PTSD risk in, 39
　suicide rates in, 39
Army Study to Assess Risk and Resilience in Service members (Army STARRS), 30, 90
Asian Americans, 16

Asymmetrical combat, 23
Atkins, D. C., 156
Atrocities, witnessing/participating in, 96
Auchterlonie, J. L., 29
Austin, K., 9, 92, 240
Autonomy, patient, 17, 123
Autonomy need, 218
Avoidance, persistent, 132
Azrael, D., 202

Bahraini, N., 66, 69, 184, 186
Barber, C., 202
Barnes, S. M., 183
Barnett, J. E., 124
Barthe, J. T., 180
Base rates, 114
Battle exhaustion, 3
Battlemind, 86
Battlemind training, 110
Baumeister, R. F., 243–244
Beardslee, W. R., 195
Beautrais, A., 206, 207
Behavioral Health Lab, 221
Behavior change interventions, 215
Behavior change theories, 216–219
    means restriction in, 220
    self-determination theory in, 216–219, 228
    systems theory in, 216, 217–218, 217f, 228
Bell, J., 96
Bell, M. R., 25, 27, 30
Belongingness, thwarted, 54–55, 54f, 64, 69, 90, 122, 244
Bender, T. W., 56–57
Beneficence, 123
Berman, A. L., 95–96, 123, 124, 128
Bernert, R. A., 123
Bertolote, J., 206, 207, 241, 243
Betthauser, L., 66, 69, 94
Bias
    acquiescence, 226
    projection, 226
Blast injury, traumatic brain injury, 28, 179
Blazer, D. G., 131
Blow, F., 93, 182
Blue Ribbon Work Group on Suicide Prevention in the Veteran Population, 90, 202, 226
Bogner, J., 180
Bohnert, A., 93
Bolhari, J., 241, 243
Bongar, B., 14, 24, 97, 107, 115–116, 118, 122, 124, 125, 197, 202, 205
Borges, G., 55, 131
Bosch, J., 68–69, 70
Bossarte, R. M., 131
Bostrom, A. G., 242, 245, 247
Botega, N. J., 241, 243
Bottom-up responsiveness, 218
Boudewyns, P., 96
Boyd, J. N., 110
Boyko, E. J., 25, 27, 30

Boyle, C. A., 103
Brandfon, S., 66
Brazaitis, K. A., 156
Brenner, L. A., 66, 69, 94, 179, 182, 183, 184, 185, 186, 203
Brett, E., 70
Brief therapies, 116–117
Britton, P. C., 59, 204
Bromet, E. J., 26, 55
Brook, J. S, 117
Brosheke, D. K., 180
Brown, D., 241, 242, 246, 248
Brown, G., 97, 98, 203
Brown, G. K., 151
Brown, J., 117
Brown, L., 202
Bryan, C. J., 59, 67, 97–98, 104, 107, 116, 182–183, 191, 192
Bueler, C. E., 183
Bullman, T., 90, 91, 92–93, 94, 104
Burdensomeness, perceived, 54, 54f, 64, 69, 90, 122
Burke, A., 184
Bush, N., 117
Bush, N. E., 6
Bush, S. S., 180
Bushido, 14
Butcher, J. N., 117

Caine, E. D., 205
Caldwell, A. B., 116, 117
Calle, E. E., 202
Campise, R., 65
CAMS SSF Stabilization Plan, 151, 153, 162f
Capability for suicide, acquired. *See* Acquired capability for suicide
Caring letters, suicide prevention, 240–250
    Caring Letters Project Pilot Study in, 245–246
    future directions in, 249–250
    implementation of, 247–249
    Motto's caring letters study on, 241–242, 243
    postcard posthospitalization caring contacts in, 242–243
    randomized controlled trial on, 246–247
    suicide risk after psychiatric treatment and, 240–241
    telephone call interventions in, 243
    theoretical basis for, 243–245
    WHO SUPRE-MISS study on, 242, 243
Caring letters study, Motto's, 241–242, 243
Castro, C. A., 29, 86
Caucasians, 16
Cha, C. B., 55
Chapman, B., 204
Chard, K. M., 183
Chen, K., 16
Chessen, C., 156
Chi, K., 16
Childhood abuse, 46, 56
Childhood adversity, 117
Children, military families with, 82
Chipps, J., 241, 242, 246, 248
Chiu, W. T., 131
Christianity, on suicide, 13, 14, 15

Chronic illness
  physician-assisted suicide with, 17
  on suicide risk, 128, 136
    in older veterans, 16, 202, 204–205
    treatment of, 216b
Chu, J., 16
Churchwell, J. C., 183
Cimbolic, P., 155
Claasen, C., 95
Clancy, Carolyn, 219
Clark, L. L., 240
Clemans, T. A., 182–183
"Client," difficulties identifying, 126–127
Client feedback, 117
Clozapine, for suicide prevention with schizophrenia, 207
Codetypes, MMPI-2, 115–117
Cognitions. *See also specific types and disorders*
  in MMPI-2 codetypes, 116
Cognitive-behavioral therapy (CBT), 6
  for PTSD, 175
  for suicide risk, 6
Cognitive impairment
  level of consciousness in, 178–180, 179t
  on suicide risk in elderly, 203
Cognitive processing therapy, 72
  for PTSD, 133
Cognitive restructuring, ABCD model, 111
Cognitive therapy, for older veterans, 207
Cohen, P., 117
Collaborative Assessment and Management of Suicidality (CAMS), 136, 147–165
  adherence framework in, 154
  CAMS tracking/updating interim sessions in, 153–154, 163f–164f
  case vignettes on
    CAMS tracking/updating interim sessions in, 154
    clinical outcome/disposition in, 155
    epilogue on, 157–158
    initial assessment in, 149–150
    initial engagement in, 148–149, 160f–161f
    patient overview in, 147–148
    SSF stabilization planning and initial treatment plan in, 153
    treatment planning in, 151–153
  clinical outcome/disposition in, 154–155, 165f
  definition of, 148
  future directions in, 157
  goals of care in, 149
  philosophy of, 149
  research on, 155–157
  Stabilization Plan in, CAMS SSF, 151, 162f
  treatment planning in, initial, 150–151
Collectivism, 80
Colpe, L., 26, 27, 30, 47, 91, 95
Columbia Suicide Severity Rating Scale, 205–206
Combat
  asymmetrical, 23
  in interpersonal-psychological theory of suicide, 57–58
Combat experience (exposure)
  acquired capability and, 53–60, 104 (*See also* Interpersonal-psychological theory of suicide (IPTS))
  risk-taking propensity and, 106
  stress of, 83–84
  on suicide risk, 90–91, 93, 122
  suicide risk assessment with, 79–86, 89–98 (*See also* Suicide risk assessment, in combat veterans)
Combat neurosis, 3
Combat-related killing, 64, 67–71
  acquired capability for suicide in, 70–71, 90–91
  frequency and nature of, 67–68
  fundamentals of, 67–68
  interpersonal-psychological theory of suicide and, 64–74 (*See also* Interpersonal-psychological theory of suicide (IPTS))
  marital and relationship problems after, 72
  perceived burdensomeness and, 69
  suicidal ideation and, 68–69, 72, 73
  suicide attempts and, 70, 73–74
  thwarted belongingness and, 69
Combat rush, 106
Combat survival driving, 104
Combat trauma. *See* Post-traumatic stress disorder (PTSD)
Community health centers, 220
Competence
  of mental health professional, 124–125
  need for, 218
Comprehensive Soldier and Family Fitness (CSF2), 32
Comprehensive Soldier Fitness (CSF) program, 191
Comstock, B., 98
Comtois, K. A., 156, 244–245, 248–249
Conditions of war. *See also specific conditions*
  access to care in, 29–30
  IEDs and traumatic brain injury in, 1–2, 23, 27–29, 173
  in OEF/OIF/OND, 27–30
  PTSD in, 29
  sleep deprivation in, 29
Conduct offense waivers, for OEF/OIF/OND recruits, 30
Conner, K. R., 204, 205, 244
Conrad, A. K., 155
Consultation, with mental health professional, 125
Contextual understanding, 127
Control over personnel, military, 128
Conwell, Y., 205
Cornette, M. M., 66, 69
Corona, C. D., 156
Corrigan, J. D., 179, 180
Cotting, D. I., 29, 70–71, 86, 106
Cottrell, L, 26–27
Covert and subintentioned suicide, veteran, 103–112. *See also* Motor vehicle accident deaths, veteran
  homicide deaths in, 103
  motor vehicle accident deaths in, 103–104
Cox, A. L., 70–71, 106
Cox, K., 26, 27, 91
Criminal histories, of OEF/OIF/OND recruits, 31
Crisis hotlines, 137
  Los Angeles Suicide Prevention Center, 229
  VA, 5, 89, 97, 146

Cukrowicz, K. C., 67, 104, 116
Culture
    differences in, 16
    military (*See* Military culture)
    of Special Operations Force, 197
Currier, D., 206, 207
Currier, G., 203

Dahlstrom, L. E., 115
Dahlstrom, W. G., 115
Danger-seeking behaviors, postdeployment, 106
Darke, S., 244
Dartmouth Atlas, 223
Dausey, D., 92
Davidson, J. R. T., 131
Death with Dignity Act, 17
Decoufle, P., 103
Deitrick, A., 203
Delaney, R. J., 103
De Leo, D., 106, 241, 243
Dementia, on suicide risk in elderly, 203
Demler, O., 131
Dependence, 218
Deployment
    of Army National Guard, OEF/OIF/OND, 39
    stress of, 82
    suicide and, in OEF/OIF/OND force, 26–27
Deployments, multiple
    of OEF/OIF/OND veterans, 24, 173
    on PTSD and suicide risk, 173, 191
Depression, 14, 96. *See also* Major depressive disorder
    mental health care for, 86
    MMPI-2 scale on, 115
    PTSD with, 132
    recognition of, physician education in, 206–207
    sleep deprivation in, 29
    on suicide risk, 92, 117
        in elderly, 203
        of OEF/OIF/OND Army National Guard, 46
        of OEF/OIF/OND veterans, 25, 57
    traumatic brain injury and, 180–181
Desai, R., 92
Desert Storm veterans
    postdeployment mortality risk in, 103–104
    vehicle-related accident deaths in, 103 (*See also* Motor vehicle accident deaths, veteran)
DeStefano, F., 103
Devore, M. D., 66
Dialectical Behavioral Therapy (DBT)
    for older veterans, 207
    for PTSD, 175
    for suicide risk, 6
Discipline, 80
Disease-centered biomedical mode, 221
Disease-centered care, 223
Dishonor, 10
Dispositions, to suicide risk, 46–47. *See also specific types*
Doane, B. M., 111

Driving, risky, 103–112. *See also* Motor vehicle accident deaths, veteran
Drug offense waivers, in OEF/OIF/OND recruits, 30
Dual identities, 126
Duberstein, P. R., 205
Duffy, S., 66
Durkheim, Émile, 11, 14–15, 109, 166–167

Earnshaw, M., 83
Egoistic suicide, 14–15
Ehler, A., 83
Elderly. *See also* Older veterans, suicide management
    suicide risk in, 41*t*, 42*t*, 43, 45, 202–206
Electroconvusive therapy (ECT), 6–7
Emotional disorders. *See also* Post-traumatic stress disorder (PTSD); *specific types*
    after deployment, 194–195
    in military children, 195
    military culture as barrier to seeking care for, 86
    with PTSD, 93–94, 96
    in Special Operations Force, 194–195, 197
    with traumatic brain injury, on suicide risk, 182, 185
    of World War II enlistee rejections, 3
Emotional health and support
    in African Americans, 16
    Comprehensive Soldier and Family Fitness for, 32
    interpersonal attachment in, 54
    military culture on, 81
    person-centered suicide prevention for, 213–231
        (*See also* Person-centered suicide prevention, primary care settings)
    of religion, 14–15
    romantic partners in, 85
    social network in, 54
    unit cohesion in, 83
Emotional reactions, to traumatic events, 84
Engberg, A. W., 94, 181
Engel, G. L., 217, 217*f*
Epidemiology, suicide. *See also specific areas*
    prevalence in, 1, 4, 10, 53, 121–122
        in Army National Guard veterans, OEF/OIF/OND, 39
        in OEF/OIF/OND veterans, 121
    of suicidal ideation, 131
    suicide attempts in, 121
    suicide completion in, 1, 4, 10, 53, 121–122
    trends in, 7
Equivocal deaths, 108–109
Ethics
    mixed-agency ethical dilemmas, 125–126
    perspectives in, 15–16
    principles of, 123
    in treatment, 121–129 (*See also* Treatment, of suicidal patients, ethical issues)
Ethnicity, 81, 84
Euthanasia, in antiquity, 11
Executive dysfunction, traumatic brain injury, suicide, and, 183–186. *See also* Traumatic brain injury (TBI)
    conceptualizing, 184–185

future directions in, 185
research on, 184
suicide and, 184–185
Exposure-based therapies, for PTSD, 132–133
Eye movement desensitization and reprocessing (EMDR), for PTSD, 133–134

False negatives, 226
False positives, 226
Families, military, 81–82
demographics of, in OEF/OIF/OND force, 24
Farberow, Norman L., 19, 20
Fatalistic suicide, 15
Fear
conditioning, 18
habituation, 55–56
two-factor theory of fear conditioning and, 18
Fear, N. T., 83
Fearlessness, 55–58
Feedback, client, 117
Fidelity, 123
Firearms
removing or restricting access to, 5, 49, 124–125
in suicide, military, 5, 59, 121–122
suicide prevention training on, 48
on suicide risk in older veterans, 203–204
Firestone, R. W., 109
Fleischmann, A., 241, 243
Flock, M. L., 103
Flores, C., 156
Fluid vulnerability theory, 47–48
Fluoxetine, for PTSD, 134
Fontana, A., 70
Foreseeability, 123, 127
Forster, J. E., 184, 186
Freud, Sigmund, 17–18, 109
Friedman, J. H., 184
Friedman, M. J., 84
Frost, L., 26–27
Fullerton, C., 26, 30, 47, 91, 95
Functional decline, on suicide risk in older veterans, 204–205

Gahm, G. A., 6, 68–69, 182
Gallaway, M. S., 27
Ganoczy, D., 66, 240
Garand, L., 203
Garcia, E., 66
Gender
on Army National Guard suicides, 45–46
of military personnel, 81
in person-centered suicide prevention, 230
on suicide risk
male, 41, 41t, 42t, 43, 46
in medieval period, 12–13
George, L. K., 131
Ghahramanlou-Holloway, M., 65, 203
Gilman, S., 26, 27, 91

Girardi, P., 109
Glassmire, D. M., 115
Goldstein, M. B., 86
Gooding, P., 91–92
Gould, M. S., 117
GR205171, for PTSD, 134
Greeks, 10–11
Greenburg, D., 29
Greene, R. L., 115–116, 118
Gregg, L., 131
Grigsby, J., 106, 108–109
Grimes, J., 182
Grinker, Roy, 3
Group suicide, Roman, 11
Gruber, M. J., 26
Grunebaum, M. F., 184
Guard. *See* National Guard
Guilt, 96
Gutierrez, P. M., 66, 69

Haas, A., 96, 167, 168, 206, 207
Haas, G. L., 184
Hacker Hughes, J., 83
Halloran, Roy, 3
Hamburger, E., 106
Hanson, Fredrick, 3
Harm, avoiding, 127, 128
Harris, J. J., 81
Harwood, J., 94
Haskell, J., 191, 192
Health system layers, 214, 214f, 217–218, 217f
"Healthy warrior effect," 27, 202
Heeringa, S., 27, 30, 47, 91, 95
Hendin, H., 96, 167, 168, 206, 207
Hernandez, A. M., 182–183
Hierarchical classification system, 80
Hierarchy of natural systems (Engel), 217, 217f
Hijjawi, S. P., 203
Hill, J., 96–97
Hippocrates, 11
Hispanic Americans, in military, 81
Historical context, 2–4, 10–20
16th–20th century, 13–15
21st century, 15–17
antiquity, 10–11
medieval period, 12–13
research lineage and accomplishments in, 17–20
Freud, Sigmund, 17–18
International Association for Suicide Prevention, 19–20
London, Perry, 18–19
Los Angeles Suicide Prevention Center, 18, 19
Mowrer, Orval Hobart, 18–19
Murray, Henry, 18
Vietnam War, 4
war syndromes in, 2
World War I, 2–3
World War II, 3–4

Hoffman, F. G., 190
Hoge, C. W., 25, 27, 29, 30, 70–71, 86, 106
Hoggatt, K., 9, 92
Holz, K.B., 105, 110
Homaifar, B. Y., 94, 179, 184, 186
Home arrival stress, 84–85
Home departure stress, 82–83
Homelessness, on suicide risk in older veterans, 205
Homicide deaths, in Vietnam vs. early war veterans, 103
Honor, 14
Hopelessness, 122
Hospital discharge, of mental health patient, suicide risk after, 240
Hotline
  crisis, 137
  Los Angeles Suicide Prevention Center, 229
  VA Crisis, 5, 89, 97, 146
Hotopf, M, 83
Huggins, J., 94, 179
Hughes, D. C., 131
Huguet, N., 203, 204
Hull, L., 83
Human Performance Program, POTFF, 193–194
Hunter, M. I., 103
Hustead, L. A., 155
Hyer, L., 96
Hyman, J., 26–27

Identities, dual, 126
Ignacio, R., 93
Ignacio, R. V., 182
Ilgen, M., 90, 93, 204
Improvised explosive devices (IEDs)
  in Iraq War, 104
  traumatic brain injury from, 23, 27–29
Independence, 218
Indicated interventions, 220, 221
Informed consent, 123–124, 127
Insomnia. *See also* Sleep disorders
  on suicidal ideation, 122
Integrated care, 220
  mental health, 213 (*See also* Person-centered suicide prevention, primary care settings)
Interactive motivational computer program (IMCP), 227
International Association for Suicide Prevention (IASP), 19–20
Interpersonal Needs Questionnaire (INQ), 65
Interpersonal-psychological theory of suicide (IPTS), 53–59, 64–74, 90–91, 244
  acquired capability for suicide in, 54f, 55, 64, 90
    combat exposure on, 104
    combat-related killing and, 70–71, 90–91
    in military and veterans, 66–67
    by military service, 55–57
  acquired capability for suicide in military and veterans in, 66–67
  clinical implications of, 71–73
  combat in, 57–58
  combat-related killing in, 67–71
    acquired capability for suicide in, 70–71, 90–91
    fundamentals of, 67–68
    marital and relationship problems after, 72
    perceived burdensomeness and, 69, 90
    suicidal ideation and, 68–69, 72, 73
    suicide attempts and, 70, 73–74
    thwarted belongingness and, 69, 90
  empirical tests of, 65–66
  future directions in, 59–60, 73–74
  limitations of, 73–74
  model of, 53–55, 54f, 64–65
  perceived burdensomeness in, 54, 54f, 64, 90, 122
  prevention and intervention implications of, 58–59
  thwarted belongingness in, 54–55, 54f, 64, 90, 122, 244
Interpersonal therapy, for suicide risk, 6
  with PTSD, 175
Intrusions, 132
Invincibility, 106
Iraq conflict veterans, 23–34. *See also* Operation Enduring Freedom (OEF) and Operation Iraqi Freedom (OIF)/Operation New Dawn (OIF/OND) veterans; *specific topics*
  IEDs on, 104
  PTSD in, 84 (*See also* Post-traumatic stress disorder (PTSD))
  stress exposure in, 53
Ireland, R., 26–27
Iverson, A. C., 83
Iverson, G. L., 180

Jacoby, A. M., 65, 155
James, L. M., 105, 110
Janis, K. E., 156
Japan, Bushido in, 14
Jaycox, L. H., 23
Jennings, K. W., 156
Jobes, D. A., 65, 94, 123, 124, 128, 147–165, 155. *See also* Collaborative Assessment and Management of Suicidality (CAMS)
Johnson, D. C., 86
Johnson, J. G., 117
Johnson, L., 65
Johnson, W. B., 124, 125–126
Joiner, T. E., Jr., 53–57, 54f, 64–65, 90, 95–96, 98, 104, 151
Joining military, developmental context of, 79–80
Jones, N. G., 81
Joshua Omvig Veterans Suicide Prevention Act of 2007, 89
June, J. D., 244–245, 248–249

Kang, H., 90, 91, 92–93, 94, 104
Kant, 15
Kaplan, M. S., 66, 203, 204
Kardiner, Abram, 166
Kasen, S., 117
Katz, I. R., 131
Kelly, P. J., 66
Kemp, J., 131, 184
Kennedy, C. H., 125–126
Keough, K. A., 110
Kessler, R., 26, 27, 30, 47, 55, 91, 95, 131
Ketamine, on suicidal ideation with PTSD, 134

Killgore, W. D. S., 106
Killgore, W. S., 70–71
Killing
　alcohol abuse and, 71, 72
　combat-related (See Combat-related killing)
　psychiatric outcomes of, 71
　PTSD and, 71, 72
Kim, H. M., 66, 240
Kim, M., 9, 92
Kimsey, B. F., 105
Kinn, J., 6
Kizer, K. W., 218
Kleespies, P., 96–97
Knight, S. J., 68–69, 70
Knight suicidality, 13
Knox, K., 95, 131, 203
Koffman, R. L., 29, 86
Koocher, G. P., 125, 126, 127
Kroke, P. E., 107
Krysinska, K., 91, 92, 93
Kung, S., 155

Language, military, 80
Lawler, E., 202
Layne, C. M., 195
LeardMann, C. A., 25, 27, 30
Leary, M. R., 243–244
Lee, S., 55
Leino, A., 16
Lento, R. M., 156
Leskela, J., 105, 110, 111
Lester, D., 91, 92, 93
Lester, P., 195
L'Estoile, Pierre de, 13
Leuty, M. E., 111
Level of consciousness (LOC)
　assessment of, 180
　in traumatic brain injury, 178–179, 179t
Lewis, R., 181
Libertarian perspective, 15–16
Lifetime-Suicide Attempt Self-Injury Count, 135
Lineberry, T., 155
Litman, Robert E., 19
Logistic regression, 49n2
London, Perry, 18–19
Loneliness. See Social isolation
Lopez-Larson, M. P., 183
Los Angeles Suicide Prevention Center (LASPC), 18, 19
Luxton, D. D., 6, 29, 68–69, 117, 240–250. See also Caring letters, suicide prevention

MacArthur, Douglas, 80
Maguen, S., 68–69, 70
Major depressive disorder. See also Depression
　in OEF/OIF veterans, 1
　PTSD with, 93, 132
Maladaptive thinking, 116
Male gender, in suicide risk, 41, 41t, 42t, 43, 46
Malley, J. C., 86
Malone, K. M., 184

Mandrusiak, M., 95–96
Mann, J. J., 184, 206, 207
Mann, R. E., 156
Marine Corps suicide rates, in OEF/OIF/OND veterans, 25–26
Marital relationships, 81–82, 122
Marital status, military personnel, 81–82
Marked alterations, 132
Marmar, C. R., 68–69, 70
Marshall, G. N., 86
Masaryk, T. G., 14
McArthur, M., 11
McCarthy, J., 9, 92, 93, 240
McCranie, E., 96
McFarland, B. H., 203, 204
McGlade, E. C., 183
McGurk, D., 29, 70–71, 86, 106
McLay, R., 241, 242, 246, 248
McRaven, William, 192
McWhorter, S. K., 105
Means restriction (counseling), 72, 98, 124–125
　behavior change theories on, 220
　effectiveness of, 59
　for firearm suicide prevention, 107 (See also Firearms)
　in older veterans, 206
Medieval period, 12–13
Melancholy, 11, 14
Menninger, K., 109
Mental health care
　barriers to, 2, 85–86
　stigma of (See Stigma, of mental health care)
Mental health conditions. See also specific disorders; specific types
　comorbid, 117–118
　hospitalization for, suicide rates after, 240
　in OEF/OIF/OND veterans, prevalence of, 1–2, 10
　on suicide risk in older veterans, 203
　suicide with, 10
Mental health practitioner
　role shifts for, unanticipated, 127
　stress of, 121
　suicidal client and, 122–123
Messer, S. C., 29, 86
Metzler, T. J., 68–69, 70
Meyer, R. G., 115, 116, 117
Microsuicides, 109
Microsystem innovations, 214–215
Military and Family Life Consultants (MFLC), SOCOM, 195
Military culture
　collectivism in, 80
　discipline in, 80
　framework of, 79
　hierarchical classification system in, 80
　joining military in, developmental context of, 79–80
　language in, 80
　as mental health care barrier, 85–86
　mission mindset in, 81
　personnel in, 81–82
　solution-focused problem-solving skills in, 81

Military culture (Contd.)
  on suicide risk assessment, 79–82
  traditions and values in, 80–81
Military sexual trauma (MST), 83–84, 139–140
Miller, M., 202
Millikan, A. M., 27
Milliken, C. S., 29
Milner, J. S., 105
Minnesota Multiphasic Personality Inventory (MMPI), 114
Minnesota Multiphasic Personality Inventory-2 (MMPI-2), 114
Minnesota Multiphasic Personality Inventory-2 (MMPI-2) suicide risk factors, 114–118
  base rate issue, low, 114
  causes and types of suicide in, 118
  client feedback in, 117
  codetypes in, 115–117
  comorbid psychiatric disorders in, 117–118
  future directions in, 114, 118
  history and fundamentals of, 114
  literature on, 114
  scales in, 115
  suicide items in, 115
  treatment considerations in, 116–117
Mirtazapine, for PTSD, 134
Mission, superordinate, 126
Mission mindset, 81
Mitchell, A. M., 203
Mitchell, M. M., 27
Mixed-agency ethical dilemmas, 125–126
Mood disorders, 122
Moods. *See also specific types*
  in MMPI-2 codetypes, 116
Moralistic perspective, 15
Morrow, C. E., 104, 116, 191, 192
Mortality risk, postdeployment
  in Desert Storm veterans, 103–104
  in Vietnam veterans, 103, 104
Motivational interviewing (MI), 136
Motor vehicle accident deaths, veteran, 103–112
  assessment and treatment of, 109–111
  driving-related suicidal ideation in, 106–107
  equivocal and subintentioned deaths in, 108–109
  future directions in, 112
  postdeployment mortality risk in, 103–104
  risky driving in, 104–106
  suicide by motor vehicle crash in, 107–108
  in Vietnam vs. early war veterans, 103
Motto, Jerome, 241–248. *See also* Caring letters, suicide prevention
Motto, J. J., 242, 245, 247
Mowrer, Orval Hobart, 18–19
Mukamal, K. J., 202
Munroe, J. F., 85, 86, 95
Murray, D., 106
Murray, H, 18
Myra Kim, H., 93
Mysliwiec, V., 29

Nademin, E., 65
Nagamoto, H., 179, 184, 186
Nagelkerke's pseudo R^2, 49n2
Nash, W. P., 195
National Guard, in OEF/OIF/OND operations
  access to care for, 29–30
  deployment of, 23–24
  mental health outcomes of, postdeployment, 29
  suicide risks and rates of, 24, 25
National Guard suicides, of Army OEF/OIF/OND veterans, 23, 39–49
  data sources and methods in, 40
  deployment of, 39
  emergent problem in, 39
  interpreting of findings on, 45–47
    age, gender, and race in, 45–46
    dispositional risk in, 46–47
    person vs. military experiences in, 45
  prevalence increase in, 39
  purpose of study on, 40
  suicide prevention implications in, 47–49
    firearm training in, 48
    follow-up protocol in, 48
    policy refocus on empirically supported risk factors in, 48–49
    screening at-risk soldiers in, 47–48
    symptom recognition training in, 48
  summary of findings on, 40–45
    suicide completer subgroups in, 43–44
    suicide rates and risk in, 39, 40–42, 40f, 41t, 49n1
    suicide risk factors in, 41t, 42–43, 42t, 49n2
    suicides in, ARNG vs. USAR, 44–45
National Guard suicides, risks and rates of, 39
National Reserve, 23–24
Navy and Marine Corps Combat and Operational Stress Control, 32
Navy suicide rates, in OEF/OIF/OND veterans, 25–26
Need-press theory, 18
Negative affectivity, 46
Negative alterations, 132
Newsom, J. T., 204
Neylan, T. C, 68–69, 70
Nielsen, A. C., 156
Nietzsche, F., 111
Nisbet, P. A., 244
Niven, A., 29
Nock, M., 30, 47, 55, 56–57, 91, 95–96
No-harm contract, 124
Noncompliance, treatment, 241
Nonmaleficence, 123
Nye, E., 96

Obligations, responsibility, 128
O'Connor, S., 156
Oetjen-Gerdes, L., 182
Older veterans, suicide management, 201–208
  assessment in, 205–206
  data issues in, 201

legal capacity in, 206
prevention/treatment strategies in, 206–208
   access to mental health services in, 207
   means restriction in, 206
   pharmacotherapy and adverse effects in, 207
   provider education in, 206–207
   psychotherapy in, 207
   VHA in, 208
prospective assessment in, 205–206
risk factors in, 202–206
   alcohol abuse in, 204
   chronic illness and functional decline in, 204–205
   firearms in, 203–204
   history of suicidal ideation/behavior in, 205
   homelessness in, 205
   identification of, 202–203
   mental health in, 203
   sleep disorders in, 204
   social isolation in, 204–205
   substance abuse in, 204
   transition periods in, 203
   in veterans, 203
suicide rates in, 201–202
suicide risk in, 41*t*, 42*t*, 43, 45, 202–206
Olfson, M., 240
Olson-Madden, J. H., 179
Operation Enduring Freedom (OEF) and Operation Iraqi Freedom (OIF)/Operation New Dawn (OIF/OND) veterans, 23–34. *See also specific topics*
Army STARRS on, 30, 90
asymmetrical combat in, 23
combat-related trauma exposure in, 57
conditions of war in, 27–30
   access to care in, 29–30
   IEDs and traumatic brain injury in, 23, 27–29
   PTSD in, 29
   sleep deprivation in, 29
deployment and suicide in
   contradictory conclusions on, 26–27
   interpersonal factors in, 27
deployments in, multiple, 24, 191
force and recruits in
   all-volunteer force in, 23–24
   criminal histories of, 31
   drug and conduct offense waivers for, 30
   family demographics of, 24
   high-risk populations in, 24
mental health condition prevalence in, 1–2, 10
National Guard and Reserve in, 23–24
OEF *vs.* OIF/OND veterans in, 23
pre-enlistment risk factors in, 30
prevalence of suicide in, 121
stress exposure in, 53
substance abuse in, 30–31, 56
suicide prevention in, 31–33
suicide rates for, 24–25
suicide rates for, by service, 25–26
Oquendo, M. A., 184

Pain tolerance, 55–58
Pan, D., 203
Panagioti, M., 91–92
Paroxetine, for PTSD, 134
Patient Aligned Care Team (PACT), 207
Patient autonomy, 17, 123
Patient-centered care, 230–231n2
Patient Health Questionnaire (PHQ-9), 135–136
Patient Protection and Affordable Care Act (PPACA), 214, 214*f*
Patton, Gen. George S., 4
Payne, C. E., 109
Peck, D. L., 108
Pedersen, C. M., 156
Perceived burdensomeness, 54, 54*f*, 64, 69, 90, 122
Perry, J. N., 117
Persistent avoidance, 132
Person-centered suicide prevention, primary care settings, 213–231
attempted *vs.* completed suicide in, 229–230
behavior change interventions in, 215
behavior change theories in, 216–219
   means restriction in, 220
   self-determination theory in, 216–219, 228
   systems theory in, 216, 217–218, 217*f*, 228
challenges of, 223–224
core propositions for, 215, 215*b*
disease- *vs.* person-centered options in, 224–228
future directions in, 229–230
gender differences in, 230
health system layers and, 214, 214*f*
integrated mental health care in, 213
less (treatment) as more (quality) in, 216*b*
macro-level initiatives in, 215
microsystem innovations in, 214–215
patients' needs and wants in, 229
person-centered prevention and ABCDE forces in, 219–220
postcard intervention in, 215
power asymmetry reduction in, 215, 223
therapeutic alliance in, 229
traditional approaches *vs.*, 220–222
   communication in, 222
   indicated interventions in, 220, 221
   integrated care in, 220
   stepped care in, 221
   targeted interventions in, 220–221
   universal prevention programs in, 220
   workplace innovations in, 222
unintended consequences in, 229
VA Primary Care-Mental Health Initiative for, 213–214
Personnel, in military culture, 81–82
Pfeiffer, P. N., 66
Pflanz, S. E., 65
Pharmacotherapy, 6. *See also specific types*
in elderly, adverse effects of, 207
for PTSD, 134

Phenelzine, for PTSD, 134
Phenomenological imperative, 229
Physical performance, POTFF, 193–194
Physician-assisted suicide, 17
Piegari, R., 131
Pietrzak, R. H., 86
Pigeon, W. R., 204
Plato, 10–11, 15
Pompili, M., 109
Possis, E., 111
Postcards. *See also* Caring letters, suicide prevention
  for person-centered suicide prevention, 215
  for posthospitalization caring contact, 242–243
Postdeployment danger-seeking behaviors, 106
Post-traumatic amnesia (PTA)
  assessment of, 180
  in traumatic brain injury, 178–179, 179t
Post-traumatic stress disorder (PTSD)
  assessment of, 135
  comorbid psychological disorders with, 132
  comorbid risk factors with, 93–94
  definition of, 131–132
  depression with, 58, 117, 132
  diagnostic criteria for, 166
  historical perspectives on, 166–167
  major depressive disorder, 93
  mental health care for, 86
  prevalence of, 131
  risky driving with, 105
  substance use disorders with, 94
  traumatic brain injury and, 94, 181
Post-traumatic stress disorder (PTSD), suicide and, 28, 57, 58, 131–132
  in Army National Guard, 39
  combat-related killing and, 69
  with combat-related PTSD, 91–93
  killing and, 71, 72
  in knights, 13
  in OEF/OIF/OND veterans, 1–2, 25, 28, 29
  risk factors for, 96, 104, 122, 131, 167
  in Vietnam War veterans, 4, 167
Post-traumatic stress disorder (PTSD) treatment, 175
Post-traumatic stress disorder (PTSD) treatment, evidence-based, 6, 131–142
  cognitive processing therapy in, 133
  with comorbid suicidality, 135–142
    assessment in, of PTSD, 135
    assessment in, of suicidality, 135–136
    barriers to, minimizing, 137
    case vignettes on, 139–142
    conceptualization in, 136–138
    with high-risk suicidality, 137–138
    with moderate suicidality, 137–138
    safety plan in, 136–137
    therapeutic rapport in, 138
    treatment specifics in, 138–139
  eye movement desensitization and reprocessing in, 133–134
  pharmacotherapy in, 134
  prolonged exposure therapy in, 132–133
  psychological comorbidities with, 132
  PTSD definition in, 131–132
  PTSD suicide risk in, 131–132
  stress inoculation training in, 134
Post-traumatic stress disorder (PTSD) treatment, for suicide risk, 166–177
  case vignettes on
    combat meaning in, 168
    protection in, 170–173
    treatment in, 169–170
    treatment in, Iraq and Afghanistan veterans, 174–175
  combat experience in, meaning of, 168
  historical context of, 166–167
  protection factors in, 170–173
  recognizing veterans at risk in, 168–169
  treatment in, 169–170
    effectiveness of, 175
    of Iraq and Afghanistan veterans, 173–175
    therapeutic relationship in, 174–175
    therapist issues in, 173–174
POTFF. *See* Preservation of the Force and Family (POTFF)
Powell, T. M., 25, 27, 30
Power asymmetries, 215, 223
Prehabilitive training program, POTFF, 193–194
Present time perspective, 110–111
Preservation of the Force and Family (POTFF), 192
  objective and development in, 192–193
  rationale of, 192
Preservation of the Force and Family (POTFF), four domains, 193–196
  history and fundamentals of, 193
  physical performance in, 193–194
  psychological performance in, 194
  social performance in, 194–195
  spiritual performance in, 195–196
Presuicidal syndrome, 19
Prevalence. *See* Epidemiology, suicide
Prevention, suicide. *See* Suicide prevention
Pridmore, S., 11
Primary care providers (PCPs)
  comprehensivist, 223
  definition of, 230n1
  suicide prevention by, 213–231 (*See also* Person-centered suicide prevention, primary care settings)
Privacy, 123
Problem-solving skills, solution-focused, 81
Professional role socialization, 221
Projection bias, 226
Prolonged exposure therapy, for PTSD, 132–133, 175
Prospective assessment, 205–206
Protective factors, 95
Provider education, for older veterans, 206–207
Psychache, 18
Psychiatric care, stigma of. *See* Stigma, of mental health care
Psychiatric disorders. *See* Mental health conditions

Psychiatric hospitalization, suicide rates after, 240
Psychodynamic therapy, 6
Psychoeducation, 5
Psychological Performance Program, POTFF, 194
Psychotherapy, 6. *See also specific types*
    for older veterans, 207
Public health initiatives, 216
    behavior change theories in, 216–219

Rabenhorst, M. M., 105
Race
    of military personnel, 81
    in military stress, 84
    in suicide risk, 42t, 43, 45–46
Rajab, M. H., 151
Rationality, of suicide, 15
Rational suicide, 17
Real Warriors initiative, 31–32
Reasonable care, 123, 124
Reger, M. A., 6, 68–69, 249–250
Regression, logistic, 49n2
Reitzschutz theory, 166
Relatedness, 218
Relativist perspective, 16
Religion, in emotional health and support, 14–15
Relova, R. M., 241, 242, 246, 248
Resilience, 191–192
Respect, 123
Responsibility obligations, 128
Ribeiro, J. D., 56–57
Ringel, Erwin, 19
Rings, J. A., 66
Risk assessment. *See* Suicide risk assessment, in combat veterans
Risk factors, suicide, 16, 122. *See also* Post-traumatic stress disorder (PTSD); *specific types*
    for Army National Guard, OEF/OIF/OND, 41t, 42–43, 42t, 49n2
    childhood adversity in, 117
    combat experience in, 90–91, 93
    empirically supported, policy focus on, 48–49
    evaluating, 95
    MMPI-2, 114–118 (*See also* Minnesota Multiphasic Personality Inventory-2 (MMPI-2) suicide risk factors)
    in older veterans, 202–206 (*See also* Older veterans, suicide management)
    predeployment, pre-enlistment, in OEF/OIF/OND, 30
    two-tier warning signs model of, 95–96
Risk management, suicide, 4–5, 96–98
Risk perception, altered, 104
Risky behavior. *See also* Motor vehicle accident deaths, veteran
    driving, 103–112 (*See also* Motor vehicle accident deaths, veteran)
    suicide risk and, 185
Roberts, L. W., 123
Rodger, L. L., 181
Romans, 11
Rosenheck, R., 70, 92
Rosenzweig, L., 111

Ross, J., 244
Rudd, M. D., 47, 56–57, 59, 90, 93, 95–98, 107, 123, 151, 182–183
Ruff, R. M., 180
Ryan, J., 29

SAD PERSONS, 136
Safety plan, 136–138
Safety Planning Intervention (SPI), 97
Salmon, Thomas, 3
Saltzman, W. R., 195
Scales, MMPI-2, 115
Schell, T. L., 86
Schneider, A. L., 179
Schoenbaum, M., 27, 91
Screening, of at-risk soldiers, 47–48
Seiden, R. H., 109
Seidlitz, L., 205
Selby, E. A., 56–57
Selective serotonin reuptake inhibitors (SSRIs), for PTSD, 134
Self-determination theory (SDT), 216–219, 228
Self-killing, 12–14
Self-medication, 94
Selzer, M. L., 109
Sepaher, I., 115–116
Sexual orientation, of military personnel, 81
Sexual trauma
    childhood, 56
    military, 83–84, 139–140
Shame, 10
Shane, S., 180
Shea, C., 155
Shell shock, 2–3
Shneidman, Edwin S., 19, 108–109, 111
Short-term psychodynamic therapy, for PTSD with suicide risk, 175
Silver, J. M., 184
Silverman, M., 95–96, 184, 186
Simpson, G. K., 183, 185
Sirotin, A., 117
Skopp, N. A., 6, 68–69, 117, 182, 249–250
Sleep deprivation, in OEF/OIF/OND veterans, 29
Sleep disorders, on suicide risk, 122
    in older veterans, 204
Sloan, L. B., 84
Smith, B., 25, 27, 30
Smith, T. C., 25, 27, 30
Smolenski, D. J., 6, 241, 242, 246, 248
Social Cognitions Scale, 48
Social isolation
    after traumatic brain injury, 181
    on suicide risk, 55, 116, 222, 224, 229, 244
        in older veterans, 204–205, 208
        with PTSD, 141
Social performance, POTFF, 194–195
Social relationships
    protective effect of, 122, 244
    in resilience, 191
Social resilience training, 191

Solution-focused problem-solving skills, 81
Southwick, S. M., 86
Special Operations Command (SOCOM), 190
Special Operations Force (SOF), 190
Special Operations Force (SOF), suicide in, 190–198
   alcohol abuse in, 191
   branch specific cultural influence in, 197
   effectiveness of, evaluating, 196
   military–civilian life gap in, 192
   POTFF four domains in, 193–196
      history and fundamentals of, 193
      physical performance in, 193–194
      psychological performance in, 194
      social performance in, 194–195
      spiritual performance in, 195–196
   POTFF objective and development in, 192–193
   POTFF rationale in, 192
   problem of, 190
   promoting from top in, 196–197
   resilience and armed forces in, 191–192
   spirituality disconnection and, 192
   stressors in, 192
   unique aspects of, 190–191
Specific norepinephrine reuptake inhibitors (SNRIs), for PTSD, 134
Speigel, J., 3
Spiegel, H., 3
Spirituality, 195–196
Spiritual Performance Program, POTFF, 195–196
Staal, M., 191, 192
Stabilization Plan, CAMS SSF, 151, 153, 162f
Stander, V. A., 105
Stanley, B., 97, 98, 151, 203
Staves, P. J., 66, 69, 94
Stein, M., 27, 30, 47, 91
Stephenson, J. A., 191, 192
Stepped care, 221
Stigma
   of moral killing, medieval, 12
   of race, in Vietnam War, 81
Stigma, of mental health care, 2, 7, 241, 247, 250
   masculine behavior and, 122
   in older veterans, 207
   in Perceived Stigma and Barriers to Care for Psychological Problems scale, 86
   psychological evaluations in, 192
   on PTSD treatment, 135, 137
   Special Operations Forces programs combating, 193, 194, 197
   on traumatic brain injury assessment, 190
Stigma, of suicide, 19
   driving-related suicide and, 106, 108, 109
Stolberg, R. A., 115
Stone, S. L., 59, 97–98, 107
Stouffer, Samuel, 3
Stresses, in combat veterans, 82–85
   combat environment, 83–84
   deployment in, 82
   home arrival, 84–85
   home departure, 82–83

Stresses, in mental health professionals, client suicide risk, 121
Stress inoculation training, for PTSD, 134
Stressors, Iraq and Afghanistan operations, 53
Strobel, W., 191
Strom, T. Q., 105, 110, 111
Subintentioned deaths, 108–109
Subintentioned suicide, veteran, 103–112. *See also* Covert and subintentioned suicide, veteran
Substance abuse and substance use disorders. *See also* Alcohol abuse
   on acquired capability for suicide, 56
   killing and, 71, 72
   in OEF/OIF/OND recruits, 30–31
   with PTSD, 94
   on suicide risk, 92
      in older veterans, 204
Suicidal behavior (suicidality), 131
   assessment of (*See* Suicide risk assessment)
   protective factors in, assessment of, 136
   in service members and veterans, 121–122
Suicidal crisis, 19
Suicidal ideation
   active and passive, 131
   combat-related killing and, 68–69, 72, 73
   communication of, lack of, 122
   driving-related, 106–107
   managing suicide risk with, 97–98
   prevalence of, 131
   on suicide completion, 95
Suicide, military, 1–7. *See also specific topics*
   barriers to care in, 2
   historical context of, 2–4 (*See also* Historical context)
   mental health conditions in, 10 (*See also* Mental health conditions; *specific types*)
   methods of, 5
   prevalence of (*See* Epidemiology, suicide)
   in Special Operations Forces, 190
Suicide attempts, 229–230
   combat-related killing and, 70, 73–74
   increase in, 53
   prevalence of, 121
Suicide by motor vehicle crash, 107–108. *See also* Motor vehicle accident deaths, veteran
Suicide completion, 229–230
   in Army National Guard, OEF/OIF/OND, 43–44
   prevalence of, 1, 4, 10, 53, 121–122
   suicidal ideation on, 95
Suicide items, MMPI-2, 115
Suicide prevention, 47–49
   by Department of Defense, for OEF/OIF/OND veterans, 31–33
   in elderly, 205–206
   empirically supported risk factors in, 48–49
   firearms and means restriction in (*See* Firearms; Means restriction)
   follow-up protocol in, 48
   future directions in, 33–34
   interpersonal-psychological theory of suicide on, 58–59
   by Los Angeles Suicide Prevention Center, 18, 19

for OEF/OIF/OND veterans, 31–33
for older veterans, 206–208
person-centered, in primary care settings, 213–231 (*See also* Person-centered suicide prevention, primary care settings)
by primary care providers, 213
with PTSD, 170–173
screening at-risk soldiers in, 47–48
Shneidman's work on, 18
symptom recognition training in, 48
Suicide Prevention Teams (SPTs), VA, 208
Suicide rates, 201–202
after psychiatric treatment, 240
in Army
historic, 40–41
in OEF/OIF/OND veterans, 25
in Army National Guard, 39
with depression, 92 (*See also* Depression)
determination of, 89–90
in males, 202
mortality ratio in, 90
in older veterans, 201–202
in Persian Gulf veterans, 104
with substance abuse, 92
in U.S., 202
Suicide rates, of OEF/OIF/OND veterans, 24–25
Air Force, 25–26
Army, 25
Marine Corps, 25–26
National Guard, 24, 39, 40–42, 40f, 41t, 49n1
Navy, 25–26
by service, 25–26
Suicide risk
in Army National Guard, 40–42, 40f, 41t, 42t, 49nn1–2
combat exposure on, 93
risky behavior and, 185
by service, 40, 40f
treatment noncompliance in, 241
Suicide risk assessment, 135–136
Army Study to Assess Risk and Resilience in Service members (Army STARRS), 30, 90
Collaborative Assessment and Management of Suicidality (CAMS), 136, 147–165 (*See also* Collaborative Assessment and Management of Suicidality (CAMS))
mental health professional competence in, 124–125
motor vehicle accident deaths and, 109–111
Suicide risk assessment, in combat veterans, 79–86, 89–98
approach to, 94–96
case examples on, 89
combat experience in, 90–91, 93
managing suicide risk in, 96–98
emergency intervention in, 98
outpatient, 97–98
military culture in, 79–82
framework of, 79
joining military in, developmental context of, 79–80
as mental health care barrier, 85–86
personnel in, 81–82
traditions and values in, 80–81

PTSD in
combat-related, 91–93
comorbid risk factors with, 93–94
risk factors related to, 96
stresses in, 82–85
combat environment in, 83–84
deployment in, 82
home arrival in, 84–85
home departure in, 82–83
suicide rate and, 89–90
Suicide risk management, 96–98
emergency intervention in, 98
initial, 4–5
outpatient, 97–98
Suicide Status Form (SSF), 136, 148–150, 160f–161f
in case vignettes, 149–152, 160f–161f
in initial treatment planning, 150–151, 160f–161f
research on, 155–156
Stabilization Plan using, 151, 153, 162f
Sullivan, G. R., 105, 107
Superordinate mission, 126
Symptom recognition training, for suicide prevention, 48
Systems theory, 216, 217–218, 217f, 228
unintended consequences and, 229

Tanielian, T., 23
Targeted interventions, 220–221
Tarrier, N., 91–92
Tarrier, T., 131
Tatarelli, G., 109
Tatarelli, R., 109
Tate, R., 183, 185
Team Strategies & Tools to Enhance Performance & Patient Safety (Team STEPPS), 221
Teasdale, T. W., 94, 181
Teesson, M., 244
Telephone call interventions, posttreatment, 243
Thanh, H. T. T., 241, 243
Theories of behavior change, 216–219
means restriction in, 220
self-determination theory in, 216–219, 228
systems theory in, 216, 217–218, 217f, 228
Therapeutic alliance, 138, 174–175, 229
Thinking, maladaptive, 116
Thomas, E. K., 241, 242, 246, 248
Thomas, J. L., 70–71, 106
Thomsen, C. J., 105
Thuras, P. D., 105, 110, 111
Thwarted belongingness, 54–55, 54f, 64, 69, 90, 122, 244
Top-down responsiveness, 218
Traditions, military cultural, 80–81
Transition periods, on suicide risk in older veterans, 203
Traumatic brain injury (TBI)
aggression and, 184
assessment of, 180
blast injury classification in, 28
in civilians, 179
classification and severity of, 179, 179t
definition of, 178–179
epidemiology of, 179

Traumatic brain injury (*Contd.*)
  from IEDs, in OEF/OIF veterans, 1–2, 23, 27–29, 173
  psychiatric disorders and, 180–181
  PTSD and, 94, 173, 181
  sequelae of, 180, 181*t*
Traumatic brain injury (TBI), mild, 179–183
  assessment of, 179, 179*t*
  sequelae of, 180, 181*t*
  suicide risk in, 181–183
Traumatic brain injury (TBI), suicide and
  executive dysfunction and, 183–186
  risk of, 28–29, 122
  with PTSD, 173
Traumatic neurosis, 166
Traumatic stress. *See also* Post-traumatic stress disorder (PTSD); *specific types*
  altruism and, 18
Traumatic war neurosis, 166. *See also* Post-traumatic stress disorder (PTSD)
Treatment, of PTSD, 175
  evidence-based, 6, 131–142 (*See also* Post-traumatic stress disorder (PTSD) treatment, evidence-based)
  for suicide risk, 166–177 (*See also* Post-traumatic stress disorder (PTSD) treatment, for suicide risk)
Treatment, of suicidal patients, 4–7
  electroconvulsive therapy in, 6–7
  follow-up in, 7
  level of care in, 5
  MMPI-2 codetypes in, 116–117
  noncompliance with, 241
  in older veterans, 201–208 (*See also* Older veterans, suicide management)
  pharmacotherapy in, 6
  psychoeducation in, 5
  psychotherapy in, 6
  risk management in, initial, 4–5
  safety plan in, 4–5, 97
  treatment plan formulation in, 5–6
  treatment planning in, 5
Treatment, of suicidal patients, ethical issues, 121–129
  "avoiding harm" in, 127, 128
  client career implications in, 127
  "client" identification difficulties in, 126–127
  competence in, 124–125
  consultation in, 125
  control and responsibility obligations in, 128
  dual identities in, 126
  embedded MHPs in, 127
  ethical principals in, 123
  future directions in, 128–129
  informed consent in, 123–124
  mental health practitioner in
    stress of, 121
    suicidal client and, 122–123
  mixed-agency ethical dilemmas in, 125–126
  role shifts in, unanticipated, 127

  suicidal behavior in, 121–122
  superordinate mission in, 126
  unique challenges of military suicidal clients in, 125–128
Trofimovich, L., 182, 240, 249–250
Two-factor theory of fear conditioning, 18

Unintended consequences, 229
Universal prevention programs, 220
Ursano, R., 30, 47, 91, 95
US Special Operations Command (SOCOM), 190
US Special Operations Force (SOF). *See* Special Operations Force (SOF)

VA Crisis Hotline, 89, 97, 146
Valenstein, M., 9, 59, 66, 92, 93, 203, 204
Values, in military culture, 80–81
Van Orden, K. A., 53–56, 64–65, 67, 244
VA Primary Care-Mental Health Initiative, 213–214
Venlafaxine, for PTSD, 134
Veracity, 123
Veterans and/or Military Crisis Line, 5
Vietnam War veterans
  chronic psychological issues in, 4
  postdeployment mortality risk in, 103–104
  PTSD and suicide risk in, 4, 167
Villarreal, E., 94
Vo, A. H., 70–71, 106
Voller, E., 105, 110

Walter, K. H., 183
Walters, E. E., 131
Walters, H., 9, 92
Warner, K., 108
War syndromes, 2
Wasserman, D., 241, 243
Waternaux, C., 184
Weigel, R., 105, 110
Welsh, G. S., 115
West, C. L., 104, 116
Wheeler, G., 29
White River Model, 221
Whole patient care, 213
WHO SUPRE-MISS study, 242, 243
Williamson, A., 244
Witte, T., 95–96
Wolfman, J. H., 184
Wood, C. W., 181
Woods, M., 96
Woodward, K., 195
World Suicide Prevention Day, 19
Worth, R. M., 103

Yurgelun-Todd, D. A., 183

Zimbardo, P. G., 110
Zivin, K., 92, 94, 240